Impact of Incomplete Metabolism on Health

Edited by

Rajesh Kumar
Maharaja Agrasen School of Pharmacy
Maharaja Agrasen University
Solan, Himachal Pradesh, India

Impact of Incomplete Metabolism on Health

Editor: Rajesh Kumar

ISBN (Online): 979-8-89881-249-2

ISBN (Print): 979-8-89881-250-8

ISBN (Paperback): 979-8-89881-251-5

Published by Bentham Science Publishers Pte. Ltd. Singapore, in collaboration with Eureka Conferences, USA.

First published in 2025.

Bentham Science Publishers Pte. Ltd.
No. 9 Raffles Place
Office No. 26-01
Singapore 048619
Singapore
Email: subscriptions@benthamscience.net

BENTHAM SCIENCE

CONTENTS

FOREWORD

In today's rapidly evolving world of medical science, the complex interplay between metabolism and health has gained increasing attention. This book titled "***Impact of Incomplete Metabolism on Health***" delves into a crucial yet often overlooked aspect of human physiology: the consequences of incomplete or impaired metabolic processes. Metabolism, the body's intricate system for converting food and other substances into energy, is essential for sustaining life. However, when these metabolic pathways are disrupted or incomplete, the effects can be far-reaching and profound, leading to a variety of health disorders.

This book offers a comprehensive exploration of how incomplete metabolism, whether due to genetic factors, environmental influences, or lifestyle choices, can adversely affect overall health. From rare metabolic disorders like phenylketonuria to more common conditions, such as diabetes, obesity, and cardiovascular disease, the book provides an in-depth analysis of how faulty metabolic processes contribute to disease development and progression. Readers will gain valuable insights into the underlying mechanisms of metabolism, the role of enzymes, and how nutrient deficiencies, toxins, and medications can exacerbate metabolic imbalances. Importantly, this book also highlights potential interventions, including dietary strategies and emerging therapies, to manage and mitigate the impact of incomplete metabolism.

Whether you are a healthcare professional, researcher, or someone seeking to understand the complexities of metabolism and its impact on health, this book will serve as an essential resource. It sheds light on the importance of early diagnosis, intervention, and personalized approaches to improve metabolic health and overall well-being.

Brijesh Kumar
Department of Pharmacology
Institute of Medical Sciences
Banaras Hindu University
Varanasi, Uttar Pradesh, India

PREFACE

Incomplete metabolism refers to the body's inability to fully break down and process nutrients or compounds, leading to the accumulation of harmful byproducts. This can be caused by genetic factors, insufficient enzyme levels, or external influences, such as medications and certain foods. Incomplete metabolic pathways may result in the accumulation of intermediate chemicals inside tissues or the circulation, which may cause abnormal cellular activities and a range of health problems.

The metabolism of drugs is also impacted by incomplete metabolism. Some drugs can cause toxicity or adverse reactions if the liver is unable to adequately metabolize them. For instance, people with CYP2D6 deficiencies may not absorb medications like codeine or antidepressants well, which might result in either a harmful build-up or an inadequate therapeutic impact.

Chronic illnesses, including obesity, diabetes, and non-alcoholic fatty liver disease, can be brought on by environmental variables that affect metabolic pathways, in addition to hereditary problems. Examples of these factors include inadequate diet and exposure to chemicals. In general, the effects of inadequate metabolism can be extensive and severe; therefore, early detection and treatment are essential to the management of many disorders.

The book contains 10 chapters. Chapter 1 discusses the introduction to metabolism. In this chapter, the book begins with deep insights into the basics behind the metabolism and production of primary and secondary metabolites. Chapter 2 explains the understanding of incomplete metabolism. Chapter 3 tells about the metabolic pathways and intermediates produced during the various metabolic reactions. Chapters 4 and 5 explain the health implications of incomplete metabolism along with case studies of patients related to incomplete metabolism. Chapter 6 highlights the diagnostic approaches and biomarkers for the detection of metabolic disorders. Chapter 7 discusses the foundation for treatment strategies and therapeutic innovations. Both chapters 8 and 9 discuss the nutritional and lifestyle interventions along with future directions and research frontiers towards the impact of incomplete metabolism. Lastly, chapter 10 discusses the role of empowering patients and caregivers in controlling metabolic disorders.

Rajesh Kumar
Maharaja Agrasen School of Pharmacy
Maharaja Agrasen University
Solan, Himachal Pradesh, India

List of Contributors

Ankit Awasthi	Chitkara University School of Pharmacy, Chitkara University, Rajpura, Punjab, India
Ashutosh Kumar	Department of Ophthalmology, University of California Los Angeles, California, USA
Anjna Sharma	Chitkara University School of Pharmacy, Chitkara University, Solan, Himachal Pradesh, India
Ashish Sharma	Washington University School of Medicine, Washington University in St. Louis, St. Louis, Missouri, United States
Amit Aggarwal	Maharaja Agrasen School of Pharmacy, Maharaja Agrasen University, Solan, Himachal Pradesh, India
Arvinder Pal Singh	Maharaja Agrasen School of Pharmacy, Maharaja Agrasen University, Solan, Himachal Pradesh, India
Ankush Goyal	Maharaja Agrasen School of Pharmacy, Maharaja Agrasen University, Solan, Himachal Pradesh, India
Ashima Thakur	Faculty of Pharmaceutical Sciences, The ICFAI University, Baddi, Himachal Pradesh, India
Ashok Kumar	Faculty of Pharmaceutical Sciences, The ICFAI University, Baddi, Himachal Pradesh, India
Brijesh Kumar	Department of Pharmacology, Institute of Medical Sciences, Banaras Hindu University, Varanasi, Uttar Pradesh, India
Devinder Kumar	Amar Shaheed Baba Ajit Singh Jujhar Singh Memorial College of Pharmacy, Bela, Ropar, Punjab, India Maharaja Agrasen School of Pharmacy, Maharaja Agrasen University, Solan, Himachal Pradesh, India
Dhirender Kaushik	Department of Pharmaceutical Sciences, Gurugram University, Gurugram, Haryana, India
Hitesh Chopra	Department of Biosciences, Saveetha School of Engineering, Saveetha Institute of Medical and Technical Sciences, Chennai, Tamil Nadu, India
Jyoti	School of Pharmaceutical Sciences, MVN University, Palwal, Haryana, India
Monika Rana	Department of Pharmacognosy, Laureate Institute of Pharmacy, Kangra, Himachal Pradesh, India
Navjot Singh Sethi	IEC University, Solan, Himachal Pradesh, India
Pankaj Kalia	Maharaja Agrasen School of Pharmacy, Maharaja Agrasen University, Solan, Himachal Pradesh, India
Pratistha Singh	Department of Ophthalmology, University of California, Los Angeles, California, USA
Prabhsharan Kaur	Department of Pharmaceutical Sciences and Technology, Maharaja Ranjit Singh Punjab Technical University, Bhatinda, Punjab, India
Parijat Pandey	Department of Pharmaceutical Sciences, Gurugram University, Gurugram, Haryana, India

Pooja Thakur	Maharaja Agrasen School of Pharmacy, Maharaja Agrasen University, Solan, Himachal Pradesh, India
Rajesh Kumar	Maharaja Agrasen School of Pharmacy, Maharaja Agrasen University, Solan, Himachal Pradesh, India
Ravinesh Mishra	School of Pharmacy and Emerging Sciences, Baddi University of Emerging Sciences & Technology, Baddi, Solan, Himachal Pradesh, India
Rahul Sharma	Amar Shaheed Baba Ajit Singh Jujhar Singh Memorial College of Pharmacy, Bela, Punjab, India
Sunil Dutt	Maharaja Agrasen School of Pharmacy, Maharaja Agrasen University, Solan, Himachal Pradesh, India
Saurav Kumar	Department of Pharmaceutical Sciences, Gurugram University, Gurugram, Haryana, India
Swati Pundir	School of Pharmaceutical Sciences, Shoolini University of Biotechnology and Management Sciences, Solan, 173212, Himachal Pradesh, India
Saurav Kumar	Department of Pharmaceutical Sciences, Gurugram University, Gurugram, Haryana, India
Shiwank Rana	Dreamz College of Pharmacy, Sundernagar, Mandi, Himachal Pradesh, India
Satish Patnaik	Stein Eye Institute, Geffen School of Medicine, University of California, Los Angeles, California, USA
Shilpa Chandel	Banasthali Vidyapith, Department of Pharmacy, Banasthali, Rajasthan, Himachal Pradesh, India
Varinder Singh	Department of Pharmaceutical Sciences and Technology, Maharaja Ranjit Singh Punjab Technical University, Bhatinda, Punjab, India
Yogendra Singh	School of Pharmaceutical Sciences, MVN University, Palwal, Haryana, India

CHAPTER 1

Insights into Metabolism

Sunil Dutt[1], **Ankit Awasthi**[2], **Rajesh Kumar**[1,*] and **Ashutosh Kumar**[3]

[1] *Maharaja Agrasen School of Pharmacy, Maharaja Agrasen University, Solan, Himachal Pradesh, India*

[2] *Chitkara University School of Pharmacy, Chitkara University Rajpura Punjab, India*

[3] *Department of Ophthalmology, University of California Los Angeles, California, USA*

Abstract: The present book chapter highlights the comprehensive epitome of metabolism, explaining the regulatory and intricate mechanisms involved. It commences with the introduction which encompasses drug metabolism pathways, enumerating linear, cyclic, and branched pathways then delves into the catabolic and anabolic pathways. The catabolic pathways explore glycolysis and the Krebs cycle, and anabolic pathways involve gluconeogenesis, protein synthesis, and lipogenesis. The further section overshadows discussion on energy in metabolism. This segment of the chapter discusses ATP as the energy currency, highlighting its structure, functions, and ATP hydrolysis. Further, it emphasizes energy transfer mechanisms that involve redox reactions. The other segments of the chapter cover the role of enzymes in metabolism, explaining catalytic functions and specificity. Afterward, enzyme regulation involves allosteric regulation and feedback inhibition. Finally, the book chapter delves the metabolism regulation, focusing on hormonal regulation, nutrient availability, and genetic-epigenetic regulation. Examining these various aspects, it is quite clear that the chapter provides a framework for understanding the complex nature of metabolism and its vital role in health and disease.

Keywords: Absorption, Acetylation, Active medications, Anabolic pathways, Catabolic pathways, Coupling reactions, Drug-metabolizing enzymes, Ecotoxicology, Elimination, Energy in metabolism, Enzymes in metabolism, Lipophilic molecules, Metabolism regulation, Metabolism, Pharmacodynamics, Pharmacokinetics, Phosphorylation, Prodrugs, Transformation, Xenobiotic.

INTRODUCTION

The study of drug metabolism, a branch of pharmacology and biochemistry, quickly advanced starting around 1950. It has significantly influenced pharmacokinetics, pharmacodynamics, ecotoxicology, and other biological fields.

* **Corresponding author Rajesh Kumar:** Maharaja Agrasen School of Pharmacy, Maharaja Agrasen University, Solan, Himachal Pradesh, India; E-mail: rajdhiman60@gmail.com

International rules increasingly require knowledge of the metabolism of medicines and other xenobiotics. The 19th century witnessed the most noteworthy achievements. The metabolism processes, such as acetylation, the synthesis of mercapturic acid, conjugation with glycine, sulfuric acid, and glucuronic acid, hydrolysis, reduction, and oxidation, were made clear [1]. The chemical sequence oxidation-conjugation created a number of pathways for different substrates and has become the cornerstone of modern xenobiochemistry [2].

Simply, medicines are small molecules that fall under the category of xenobiotics, or compounds that are not naturally occurring in the human body. Nonetheless, a variety of endogenous compounds, such as steroids and hormones, are occasionally referred to as drugs and are used to treat certain medical conditions. The term "metabolism" refers to the process by which an enzyme changes a chemical's moiety. The most well-known enzymes involved in drug metabolism are the reductases, hydrolases, and oxidases that make up the cytochrome P450s (CYP450s). The primary objective of metabolism is the elimination of endogenous and/or exogenous chemicals from the body [3].

In the metabolic process, there will be the conversion of lipophilic molecules into hydrophilic ones or will facilitate elimination. On the other hand, the drug-metabolizing enzymes transform compounds into their pharmacologically active form in many conditions. Prodrugs are pharmaceutically inactive drugs designed to treat absorption and bioavailability difficulties. They pass through a process of transformation in the body subsequent to absorption to become active medications. To get around ampicillin's low bioavailability, pivampicillin is produced as a prodrug that has the potential to hydrolyze into ampicillin after being absorbed into the bloodstream. A significant instance of a prodrug is the use of mycophenolate mofetil to improve the oral bioavailability of mycophenolic acid [4].

The byproducts of metabolism that are called metabolites can either be biologically active or inactive. The smooth endoplasmic reticulum (ER), mitochondria, and small intestinal epithelia of hepatocytes are the primary locations for five CYP450 enzymes that are crucial for drug removal. To a lesser extent, they are also present in the proximal tubules of the kidneys. The significance and essential functions of conjugating enzymes and drug transporters are becoming increasingly clear to humans. Modifications to these pathways may have an impact on a drug's pharmacokinetics and pharmacodynamics. Drugs interact with these pathways during their absorption, distribution, metabolism, and excretion [5].

DRUG METABOLISM PATHWAYS

There are various and diverse ways that drug-food substances can be metabolized or biotransformed, including through the chemical processes of hydration, reduction, oxidation, hydrolysis, conjugation, and condensation. Studying these pathways is critical since a drug's metabolism might reveal whether or not it has pharmacological or toxicological effects. Phase I, also known as functionalization reactions, and Phase II, sometimes known as conjugative reactions, are the two stages that generally makeup drug metabolism (Table **1**). The chemical processes are often connected to drug metabolism in phases I and II [6].

Table 1. Phase I and Phase II reactions.

Phase I	Phase II
Oxidation	Glucuronidation/glucosidation
Reduction	Sulfation
Hydrolysis	Acetylation
Dethioacetylation	Methylation
Isomerisation	Glutathione conjugation
Hydration	Amino acid conjugation

On the other hand, the drug and the food go through the drug metabolism process. It can undergo different types of pathways based on their structural flow and functionality. Here's an explanation of these pathways:

Linear Pathway

In linear pathways, the drug undergoes a series of sequential biochemical transformations, with each step dependent on the previous one.

By converting acetaldehyde to acetate, mammals metabolize ethanol. By conversion to acetyl CoA, acetate is absorbed by the general metabolic pool. Although other organs subsequently metabolize the majority of the acetate, the liver is the primary organ that converts ethanol to acetate. Alcohol Dehydrogenase (ADH) is the enzyme that is principally responsible for converting ethanol to acetaldehyde by catalyzing the following reaction: acetaldehyde + NADH + H^+ - ethanol + NAD^+. NAD^+ is an essential coenzyme that accepts electrons during the oxidation of ethanol to acetaldehyde (Fig. **1**).

It is reduced to NADH in the process. The availability of NAD^+ is a limiting factor in ethanol metabolism. Since NAD^+ is required for glycolysis, the citric acid cycle, and the electron transport chain, extensive ethanol metabolism can lead to a depletion of NAD+, disrupting these critical metabolic processes. The liver enzymes found in mammals are soluble proteins with a molecular weight of 80,000 located in the cytosolic compartment [7]. On the other hand, NADH is generated as a byproduct of the ADH-catalyzed reaction. The accumulation of NADH relative to NAD^+ alters the cellular redox state, leading to several metabolic consequences. Increased NADH levels lead to various consequences. For example, high levels of NADH inhibit gluconeogenesis, which can lead to hypoglycemia, particularly in fasting individuals. The increased $NADH/NAD^+$ ratio favors the conversion of pyruvate to lactate (lactic acidosis) and oxaloacetate to malate, reducing the availability of substrates for the citric acid cycle and gluconeogenesis. Acetaldehyde is converted into Acetate (*via* Aldehyde Dehydrogenase, ALDH), and acetate is converted into acetyl-CoA (*via* Acetyl-CoA synthetase). Finally, Acetyl-CoA enters the Krebs cycle. There are also multiple molecular forms of aldehyde dehydrogenase (ALDH) in the human liver, but only class I and class II isozymes, encoded by ALDH1 (or ALDH1A1 in the new nomenclature) and ALDH2, respectively, are thought to be involved in acetaldehyde oxidation [8].

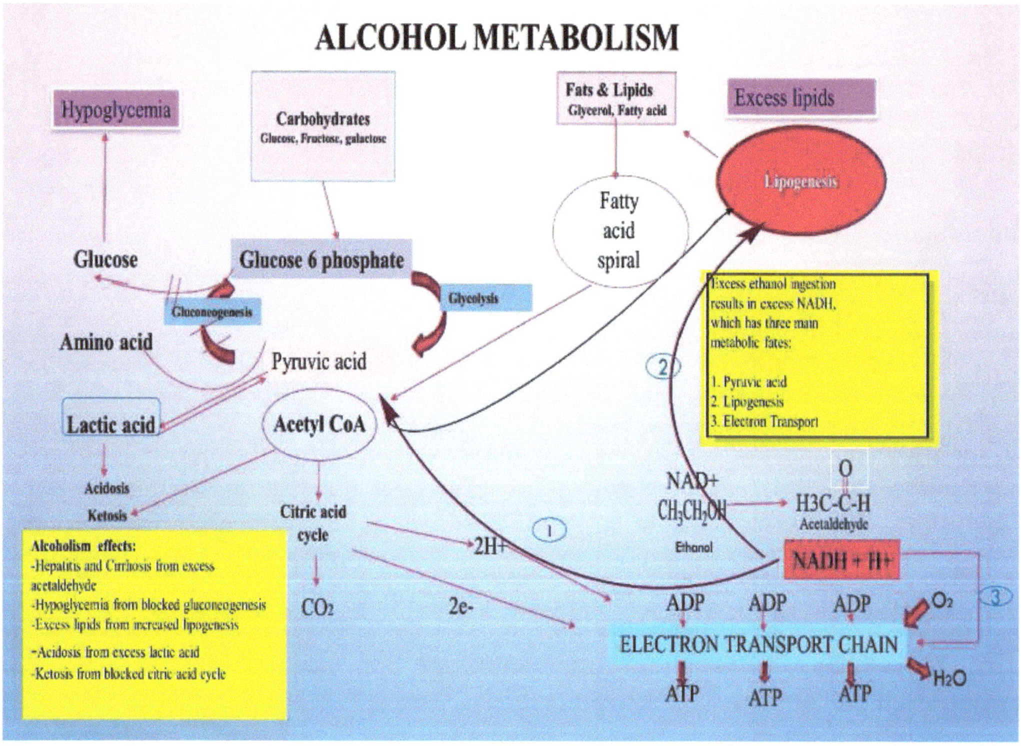

Fig. (1). Process of alcohol metabolism.

Cyclic Pathways

Cyclic pathways involve a series of reactions that regenerate the initial reactant, forming a cycle.

Krebs Cycle (Citric Acid Cycle)

Although not directly a drug metabolism pathway, the Krebs cycle is involved in the metabolism of drug metabolites that enter as intermediates (*e.g.*, acetyl-CoA from ethanol metabolism). TCA cycle derivatives like citrate and alpha-ketoglutarate, which are hypothesized to be involved in the synthesis of coupling factors, may act as mitochondrion-derived regulators of glucose-stimulated insulin secretion. In addition to being crucial for ATP production, the mitochondrion in general—and the TCA cycle in particular—serve as major metabolic hubs that enable fuel oxidation and the generation of lipid and protein building blocks through cataplerosis. Both glutamate and malonyl-CoA play key roles in these processes.

Approximately 50% of pyruvate is oxidized in beta cells to acetyl-CoA by pyruvate dehydrogenase. Pyruvate dehydrogenase is a foremost site of modulation as, among other effectors, the enzyme is activated by elevation of mitochondrial Ca^{2+} and, conversely, its activity is reduced upon exposures to either excess chronic high glucose Oxaloacetate or fatty acids produced by the anaplerotic enzyme pyruvate carboxylase, condenses with acetyl-CoA forming citrate, which undergoes stepwise decarboxylation and oxidation yielding alpha-ketoglutarate (Fig. **2**). The TCA cycle is ended up *via* succinate, fumarate, and malate, in turn producing oxaloacetate. The fate of alpha-ketoglutarate is influenced by the redox state of mitochondria [9].

Branched Pathways

Amino Acid Biosynthesis

In mammals, amino acid biosynthesis consists of a series of enzyme-catalyzed processes which transform precursors into amino acids. These precursors often come from key metabolic routes, including glycolysis, the citric acid cycle, and the pentose phosphate pathway. The synthesis of proteins, the generation of neurotransmitters, and other essential processes all depend on the amino acids. It is an essential process that relies on the availability of particular metabolic intermediates and enzymes is the production of amino acids in mammals [10, 11]. Essential amino acids must be obtained from nutrition, whereas animals *via* certain mechanisms, can synthesize non-essential amino acids. Because the routes for the production of amino acids are correlated with vital metabolic activities, cells are able to create these essential molecules as needed. It is essential to comprehend these pathways in order to understand nutrition, metabolism, and associated illnesses. Aromatic amino acids (AAAs) include tryptophan, phenylalanine, and tyrosine.

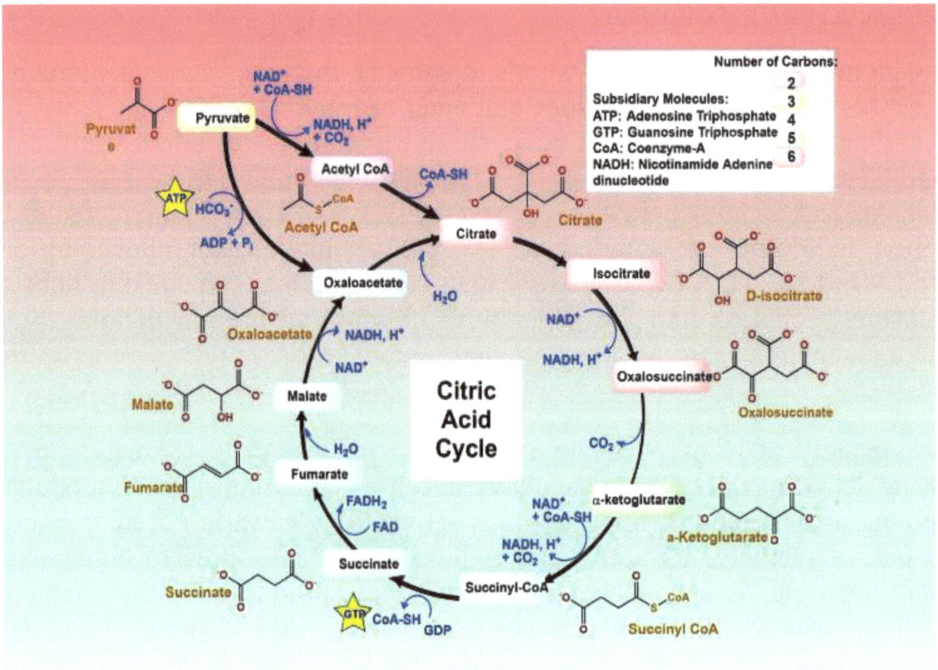

Fig. (2). Process of citric acid cycle.

Serine, glycine, and cysteine are amino acids derived from 3-phosphoglycerate. Alanine, synthesized from pyruvate, is an intermediate in glycolysis. Asparagine and aspartate are synthesized from oxaloacetate or glutamate, glutamine, proline, and arginine derived from α-ketoglutarate or are the intermediate of the citric acid cycle. Histidine is an intermediate in the pentose phosphate pathway derived from ribose-5-phosphate.

In plants, these amino acids are precursors to many natural products, including pigments, alkaloids, hormones, and components of cell walls. The shikimate pathway, which receives around 30% of photosynthesis-fixed carbon in vascular plants, is the source of all three AAAs. The AAAs are vital parts of human diets since animal lineages no longer possess their biosynthetic pathways, and the enzymes needed to synthesize them have been targeted for use in herbicide research [9, 12].

Non-essential Amino Acids

Animals are able to synthesize the following amino acids that are 9 in number; phenylalanine-derived, which is conditionally essential, are tyrosine, alanine, asparagine, aspartate, glutamate, glutamine, glycine, proline, and serine.

Conditionally Essential Amino Acids

The body normally synthesizes these amino acids, but in some situations, such as disease or stress, you may need to add more of them to your diet: phenylalanine-derived tyrosine, methionine-derived cysteine, and arginine.

IMPORTANT BIOSYNTHETIC ROUTES FOR AMINO ACIDS THAT ARE NOT ESSENTIAL

Alanine Biosynthesis

Pyruvate is the precursor for the alanine pathway. Through a transamination reaction, alanine is synthesized, or an amino group from glutamate is transferred to pyruvate, producing alanine and α-ketoglutarate. Pyruvate+Glutamate→Alanine+α-Ketoglutarate in the presence of Enzyme Alanine aminotransferase (ALT).

Tyrosine and Phenylalanine Synthesis

Tyrosine or phenylalanine can be synthesized *via* two distinct processes that can be formed from the amino acid precursor chorismate. Tyrosine or phenylalanine, the pathway's final product, feedback blocks it, regulating the process and guaranteeing balanced synthesis in accordance with cellular requirements. Unlike the Trp pathway, the plant Phe and Tyr pathways are still relatively unexplored, and the genes encoding the pathway enzymes have only recently been identified. Chorismate, the end product of the shikimate pathway, serves as the precursor for phenylalanine and tyrosine. CM transforms chorismate into prephenate, which can then be transformed *via* two different routes to Phe and Tyr [13, 14]. Another process, called the arogenate pathway, starts with prephenate being transaminated to L -arogenate (formerly called pretyrosine). Arogenate dehydratase (ADT) or arogenate dehydrogenase (ADH) catalyzes the subsequent dehydrogenation/decarboxylation to Tyr or Phe, respectively. These processes occur in reverse order in the alternative pathway, known as the phenylpyruvate or 4-hydroxyphenylpyruvate pathway. In this pathway, prephenate first undergoes dehydration and decarboxylation by prephenate dehydratase (PDT) to form phenylpyruvate, or dehydrogenation and decarboxylation by prephenate dehydrogenase (PDH) to form 4-hydroxyphenylpyruvate. These intermediates are then transaminated to produce phenylalanine (Phe) and tyrosine (Tyr), respectively [13, 15].

Fatty Acid Metabolism

Acetyl-CoA as a Central Metabolite: Acetyl-CoA is a critical metabolite that may enter several routes, including the fatty acid synthesis pathway lipogenesis, the Krebs cycle for energy generation, and ketogenesis for the synthesis of ketone bodies. Acetyl-CoA's ability to be utilized for energy generation or stored as fat is determined by hormonal control, which includes the actions of insulin and glucagon.

Pentose Phosphate Pathway (PPP)

Ribose-5-Phosphate *vs.* NADPH Production: The PPP can branch into two different pathways: one leading to the production of ribose-5-phosphate (important for nucleotide synthesis) and another leading to the production of NADPH (essential for reductive biosynthesis and maintaining redox balance). The pentose phosphate pathway (PPP) is a glucose-oxidizing system that generates nicotinamide adenine dinucleotide phosphate (NADPH) and ribose 5-phosphate in tandem with higher glycolysis. While NADPH is important in redox homeostasis and promotes biosynthetic activities such as the production of proline, fatty acids, cholesterol, tetrahydrofolate, and deoxyribonucleotides, ribose 5-phosphate is needed for nucleotide synthesis. The PPP is a two-branch route that breaks down glucose. Glucose 6-phosphate is the input of the oxidative branch (oxPPP), and it is also necessary for glycolysis. The oxPPP produces two molecules of NADPH while converting glucose 6-phosphate to ribose 5-phosphate. Glycolytic intermediates and ribose 5-phosphate are reversibly interconverted by the non-oxidative branch (non-oxPPP). When under unstressed conditions, the PPP shows a substantially lower flow than glycolysis in the majority of healthy mammalian cells [16, 17].

The demand for ribose-5-phosphate *vs.* NADPH determines the direction of the pathway. For instance, if the need for NADPH is high, more glucose-6-phosphate will be diverted into the oxidative phase of the PPP [18, 19].

Purine Metabolism

Adenine and Guanine Nucleotide Synthesis: The intermediate inosine monophosphate (IMP), which may be changed into either adenosine monophosphate (AMP) or guanosine monophosphate (GMP), is where the purine nucleotide biosynthesis pathway splits off. Feedback inhibition strictly controls the synthesis of both AMP and GMP, guaranteeing that the two nucleotides are generated in equal proportions [20, 21].

Glycolysis and Gluconeogenesis

Glucose-6-phosphate can be sent to the pentose phosphate route for the synthesis of NADPH and ribose-5-phosphate during glycolysis, or it can proceed along the glycolytic pathway to generate energy. The demands of the cell determine how these routes are balanced; for example, insulin and other hormones stimulate glycolysis and energy generation, while a cell's need for redox balance or nucleotide synthesis might boost the flow *via* the pentose phosphate pathway. The liver and kidneys are the sites of glucose biosynthesis (Fig. **3**).

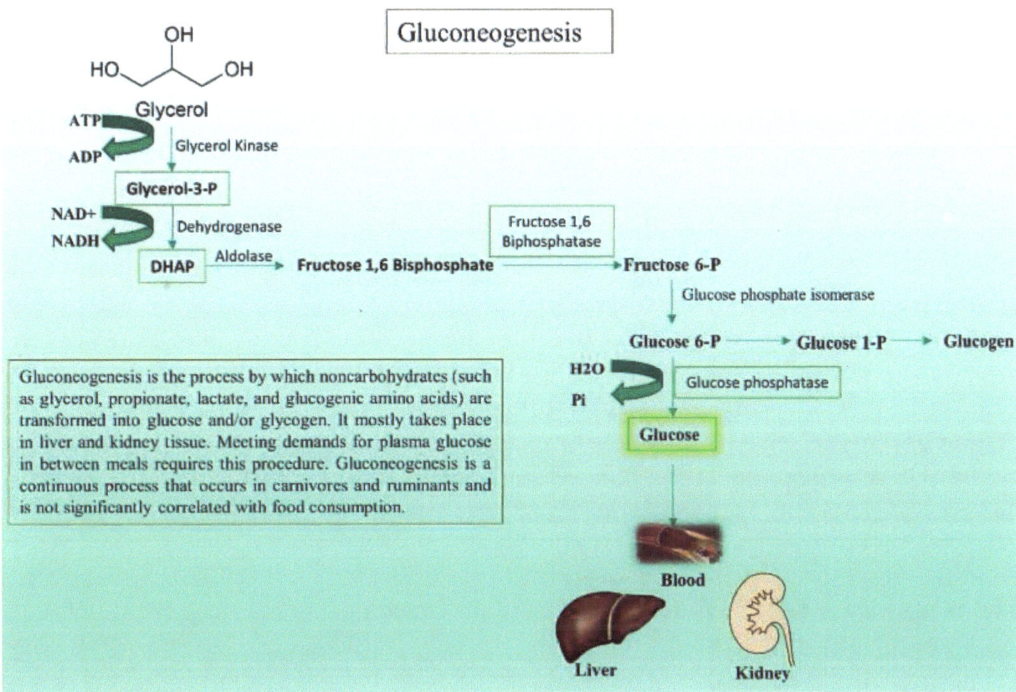

Fig. (3). Process of gluconeogenesis.

The process of gluconeogenesis provides the plasma glucose required in between meals. The diabetogenic hormones (glucagon, growth hormone, adrenaline, and cortisol) promote glucose synthesis. Certain amino acids, glycerol, lactate, and propionate are examples of gluconeogenic substrates. In gluconeogenesis, the rate-limiting process is catalyzed by PEP carboxykinase. During gluconeogenesis, hydrocarbons are transferred from pyruvate to PEP *via* the dicarboxylic acid shuttle. Since ruminant and carnivorous animals constantly produce glucose, they are not required to store glycogen in their liver cells. Ala is the most common amino acid that is transferred from muscle to the liver during exercise and fasting. A (small) gluconeogenic substrate is b-aminoisobutyrate, which is produced when pyrimidines degrade [22, 23].

ENERGY IN METABOLISM

Energy plays a pivotal role in maintaining a healthy metabolism inside living beings. It circumscribes all the essential chemical processes within the living organisms to perpetuate life. Most importantly, the role of energy in metabolism includes energy conversion, cellular processes, basic metabolic rate (BMR), bioenergetics, photosynthesis, and respiration [24, 25]. In the energy conversion process, metabolism utilizes energy obtained from food (proteins, carbohydrates, and fats) and light, which is stored within the cells. It then breaks these substances down into simple molecules and produces energy.

Furthermore, the obtained energy can be utilized by microorganisms to perform basic cellular functions such as maintaining body temperature, muscle contraction, and transmission of nerve impulses [26, 27]. In addition to this, the obtained energy is also utilized to maintain BMR. Whereas, bioenergetics describe the transformation and flow of energy within the systems. It comprises all the chemical reactions needed to consume or release energy, which are essential for breaking and building complex molecules. Besides this, plants obtain energy from the sunlight and convert it to usable forms *via* cellular respiration. In addition to this, the various steps involved in the process of energy metabolism include glycolysis, pyruvate oxidation, citric acid cycle, electron transport chain and oxidative phosphorylation. These steps collectively help in the confirmation that energy from the source is converted into adenosine triphosphate (ATP) [28, 29].

ATP is also known as the energy currency of the cells. This is because it helps in the storage and transport of chemical energy within cells to perform life-sustaining cellular processes. The structure of ATP includes adenine (nitrogenous base), ribose, and three phosphate groups, which are connected *via* high-energy bonds. These high-energy bonds help in the storage of energy within the ATP, especially *via* linkage between the beta and gamma phosphates [30]. This stored energy is released with the help of ATP hydrolysis. It is the process in which ATP is broken down into inorganic phosphate (Pi) and ADP (adenosine diphosphate) and releases energy.

The various steps involved in ATP hydrolysis include nucleophilic attack, bond cleavage, proton transfer, release of energy, and stabilization of products [31, 32].

-Nucleophilic attack: Water bounds with gamma phosphates group of the ATP. This mechanism takes place in the presence of ATPase enzymes.

-Bond cleavage: After nucleophilic attack, there is a cleavage of beta and gamma phosphoanhydride bonds [33]. This results in the formation of Pi and ADP. Proton transfer: During bond cleavage, there is a transfer of a proton from the attacking water molecule to the surrounding base. This helps in the stabilization of the transition state.

-Release of energy: The lysis of the phosphoanhydride bond exhibits a high amount of energy, which the cell harnesses to perform various functions such as biosynthesis, muscle contraction, and active transport.

- Stabilization: ADP and Pi are finally stabilized through hydration and interactions with the other molecules present in the cells. This step ensures that the released energy is effectively used and products can be utilized further to regenerate ATP. The advantages of ATP hydrolysis include controlled energy supply, regulation of metabolic pathways, stability/reversibility, heat production, and electrochemical gradients [34, 35].

-Similarly, the mechanism of energy transfer includes coupling reactions, phosphorylation, and electron transfer chain (ETC).

-Coupling reaction is a linkage between endergonic reaction with exergonic reaction. The energy emitted from the exergonic reaction is utilized to initiate the endergonic reaction. The mechanism of coupling reaction involves various steps, which include ATP hydrolysis, phosphorylation and common intermediate [36].

-ATP hydrolysis is an energy-releasing process which is already discussed in the aforementioned sections.

-Phosphorylation: During ATP hydrolysis, the obtained Pi is relocated to another molecule, a process termed phosphorylation. This process frequently varies the activity of the target molecule and enables them to initiate further reactions [37].

-Common Intermediate: During the coupling reaction, there is a formation of a common intermediate. For instance, synthesis of glutamine from ammonia and glutamate, ATP is required to make an intermediate. This further reacts with ammonia and form glutamine. A few examples of coupling reactions are:

(a) Protein synthesis: The formation of peptide bonds among amino acids is an endergonic process. Additionally, coupled with guanosine triphosphate (GTP), a similar molecule to ATP with high energy [38]. The other example is:

(b) Glucose Phosphorylation: In this endergonic reaction glucose is phosphorylated to form glucose-6-phosphate and ADP. The coupling of this reaction is due to the exergonic ATP hydrolysis [39].

(c) Active Transport: The movement of ions through the cell membrane from lower concentration to high concentration, *e.g.*, sodium potassium pump.

This process takes place with the help of ATP hydrolysis [40]. Most importantly, the advantage of coupling reactions lies in their efficiency and regulatory role:

-Efficiency: The energy released from exergonic reactions is not wasted but instead utilized for various cellular processes.

-Regulation: Coupling helps effectively regulate metabolic pathways, ensuring that energy is available where and when it is needed.

-Homeostasis: It contributes to maintaining homeostasis at the cellular level by balancing energy utilization and production, which also supports temperature stability [41, 42].

-Phosphorylation: It is the process *via* which energy is transferred within the cells. This process takes place through oxidative and substrate-level phosphorylation. The mechanism of oxidative phosphorylation involves electron transfer chain (ETC) initiation, electron transfer, proton pumping, formation of an electrochemical gradient, ATP synthesis, and oxygen as the final electron acceptor [43, 44].

(a)*ETC*: In this step, there is a formation of FADH2 and NADH from citric acid and glycolysis. These donate electrons to the ETP located in the inner membrane of mitochondria [45].

(b) Electron transfer: In this, the electrons are passed from protein complexes I to IV. Once the electrons pass from these complexes, they produce energy.

(c) Proton pumping: The released energy is further used to drive H^+ ions (pump protons) from the mitochondrial matrix to intermembrane space and create a proton gradient.

(d) Formation of electrochemical gradient: In this step, electrochemical potential is generated across the intermembrane space, which is also known as proton motive force.

(e) ATP synthesis: In this step, H^+ ions return to the mitochondrial matrix with the help of ATP synthase. It is a complex that utilizes energy from this flow. Finally, results in the formation of Pi and ATP from ADP [46, 47].

(f) Oxygen as final electron acceptor: In the end stage of*ETC*, oxygen accepts electrons and merges up with protons and form water. This final step is very essential for the electron flow continuation *via ETC*. Additionally, in substrate-level phosphorylation, the phosphorylated intermediate transfers high-energy phosphate group. This results in the direct formation of ATP from ADP. This process is very crucial for the citric acid cycle and glycolysis. During the citric cycle, there is a formation of succinyl-CoA to succinate from succinyl-CoA synthetase. Simultaneously, results in the generation of ATP/GTP from GDP/ADP and Pi. These reactions are very important for cells, particularly under the presence of anaerobic conditions or in the cells, which lacks mitochondria. As they offer a rapid source of ATP in the absence of oxygen. Similarly, in glycolysis the limiting factor is enzyme phosphoglycerate kinase. Further, catalyzes the phosphate group turn over from 1,3-bisphosphoglycerate to ADP, casting ATP and 3-phosphoglycerate. In this process, another rate limiting step is catalyzed *via* pyruvate kinase, where ADP accepts phosphate group from phosphoenolpyruvate (PEP), forming ATP and pyruvate [48, 49].

ETC: It is a cluster of protein complexes and various other molecules that are embedded in the inner region of the mitochondrial membrane. The mechanism of*ETC* is already discussed in the subsequent sections.

ENZYMES IN METABOLISM

Enzymes are specialized proteins as they are made up of amino acids. Moreover, they have complex structures atypical of proteins, enabling them to perpetuate catalytic role in metabolism. The different enzymes that play critical roles in metabolism include hexokinase, phosphofructokinase, pyruvate kinase, lactate dehydrogenase, pyruvate dehydrogenase, citrate synthase, aconitase, isocitrate dehydrogenase, α-Ketoglutarate dehydrogenase, succinyl-CoA synthetase, succinate dehydrogenase, fumarase, malate dehydrogenase, ATP synthase, glucose-6-phosphatase, fructose-1,6-bisphosphatase, phosphoenolpyruvate carboxykinase, glycogen phosphorylase, glycogen synthase, lipase, acetyl-CoA carboxylase, fatty acid synthase, carnitine acyltransferase, HMG-CoA reductase, aminotransferase, glutamine synthetase, arginase, ornithine transcarbamylase, NADH dehydrogenase and cytochrome c oxidase. The role of these aforementioned enzymes in metabolism is discussed below.

Hexokinase is one of the vital enzymes in cellular metabolism and primarily participates in the glycolytic pathway. It helps in the catalyzation of the phosphorus and results in the formation of glucose-6-phosphate from glucose. This is the initial step of glycolysis and is essential because it entraps glucose within cells. So that it cannot diffuse back and be available to cells to proceed metabolic processes further. The aforementioned conversion of glucose *via* hexokinase also helps in attaining a concentration gradientthereby promoting continuous glucose uptake in the cells. Further, this ensures glucose utilization for the production of energy and is stored as a glycogen. Additionally, the resulting glucose-6-phosphate regulates hexokinase activity to maintain optimal glycogen accumulation and ensure the necessary production of energy. Furthermore, hexokinase has different isoenzymes, *e.g.*, hexokinase I-IV, which also plays an important role in specific regulatory mechanisms of metabolism [50, 51]. Phosphofructokinase is also a regulatory enzyme for glycolytic pathway progression, having a key role in metabolism. It helps in the catalyzation of the fructose-6-phosphate and results in the formation of fructose-1,6-bisphosphate. This step helps in the lysis of glucose to produce energy, has high regulation and serves as a critical control factor in glycolysis. In addition to this, it is inhibited allosterically by citrate and ATP. This step keeps an eye on the energy levels. Once they increase, is down downregulated *via* glycolysis and slow down. Conversely, its activation occurs through ADP and AMP and provides signals when there is a decrease in energy levels. Hence, promotes glycolysis acceleration and upregulates the production of ATP. Besides this, one of the important isoforms of phosphofructokinase is phosphofructokinase present in liver tissues and muscles and helps in the regulation of tissue-specific glycolysis. Most importantly, the role of phosphofructokinase is to regulate the balance in cells for energy demand and supply [52]. Therefore, makes it a central player in the regulation of metabolism. Similarly, pyruvate kinase helps in the catalyzation of the final step of glycolysis. Its work is to promote the transfer of phosphate group to form ADP from phosphoenolpyruvate. Further, it results in the production of one molecule of ATP and pyruvate. This is an irreversible process, which keeps control over the glycolysis process and, ensures efficient ATP production important for cellular energy. Moreover, its isoforms are pyruvate kinase I and II, which play a critical role in regulatory mechanisms [53, 54]. For example, pyruvate kinase II helps in the formation of dimers and tetramers. Dimers are fewer and Tetramers are highly active to allow fine-tuned regulation of glycolysis as per the cell's metabolic needs. Additionally, this enzyme is inhibited by ATP and alanine. Allosterically, activated by fructose-1,6-bis bisphosphate thereby reflecting the status of cells' energy and acts on glycolysis when cells require energy. Hence, it is clear that pyruvate kinase is a key point controller in cells for energy production and metabolic flux [55, 56]. Lactate dehydrogenase takes place under anaerobic conditions and helps in cellular metabolism. It helps in the catalyzation of pyruvate conversion into lactate. Furthermore, simultaneously converts NADH to NAD^+. This process is required when oxygen levels are low because it helps in the continuation of glycolysis *via* NAD^+ regeneration. This results in the production of ATP. For instance, when we do intense exercise and oxygen is scarce. Then, lactate dehydrogenase helps in the energy production within muscle cells, albeit less efficiently than oxidative phosphorylation. Afterward, the produced lactate is converted back to pyruvate in the liver. Then it enters to Krebs cycle to produce energy when there is availability of oxygen again [57].

Pyruvate dehydrogenase: It is one of the important enzymes in cellular metabolism. It acts as a bridge between the citric acid cycle and glycolysis. It helps in the catalyzation of glycolysis-derived pyruvate acetyl-CoA. Further, it produces energy by entering into the citric acid cycle. Most importantly, this step provides a linkage between the glycolysis aerobic process with the citric acid cycle aerobic process and oxidative phosphorylation. Furthermore, from this step, pyruvate dehydrogenase promotes complete glucose oxidation. Hence, it results in the maximum yield of ATP. In addition to this, the obtained acetyl-CoA is used in fatty acid synthesis, highlighting PDH's role in biosynthetic pathways and energy production [58].

Citrate synthase: It plays a critical role in the Krebs cycle and catalyzes the primary step in the cycle. In this step oxaloacetate combines with acetyl-CoA and forms citrate. This step is crucial as it facilitates the cycle and results in ATP production. In addition to this, citrate synthase is present inside the mitochondrial matrix and acts as a pace-making enzyme. This is because its functions affect the rate of the entire Krebs cycle. Furthermore, it also confirms that acetyl groups derived from proteins, carbohydrates, and fats are efficiently utilized for energy production. Additionally, the produced citrate in this process can be transported out of the mitochondria to serve as a precursor for cholesterol and fatty acids [59].

Aconitase: It helps in the catalyzation of citrate isomerization to isocitrate through cis-aconitate. This step is important for ATP production, the cell's primary energy currency. Moreover, aconitase contains a cluster of sulfur that directly comes in contact with the substrate, distinguishing it from different iron-sulfur proteins that function as electron carriers. Additionally, aconitase also helps in iron homeostasis and under iron-depleted conditions, it behaves as an iron-responsive element-binding protein and regulates the gene expression involved in iron metabolism [60, 61].

Isocitrate dehydrogenase (IDH): They are essential enzymes involved in various metabolic processes, including the Krebs cycle and energy production. They catalyze the conversion of isocitrate to α-ketoglutarate (α-KG), NAD(P)H, and

CO_2. IDHs help maintain cellular balance by regulating respiration, energy production, redox balance, promoting anaplerosis, and influencing gene expression through epigenetic mechanisms [62]. For example, a study conducted by Lee *et al.* observed that mice on a high-fat diet with a lack of IDH2 caused weight gain and reduced mitochondrial function due to high ROS levels [63]. α-Ketoglutarate dehydrogenase plays a vital role in the Krebs cycle by converting α-ketoglutarate to succinyl-CoA and producing NADH. This reaction supports energy production, amino acid metabolism, and cellular redox balance. In the brain, KGDH is important for cognitive function and protection against neurodegenerative diseases. Proper KGDH activity helps manage oxidative stress and prevents various metabolic and cellular disorders [64, 65]. Succinyl-CoA synthetase (SCS) is a mitochondrial enzyme essential for several metabolic processes. It facilitates the citric acid cycle by integrating amino acids, fatty acids, and propionate and generating high-energy phosphates. SCS also plays a role in ketone body biosynthesis and is crucial for heme synthesis. Additionally, it is involved in protein succinylation and is linked to mtDNA depletion. Additionally, SCS can produce ATP through substrate-level phosphorylation even without oxygen by converting succinyl-CoA to succinate, thus influencing various metabolic pathways [66, 67]. Fumarase is essential in metabolism for converting fumarate to malate in the citric acid cycle, a process critical for ATP production and metabolic balance. Proper fumarase function supports cellular energy and biosynthetic pathways. Mutations in the FH gene, which encodes fumarase, can disrupt this conversion, leading to impaired energy production and health issues such as developmental problems in the brain and increased risk of cancers like hereditary leiomyomatosis and renal cell cancer (HLRCC) [68].

Malate dehydrogenase: It is an important enzyme involved in the citric acid cycle and the malate-aspartate shuttle. It facilitates the reversible reaction where malate is converted to oxaloacetate, with NAD^+ serving as a cofactor. This reaction produces oxaloacetate, NADH, and a proton.

MDH exists in two main forms: cytosolic MDH, which supports the malate-aspartate shuttle to transfer reducing equivalents into the mitochondria, and mitochondrial MDH, which is crucial for the citric acid cycle and energy production [69]. ATP synthase is an enzyme in the inner mitochondrial membrane (or thylakoid membrane in chloroplasts) that produces ATP, the cell's main energy source. It works by using the flow of protons across the membrane to drive the rotation of its components, which then convert ADP and phosphate into ATP [70]. Glucose-6-phosphatase is an enzyme primarily present in the liver and kidneys. It converts glucose-6-phosphate into glucose and inorganic phosphate. This reaction is crucial for gluconeogenesis and glycogenolysis, processes that generate glucose for the bloodstream. By removing the phosphate group from glucose-6-phosphate, glucose-6-phosphatase helps maintain blood sugar levels, especially during fasting or vigorous exercise [71, 72].

REGULATION IN METABOLISM

Metabolic regulation is also known as another complex and dynamic signal transduction system that is important for cells to preserve homeostasis. These are favourable conditions vital for the proper running of the organism within the internal context [73]. Metabolic regulation is, therefore, very strict at several levels, with an overwhelming concern for the kinds of enzymes and other proteins that are employed in the metabolic pathways at the genetic level. Suppose a specific amount of an enzyme is required [74]. This can be achieved because genes are regulated and can be upregulated or downregulated in response to signals such as hormones, nutrients, or stress. For example, when blood glucose concentration is elevated, insulin, a hormone secreted by the pancreas, stimulates the activation of lipogenic and glycogenic genes to ensure that the excess glucose is stored as glycogen or fat [65, 75]. Regulation at the enzymatic level is perhaps the most illustrative of the ability of cells to change. It entails fine-tuning the reactions and the enzymes that catalyze these reactions in these pathways. They can be turned on or off by molecules that bind to them in a way that is not at the active site [76], for example, by allosteric substances or through chemical reactions like phosphorylation. Yet the ability to adapt at this pace is critical for allowing cells to maintain their function, such as increasing energy production as is needed during exercise or reducing energy consumption during fasting [77]. At a subcellular level, compartmentalization of metabolic pathways in various organelles, such as mitochondria for energy metabolism, enables rigid regulation and control of processes. Cells have also means of communicating with one another and the whole body through signaling molecules, which include hormones; these modify some metabolic processes to match the energy state or growth requirements of the organism and conditions prevailing in the environment [78].

Hormonal Regulation

Hormonal regulation is one of the main aspects of metabolic regulation; hormones are considered chemical messengers that control and adapt the functioning of organs and tissues to maintain homeostasis in the organism. The hormone propagated by the beta cells in the pancreas, commonly known as insulin, is influential in this process [79, 80]. It is secreted when blood glucose concentration is high, for instance, after a meal. It stimulates the uptake of glucose by the cells, especially the liver, skeletal muscle, and fat cells, and it stimulates the conversion of glucose to glycogen for storage or fat in adipose tissue. Insulin also suppresses the breakdown of glycogen into glucose (glycogenolysis), as well as the synthesis of glucose from other substrates (gluconeogenesis) so that blood glucose concentrations do not rise [81, 82].

On the other hand, glucagon, which is secreted from the alpha cells of the pancreas, tells the body that glucose levels are low during fasting or exercise. It triggers the liver to release glycogen into glucose and supply the required energy in the bloodstream, especially in important organs such as the brain and muscles [83]. This hormone also promotes glycogen synthesis from amino acids and other precursors, such as gluconeogenesis. The impact of adrenaline, which is secreted by the adrenal glands when the body is stressed or physically engaged, is even stronger because adrenaline encourages glycogenolysis and lipolysis – the processes of burning glycogen and fats – to get a boost of energy as early as possible [84, 85].

These hormones collectively function to maintain very close and strict control of blood glucose levels so that energy can always be provided when required. These hormones work somewhat reciprocally when one of them is being secreted and the other is inhibited in order to avoid rapid changes in blood glucose [86]. These hormones are widely involved in the maintenance of the metabolic balance and play a crucial role in changing the metabolic need for feeding, fasting, stress, or any other condition. Suppose the regulation of hormones is not kept in check, as is seen in diabetic cases. In that case, a fine balance of the glucose and energy supply can be hugely affected, thereby causing many complications [87].

NUTRIENT AVAILABILITY

Nutrient signals are a core component used by cells and tissues directly to control metabolism and thus are a primary prerequisite to achieving metabolic control. The body's metabolic processes are extremely sensitive to the differing quality and quantity of nutrients to work within a given situation at different periods, for example, eating, starvation, or exercise [88]. This dynamic regulation is important to maintain energy balance and, thus, cellular functions while preventing dysfunction that results from nutrient excess and deficiency.

Fed State: Glucose Abundance and Metabolic Shifts

In the fed state, when the concentration of macronutrients, especially glucose, is high, the body switches energy-producing pathways to storage and biosynthetic ones. When one has taken cooked grains, which are rich in carbohydrates, the level of glucose in the blood increases, and this acts as a signal for the pancreas gland to release insulin [89, 90]. Glucose, the sugar that is mainly obtained from the food we take, is absorbed into the cells with the help of insulin, the hormone mainly secreted by the pancreas. In the liver, it stimulates glycogenesis, the conversion of excess glucose into glycogen for storage in the body. Glycogen is in the form of a storage form of energy which can easily be retrieved whenever there is a low UTI of glucose [91]. In the same regard, insulin suppresses gluconeogenesis, which is the synthesis of glucose from any other material besides carbohydrates. It offers a check to glycogenolysis, the breakdown of glycogen into glucose [92]. In the muscle cells, the glucose is converted to glycogen, and this represents a store of energy that is used during contraction. In the adipose tissue, as shown in the equation, insulin induces the transformation of excess glucose into fatty acids that are then converted into triglycerides, which are the most stored fats. This is a process known as lipogenesis, which helps store any excess energy that has not been used in one's body [93]. Further, the fed state is accompanied by a higher rate of glycolysis; this is a crucial metabolic pathway that is involved in breaking down glucose to derive ATP –the 'energy currency' of cells. Glycolysis is most intense in tissues with a high energy turnover rate, such as the brain and red blood cells, where glucose is the exclusive source of energy [94]. Pyruvate and lactate, which are the end products of glycolysis, can also be utilized in other metabolic pathways, making a clear connection between the body's total energy regulation.

Fasting State: Metabolic Adaptations and Energy Mobilization

For instance, during a fasting period or when there is a long interval between meals, the body has to draw strategies to ensure an orderly provision of energy, especially to organs that work round the clock, such as the brain or heart [95]. When blood glucose levels start to drop, insulin secretion declines, and another hormone secreted by the pancreas, glucagon, rises. The main action of glucagon in the context of the fasting state is to help regulate blood glucose concentrations by encouraging the breakdown of glycogen and gluconeogenesis in the liver [96]. If glycogen reserves are depleted—after approximately 12–24 h of fasting—the body turns its attention to fats. Lipolysis starts in the adipose tissue through the hydrolysis of triglycerides into free fatty acids and glycerol [97]. These fatty acids are liberated in the bloodstream and transported by the different tissues, where beta-oxidation in the mitochondria yields ATP. The liver also transforms part of the fatty acids into ketone bodies, another source of energy that the brain and other body parts can use in case of a lack of glucose. There is a shift to ketogenesis or the production of ketone bodies, especially during fasting, because it also offers an essential energy source, with little demand on gluconeogenesis from amino acids, hence sparing muscles [98]. This metabolic flexibility that enables human beings to switch from burning glucose to fats and ketones is one of the ways that people survive cases of famine. However, this state must be well-controlled because high levels of ketone can result in a dangerous condition known as ketoacidosis, which is caused by a very acidic pH in the blood [99].

Protein Metabolism: Adaptation and Conservation

During both fed and fasting states, amino acids—the constituents of proteins—are important regulators of metabolism. During the fed state, amino acids from ingested protein are employed in this process for the growth, repair, and formation of enzymes and hormones [100]. Amino acids, some of which are taken up by cells that need to make new protein molecules, and any excess amino acid can be processed as follows: deaminated; the resulting carbon skeleton is used in ATP production or is converted into glucose *via* gluconeogenesis pathways or fatty acids [101]. If the animal is actively fasting, then it should focus on strategies for conserving protein. First, amino acids from muscle protein will be utilized as the substrate for gluconeogenesis to maintain blood glucose concentration [102]. Nonetheless, as fasting is prolonged, the body starts to use less protein catabolism and preserves muscle mass more. The preferred operation of the liver shifts towards beta-oxidation, ketogenesis, which is characterized as decreasing glucose reliance and, therefore, saving amino acids from fuel consumption [103]. The conservation of protein is essential to retain muscle performance during extended periods without nourishment.

Hormonal and Cellular Signaling: Integration and Regulation

A network of hormonal signals regulates nutrient availability in the body. Insulin and glucagon serve as primary responses. In addition to limiting glucose, insulin also affects lipid and protein metabolism. It stimulates anabolic or energy-conserving reactions. On the other hand, glucagon induces catabolic processes. It breaks down stored energy to keep glucose levels stable while fasting [103, 104].

At the cellular level, nutrient sensing pathways, including the mTOR (mechanistic target of rapamycin) pathway, serve as important integrators for signals on surrounding nutrients and energy status. For instance, amino acids in combination with glucose activate the mTOR pathway. This leads to promoting protein synthesis and cell growth. On the other hand, insufficient nutrients activate pathways like those that involve AMPK (AMP-activated protein kinase) [105]. These pathways direct cells to conserve energy and promote alternative sources of cell fuel. Flexibility to use sugars, fats and proteins depending on available nutrients is crucial for both survival and well-being. This sophisticated cross-talk between hormonal signals, nutrient sensing pathways, and metabolic processes allows cells and tissues to adjust their function under different physiological conditions. These adjustments maintain energy balance or systemic functional demands. Alterations in this balance, as seen through the lens of metabolic disorders such as diabetes, can have pronounced effects on health. This emphasizes the need for fine nutrient control related to optimal metabolic health [106].

GENETIC AND EPIGENETIC REGULATION

Genetic and epigenetic regulation play crucial roles in metabolism. They alter cells' metabolic pathways to ensure that they can adapt to various internal and external stimuli. These regulatory mechanisms control the expression of genes involved in metabolic processes, allowing cells to meet their energy demands, respond to environmental changes, and maintain homeostasis [107].

Genetic Regulation in Metabolism

Genetic regulation refers to the control of gene expression at the transcriptional level, where genes encoding enzymes and other proteins involved in metabolism are turned on or off depending on the cell's needs. This regulation is primarily mediated by transcription factors, which are proteins that bind to specific DNA sequences near the genes they regulate [108].

Transcription Factors

Transcription factors are crucial in the genetic regulation of metabolism. For example, the transcription factor SREBP-1c (Sterol Regulatory Element-Binding Protein-1c) is activated in response to insulin and promotes the expression of genes involved in lipid synthesis [109]. This allows the cell to produce fatty acids and triglycerides when nutrients are abundant. Another important transcription factor is PPARα (Peroxisome Proliferator-Activated Receptor Alpha), which is activated during fasting and promotes the expression of genes involved in fatty acid oxidation, enabling the body to utilize fats as an energy source [110].

Gene Promoters and Enhancers

The regions of DNA known as promoters and enhancers are critical in determining when and where genes are expressed. Promoters are located near the start of a gene and provide a binding site for RNA polymerase and transcription factors [111]. Enhancers can be located further away from the gene and help increase the efficiency of transcription. These elements ensure that metabolic genes are expressed in response to the right signals, such as nutrient availability, hormonal changes, or stress [112].

Nutrient-Sensitive Pathways

Nutrient availability directly influences genetic regulation through nutrient-sensitive signalling pathways. For example, the mTOR (mechanistic target of rapamycin) pathway is activated in the presence of sufficient nutrients, particularly amino acids [113]. It promotes cell growth and protein synthesis by increasing the expression of genes involved in these processes. Conversely, when nutrients are scarce, the AMPK (AMP-activated protein kinase) pathway is activated, leading to the downregulation of anabolic processes and the upregulation of catabolic pathways to conserve energy [114].

EPIGENETIC REGULATION IN METABOLISM

Epigenetic regulation refers to changes in gene expression that do not involve alterations in the DNA sequence but are instead mediated by chemical modifications to DNA or histone proteins, which package DNA in the cell nucleus. These modifications can be influenced by environmental factors, such as diet, stress, and exercise, and play a significant role in metabolic regulation [115].

DNA Methylation

DNA methylation involves the addition of a methyl group to the cytosine base in DNA, typically at CpG sites (regions of DNA where a guanine follows a cytosine). Methylation usually represses gene expression by preventing the binding of transcription factors to the DNA [116]. In metabolism, DNA methylation can regulate the expression of key metabolic genes. For example, hypermethylation of the promoter region of the POMC (Proopiomelanocortin) gene, which is involved in appetite regulation, can lead to reduced expression and contribute to obesity.

Histone Modification

Histones are proteins around which DNA is wrapped, forming chromatin. Chemical modifications to histones, such as acetylation, methylation, phosphorylation, and ubiquitination, can alter the structure of chromatin, making it either more or less accessible to transcription machinery [117]. Histone acetylation, for example, typically makes chromatin more open and accessible, promoting gene expression. In the context of metabolism, histone modifications can regulate genes involved in processes such as glucose metabolism, fatty acid oxidation, and mitochondrial function [118]. For instance, histone deacetylases (HDACs), which remove acetyl groups from histones, can repress the expression of genes involved in energy expenditure, contributing to metabolic diseases like obesity and type 2 diabetes [119].

Non-Coding RNAs

Non-coding RNAs, such as microRNAs (miRNAs) and long non-coding RNAs (lncRNAs), also play a role in epigenetic regulation by influencing the stability and translation of mRNAs or by interacting with chromatin to affect gene expression [119]. miRNAs can bind to the mRNA of metabolic genes, leading to their degradation or the inhibition of their translation. For example, miR-33a/b is known to regulate cholesterol and lipid metabolism by targeting genes involved in cholesterol efflux and fatty acid oxidation [120].

Environmental Influences

Epigenetic modifications are highly responsive to environmental factors, including diet, exercise, and exposure to toxins. For instance, a high-fat diet can lead to epigenetic changes that affect the expression of genes involved in lipid metabolism, potentially leading to obesity and insulin resistance [121]. Conversely, regular physical activity has been shown to induce beneficial epigenetic changes, such as increased expression of genes involved in mitochondrial function and energy metabolism [122].

INTERACTION BETWEEN GENETIC AND EPIGENETIC REGULATION

The interaction between genetic and epigenetic regulation in metabolism is a highly intricate process that involves the interplay of DNA sequences and chemical modifications that influence gene expression. Genetic regulation is primarily governed by the sequences of DNA and the activity of transcription factors that bind to specific regions of the genome to either activate or repress gene expression [123]. These transcription factors are essential for turning genes on or off in response to various signals, such as hormonal changes, nutrient availability, or stress, thereby directly influencing metabolic pathways.

However, the regulation of metabolism extends beyond the genetic code itself. Epigenetic modifications such as DNA methylation, histone modification, and the action of non-coding RNAs add a layer of control that can fine-tune gene expression without altering the underlying DNA sequence [124]. These epigenetic changes can influence the accessibility of the DNA to transcription factors, thereby modulating how genes are expressed in response to internal and external cues. For instance, DNA methylation can silence genes by preventing transcription factors from binding to their target sequences. At the same time, histone acetylation can open up the chromatin structure, making genes more accessible for transcription [125].

The interaction between genetic and epigenetic regulation is not one-way; it is a dynamic and reciprocal relationship. Transcription factors involved in genetic regulation can recruit enzymes that modify histones or DNA, effectively linking genetic control to epigenetic changes [126]. For example, a transcription factor might recruit a histone acetyltransferase (HAT) enzyme that adds acetyl groups to histones, leading to a more open chromatin structure and increased gene expression. Conversely, epigenetic modifications can affect the binding of transcription factors to DNA. Methylation of a gene promoter, for instance, can block the binding of a transcription factor, thereby repressing gene expression [127].

This interplay is particularly important during key developmental stages when epigenetic modifications can establish long-term patterns of gene expression that influence an individual's metabolic profile. For example, during early development, epigenetic changes can set a metabolic program that determines how genes involved in energy metabolism are expressed throughout life [128]. This process, often referred to as "metabolic programming," can have lasting effects on an individual's susceptibility to metabolic disorders such as obesity, diabetes, and cardiovascular disease. Environmental factors, including diet, stress, and exposure to toxins, can also influence these epigenetic modifications, further impacting how genetic and epigenetic mechanisms interact to regulate metabolism.

Lastly, the interaction between genetic and epigenetic regulation in metabolism is a complex and dynamic process that involves a constant dialogue between DNA sequences and chemical modifications. This interaction allows for a highly adaptable system of metabolic control that can respond to a wide range of internal and external signals, ensuring that the body's metabolic processes are finely tuned to meet its needs. Understanding this interplay is crucial for unraveling the mechanisms underlying metabolic diseases and developing targeted therapies that can modify gene expression patterns to restore metabolic balance.

CONCLUSION

The interaction between genetic and epigenetic regulation in metabolism is a dynamic and interconnected process that controls how metabolic genes are expressed in response to various signals. Genetic regulation involves the activity of transcription factors, which are proteins that bind to specific DNA sequences to either activate or repress the transcription of genes involved in metabolic pathways. These transcription factors respond to signals such as hormones, nutrients, and environmental stressors, ensuring that genes are expressed at the right time and in the right amounts to meet the cell's metabolic needs.

Epigenetic regulation adds a layer of control by modifying the DNA or histone proteins around which DNA is wrapped without altering the DNA sequence itself. These modifications, such as DNA methylation and histone acetylation, influence how tightly DNA is packaged and, consequently, how accessible it is to the transcriptional machinery. For example, DNA methylation typically silences gene expression by preventing transcription factors from binding to their target sites. At the same time, histone acetylation generally promotes gene expression by loosening the chromatin structure, making the DNA more accessible. The interaction between these mechanisms is crucial for fine-tuning metabolism. Transcription factors can recruit enzymes that add or remove epigenetic marks, directly linking genetic signals to epigenetic changes. Conversely, epigenetic modifications can affect the ability of transcription factors to bind to DNA, thereby influencing gene expression. This intricate interplay allows cells to adapt their metabolic activities to changes in nutrient availability, energy demands, and environmental conditions. Moreover, these regulatory processes are responsive to external factors, meaning that diet, exercise, and other lifestyle choices can lead to lasting epigenetic changes that influence metabolic health throughout life.

REFERENCES

[1] A. Conti, and M.H. Bickel, "History of Drug Metabolism: Discoveries of the Major Pathways in the 19th Century", *Drug Metab. Rev.,* vol. 6, no. 1, pp. 1-50, 1977. [http://dx.doi.org/10.3109/03602537708993764]

[2] Y. Li, Q. Meng, M. Yang, D. Liu, X. Hou, L. Tang, X. Wang, Y. Lyu, X. Chen, K. Liu, A.M. Yu, Z. Zuo, and H. Bi, "Current trends in drug metabolism and pharmacokinetics", *Acta Pharm. Sin. B,* vol. 9, no. 6, pp. 1113-1144, 2019. [http://dx.doi.org/10.1016/j.apsb.2019.10.001] [PMID: 31867160]

[3] J. Zhang, R. Wang, and S.A. Pereira, *Drug metabolism and transport: the frontier of personalized medicine.* vol. 14. , 2023.*Front. Pharmacol.,* vol. 14. , 2023. [http://dx.doi.org/10.3389/978-2-8325-3067-2]

[4] A. Najjar, and R. Karaman, "The prodrug approach in the era of drug design", *Expert Opin. Drug Deliv.,* vol. 16, no. 1, pp. 1-5, 2019. [http://dx.doi.org/10.1080/17425247.2019.1553954] [PMID: 30558447]

[5] N. Isoherranen, and K.E. Thummel, "Drug metabolism and transport during pregnancy: how does drug disposition change during pregnancy and what are the mechanisms that cause such changes?", *Drug Metab. Dispos.,* vol. 41, no. 2, pp. 256-262, 2013. [http://dx.doi.org/10.1124/dmd.112.050245] [PMID: 23328895]

[6] G. G. Gibson and P. Skett, "Pathways of drug metabolism," In: *Introduction to Drug Metabolism,* Boston, MA: Springer US, 1994, pp. 1–34.

[7] R.R. Brooks, and E.D. Wither, "Nickel accumulation by Rinorea bengalensis (Wall.) O.K.,", *J. Geochem. Explor.,* vol. 7, pp. 295-300, 1977. [http://dx.doi.org/10.1016/0375-6742(77)90085-1]

[8] V.A. Ramchandani, W.F. Bosron, and T.K. Li, "Research advances in ethanol metabolism", *Pathol. Biol. (Paris),* vol. 49, no. 9, pp. 676-682, 2001. [http://dx.doi.org/10.1016/S0369-8114(01)00232-2] [PMID: 11762128]

[9] O.E. Owen, S.C. Kalhan, and R.W. Hanson, "The key role of anaplerosis and cataplerosis for citric

acid cycle function", *J. Biol. Chem.,* vol. 277, no. 34, pp. 30409-30412, 2002.
[http://dx.doi.org/10.1074/jbc.R200006200] [PMID: 12087111]

[10] Y. Kamiyama, "Effect of obstructive jaundice on the electrophysiological characteristics of the gastric mucosa and gastric acid secretion in rats", *Nippon Geka Gakkai Zasshi,* vol. 91, no. 4, pp. 464-471, 1990.
[PMID: 2377162]

[11] A. Singh, L.B. Prasad, K. Shiv, R. Kumar, and S. Garai, "Synthesis, characterization, and in vitro antibacterial and cytotoxic study of Co(II), Ni(II), Cu(II), and Zn(II) complexes of N-(--methoxybenzyl) N-(phenylethyl) dithiocarbamate ligand", *J. Mol. Struct.,* vol. 1288, p. 135835, 2023.
[http://dx.doi.org/10.1016/j.molstruc.2023.135835]

[12] A. Singh, R. Kumar, K. Shiv, S.K. Pandey, M.K. Bharty, R.J. Butcher, and L.B. Prasad, "Synthesis, crystal structure and screening for anticonvulsant and antianxiety activities of three new Ni(II), Cu(II), and Zn(II) dithiocarbamate complexes", *J. Mol. Struct.,* vol. 1298, no. P2, p. 137052, 2024.
[http://dx.doi.org/10.1016/j.molstruc.2023.137052]

[13] B. K. Singh, "The biosynthesis of tryptophan, tyrosine, and phenylalanine from chorismate," In: *Plant Amino Acids*, T. Aftab and K. R. Hakeem, Eds. Springer, 2020, pp. 185–218.

[14] A. Singh, R. Kumar, R. Patel, R.N. Gautam, M.K. Bharty, and L.B. Prasad, "Dithiocarbamate-based novel anti-histaminic agents: synthesis, characterization, crystal structure and thermal study", In: *Dalt. Trans*, vol. 53, no. 33, pp. 14077–14088, 2024..
[http://dx.doi.org/10.1039/D4DT01777C]

[15] R. Garg, K. Dobhal, B. Kumar, and A. K. Gupta, "COVID-19: a challenge towards the sustainability of health in platform era," NeuroQuantology, vol. 20, no. 11, pp. 2503–2519, 2022.
[http://dx.doi.org/10.14704/NQ.2022.20.11.NQ66250]

[16] T. TeSlaa, M. Ralser, J. Fan, and J.D. Rabinowitz, "The pentose phosphate pathway in health and disease", *Nat. Metab.,* vol. 5, no. 8, pp. 1275-1289, 2023.
[http://dx.doi.org/10.1038/s42255-023-00863-2] [PMID: 37612403]

[17] A. Kumar, B. Kumar, R. Kumar, A. Kumar, M. Singh, V. Tiwari, A. Trigunayat, P. Paul, and P. Singh, "Acute and subacute toxicity study of ethanolic extract of Calotropis procera (Aiton) Dryand flower in Swiss albino mice," Phytomedicine Plus, vol. 2, no. 2, 2022.
[http://dx.doi.org/10.1016/j.phyplu.2022.100224]

[18] A. Stincone, A. Prigione, T. Cramer, M. M. C. Wamelink, K. Campbell, E. Cheung, V. Olin-Sandoval, N.-M. Grüning, A. Krüger, M. T. Alam, M. A. Keller, M. Breitenbach, K. M. Brindle, J. D. Rabinowitz, and M. Ralser, "The return of metabolism: biochemistry and physiology of the pentose phosphate pathway," Biol. Rev. Camb. Philos. Soc. vol. 90, no. 3, pp. 927-963, 2015
[http://dx.doi.org/10.1111/brv.12140.The] [PMID: 25243985]

[19] R. Dhiman, "Novel LC-MS/MS bioanalytical method validation for gabapentin in human plasma in accordance with current regulatory guidances", *Int. J. Pharm. Chem. Sci.,* vol. 4, 2015no. 4, pp. 462-468. Available from: http://www.ijpcsonline.com/files/08-10-15/07-1035.pdf

[20] B.A. Moffatt, and H. Ashihara, "Purine and pyrimidine nucleotide synthesis and metabolism", *The Arabidopsis Book,* vol. 1, p. e0018, 2002.
[http://dx.doi.org/10.1199/tab.0018] [PMID: 22303196]

[21] H. Pandey, S. Srivastava, R. Kumar, and Y. B. Tripathi, "Preclinical acute and repeated dose toxicity of Pueraria tuberosa (PTWE) on Charles Foster rats," Int. J. Pharm. Sci. Res., pp. 4572–4581, 2018. Available from: https://ijpsr.com/bft-article/preclinical-acute-and-repeated-dose-toxicity-of--ueraria-tuberosa-ptwe-on-charles-foster-rats/

[22] L. R. Engelking, "Gluconeogenesis," In: *Textbook of Veterinary Physiological Chemistry*, 3rd ed., Academic Press, 2015, pp. 225–230.
[http://dx.doi.org/10.1016/B978-0-12-391909-0.50037-2]

[23] R.R. Wolfe, "The underappreciated role of muscle in health and disease", *Am. J. Clin. Nutr.,* vol. 84, no. 3, pp. 475-482, 2006.
[http://dx.doi.org/10.1093/ajcn/84.3.475] [PMID: 16960159]

[24] J. Yang, H. Ueharu, and Y. Mishina, "Energy metabolism: A newly emerging target of BMP signaling in bone homeostasis", *Bone,* vol. 138, p. 115467, 2020.
[http://dx.doi.org/10.1016/j.bone.2020.115467] [PMID: 32512164]

[25] R. Dhiman, D. Kumar, B. Kumar, and B.L. Pandey, "Quantitative determination of atorvastatin, ortho-hydroxy atorvastatin, para-hydroxy atorvastatin in human plasma using rosuvastatin as internal standard by LC-MS / MS", *Int. J. Pharm. Chem. Sci.,* vol. 4, no. 4, pp. 487-500, 2016.

[26] C.L. Tan, and Z.A. Knight, "Regulation of Body Temperature by the Nervous System", *Neuron,* vol. 98, no. 1, pp. 31-48, 2018.
[http://dx.doi.org/10.1016/j.neuron.2018.02.022] [PMID: 29621489]

[27] M. Singh, B. Kumar, A. K Singh, C. Shekhar Azad, M. K Yadav, A. Kumar, R. Kumar, A. Kumar, and P. Chaudhary, "Pharmacokinetic screening to estimate the drug likeliness characteristics of selected herbal anticancer drugs", *RJPT,* vol. 16, no. 7, pp. 3422-3426, 2023.
[http://dx.doi.org/10.52711/0974-360X.2023.00566]

[28] M.S. Saghiv, and M.S. Sagiv, "Metabolism", In: *Basic Exercise Physiology: Clinical and Laboratory Perspectives.* Springer International Publishing: Cham, 2020, pp. 33-147.
[http://dx.doi.org/10.1007/978-3-030-48806-2_2]

[29] B. Kumar, R. Kumar, H. Patel, and A. Kumar, "Development and validation of an RP-HPLC method for simultaneous estimation of amlodipine besylate and indapamide in bulk and tablet dosage form", Indo Am. J. Pharm. Sci., vol. 4, no. 11, pp. 4278-4287, 2017. DOI: 10.5281/zenodo.1057601

[30] J. Dunn, and M.H. Grider, "Physiology. Adenosine Triphosphate", *StatPearls*, StatPearls Publishing, 2023.

[31] J.C. Fontecilla-Camps, "The Complex Roles of Adenosine Triphosphate in Bioenergetics", *ChemBioChem,* vol. 23, no. 10, p. e202200064, 2022.
[http://dx.doi.org/10.1002/cbic.202200064] [PMID: 35353443]

[32] B. Kumar, R. Kumar, A. Kumar, and R. Patel, "Simultaneous RP-HPLC method development and validation of fenofibrate and rosuvastatin calcium in bulk and tablet dosage form," *Indo Am. J. Pharm. Sci.*, vol. 4, no. 11, pp. 4288-4297, 2017.

[33] M. Ogrizek, M. Janežič, K. Valjavec, and A. Perdih, "Catalytic mechanism of ATP hydrolysis in the ATPase domain of human DNA topoisomerase IIα," *J. Chem. Inf. Model.*, vol. 62, no. 16, 2022.

[34] O. Brylski, P. Shrestha, P. Gnutt, D. Gnutt, J. W. Mueller, and S. Ebbinghaus, "Cellular ATP levels determine the stability of a nucleotide kinase," Front. Mol. Biosci., vol. 8, 2021.

[35] K. Ajay, N. Manmath Kumar, K. Brijesh, K. Ashutosh, K. Rajesh, K. Vikas, and S. Ambrish Kumar, "Toxicity (acute and subacute) assessment and in-vivo antiurolithiatic activity of ethanolic extract of Caesalpinia bonducella seed in albino Wistar rat", *J. Appl. Pharm. Sci.,* vol. 12, no. 2, pp. 187-197, 2022.
[http://dx.doi.org/10.7324/JAPS.2022.120220]

[36] N.S. Chandel, "Basics of Metabolic Reactions", *Cold Spring Harb. Perspect. Biol.,* vol. 13, no. 8, p. a040527, 2021.
[http://dx.doi.org/10.1101/cshperspect.a040527] [PMID: 34341010]

[37] S. Nath, "Phosphorus chemistry at the roots of bioenergetics: ligand permutation as the molecular basis of the mechanism of ATP synthesis/hydrolysis by FOF1-ATP synthase", *Molecules,* vol. 28, no. 22, p. 7486, 2023.
[http://dx.doi.org/10.3390/molecules28227486] [PMID: 38005208]

[38] N.S. Chandel, "Amino Acid Metabolism", *Cold Spring Harb. Perspect. Biol.,* vol. 13, no. 4, p.

a040584, 2021.
[http://dx.doi.org/10.1101/cshperspect.a040584] [PMID: 33795250]

[39] P.L. Folch, M.M.M. Bisschops, and R.A. Weusthuis, "Metabolic energy conservation for fermentative product formation", *Microb. Biotechnol.,* vol. 14, no. 3, pp. 829-858, 2021. [http://dx.doi.org/10.1111/1751-7915.13746] [PMID: 33438829]

[40] S.M. Rapoport, Medical biochemistry, Veb Verlag Volk Gesundh: Berlin, 1975.

[41] M.J. Buskes, and M.J. Blanco, "Impact of cross-coupling reactions in drug discovery and development", *Molecules,* vol. 25, no. 15, p. 3493, 2020. [http://dx.doi.org/10.3390/molecules25153493] [PMID: 32751973]

[42] R. Kumar, B. Kumar, and A. Kumar, "Stability indicating method development and validation of abiraterone in human plasma by using LC-MS/MS", In: *Asian J. Pharm.* vol. 3. , 2017, no. 5, pp. 12-26.

[43] C. Cardenas, A. Lovy, E. Silva-Pavez, F. Urra, C. Mizzoni, U. Ahumada-Castro, G. Bustos, F. Jaña, P. Cruz, P. Farias, E. Mendoza, H. Huerta, P. Murgas, M. Hunter, M. Rios, O. Cerda, I. Georgakoudi, A. Zakarian, J. Molgó, and J.K. Foskett, "Cancer cells with defective oxidative phosphorylation require endoplasmic reticulum–to–mitochondria Ca 2+ transfer for survival", *Sci. Signal.,* vol. 13, no. 640, p. eaay1212, 2020.
[http://dx.doi.org/10.1126/scisignal.aay1212] [PMID: 32665411]

[44] R. Kumar, B. Kumar, and A. Kumar, "Pharmacokinetic profile of antitubercular drugs and their clinical applications in therapeutic drug monitoring", *Asian J. Pharm.,* vol. 3, 2017no. 5, pp. 145-154. Available from: http://ajpp.in/uploaded/p71.pdf

[45] A. Verma, A. Basak, K. Sampath, and K. Y. Karuna, “Frontier in Biochemistry, Physiology and Food Science,” In: Frontier in Biochemistry, Physiology and Food Science, vol. 1, Cape Comorin Publisher, 2023, pp. 85–101.

[46] B. Alberts, A. Johnson, J. Lewis, M. Raff, K. Roberts, and P. Walter, Molecular Biology of the Cell, 4th ed., New York, USA: Garland Science, 2002.

[47] A. Kumar, "Assessment of pain-ameliorative potential of ethanolic extract of calotropis procera and its active components characterization by hptlc and gc-ms analysis", *In: Int. J. Pharm. Sci. Res*, vol. 13, no. 9, pp. 581-586, 2022.

[48] J. W. Pelley, “Citric acid cycle, electron transport chain, and oxidative phosphorylation,” In: Elsevier’s Integrated Biochemistry, Elsevier, 2007, pp. 55–63.
[http://dx.doi.org/10.1016/B978-0-323-03410-4.50013-4]

[49] R.D. Vijaya Durga, "Determination of S-amlodipine in human plasma using tizanidine as internal standard by LC/MS/MS method", *J. Bioanal. Biomed.,* vol. 7, no. 3, pp. 91-96, 2015.
[http://dx.doi.org/10.4172/1948-593X.1000130]

[50] D. Guo, Y. Meng, X. Jiang, and Z. Lu, "Hexokinases in cancer and other pathologies", *Cell Insight,* vol. 2, no. 1, p. 100077, 2023.
[http://dx.doi.org/10.1016/j.cellin.2023.100077] [PMID: 37192912]

[51] A. Kumar, M.K. Nandi, B. Kumar, A. Kumar, R. Kumar, A.K. Singh, and M. Singh, "Antiurolithiasis, antioxidant, anti-inflammatory, analgesic, and diuretic activity of ethanolic extract of seeds of Caesalpinia bonducella", *Int. J. Pharm. Investig.,* vol. 11, no. 3, pp. 306-311, 2021. [http://dx.doi.org/10.5530/ijpi.2021.3.54]

[52] A.D. Nguyen, G. Nam, D. Kim, and E.Y. Lee, "Metabolic role of pyrophosphate-linked phosphofructokinase pfk for C1 assimilation in Methylotuvimicrobium alcaliphilum 20Z", *Microb. Cell Fact.,* vol. 19, no. 1, p. 131, 2020.
[http://dx.doi.org/10.1186/s12934-020-01382-5] [PMID: 32546161]

[53] W.J. Israelsen, and M.G. Vander Heiden, "Pyruvate kinase: Function, regulation and role in cancer", *Semin. Cell Dev. Biol.,* vol. 43, pp. 43-51, 2015.

[http://dx.doi.org/10.1016/j.semcdb.2015.08.004] [PMID: 26277545]

[54] R. Dhiman, V. Durga, K. Jayasankar, and S. Raj, "Solid phase extraction: Potential carbocisteine bioanalysis using carbocisteine 13c3 stable isotope: LC-MS/MS technique with pharmacokinetics on healthy human subjects", *Pharm. Lett.,* vol. 7, no. 8, pp. 126-134, 2015.

[55] S. Saad, G. Cereghetti, Y. Feng *et al.*, "Reversible protein aggregation is a protective mechanism to ensure cell cycle restart after stress," *Nat. Cell Biol.*, vol. 19, no. 10, pp. 1202–1213, 2017. [http://dx.doi.org/10.1038/ncb3600]

[56] R. Dhiman, V. Durga, and A. J. Rio, "Bioanalytical method for lornoxicam determination in human plasma by using piroxicam as internal standard by LC-MS/MS", *J. Chem. Pharm. Res.,* vol. 7, no. 8, pp. 54-60, 2015.

[57] X. Li, Y. Yang, B. Zhang, X. Lin, X. Fu, Y. An, Y. Zou, J.X. Wang, Z. Wang, and T. Yu, "Lactate metabolism in human health and disease", *Signal Transduct. Target. Ther.,* vol. 7, no. 1, p. 305, 2022. [http://dx.doi.org/10.1038/s41392-022-01151-3] [PMID: 36050306]

[58] S. Park, J.H. Jeon, B.K. Min, C.M. Ha, T. Thoudam, B.Y. Park, and I.K. Lee, "Role of the pyruvate dehydrogenase complex in metabolic remodeling: Differential pyruvate dehydrogenase complex functions in metabolism", *Diabetes Metab. J.,* vol. 42, no. 4, pp. 270-281, 2018. [http://dx.doi.org/10.4093/dmj.2018.0101] [PMID: 30136450]

[59] K. Sumi, Y. Hatanaka, R. Takahashi, N. Wada, C. Ono, Y. Sakamoto, H. Sone, and K. Iida, "Citrate synthase insufficiency leads to specific metabolic adaptations in the heart and skeletal muscles upon low-carbohydrate diet feeding in mice", *Front. Nutr.,* vol. 9, p. 925908, 2022. [http://dx.doi.org/10.3389/fnut.2022.925908] [PMID: 35873436]

[60] O.V. Lushchak, M. Piroddi, F. Galli, and V.I. Lushchak, "Aconitase post-translational modification as a key in linkage between Krebs cycle, iron homeostasis, redox signaling, and metabolism of reactive oxygen species", *Redox Rep.,* vol. 19, no. 1, pp. 8-15, 2014. [http://dx.doi.org/10.1179/1351000213Y.0000000073] [PMID: 24266943]

[61] R. Dhiman, B. Kumar, D. Kumar, and A. Kumar, "Liquid-liquid extraction method for ziprasidone (zrs) bioanalysis by using zrs-d8 (stable isotope) as internal standard", *J. Chem. Pharm. Res.,* vol. 8, no. 3, pp. 722-727, 2016.

[62] K. White, and S. Someya, "The roles of NADPH and isocitrate dehydrogenase in cochlear mitochondrial antioxidant defense and aging", *Hear. Res.,* vol. 427, p. 108659, 2023. [http://dx.doi.org/10.1016/j.heares.2022.108659] [PMID: 36493529]

[63] J.-H. Lee, Y. Go, D.-Y. Kim, S. H. Lee, O.-H. Kim, Y. H. Jeon, T. K. Kwon, J.-H. Bae, D.-K. Song, I. J. Rhyu, I.-K. Lee, M. Shong, B.-C. Oh, C. Petucci, J.-W. Park, T. F. Osborne, and S.-S. Im, "Correction: isocitrate dehydrogenase 2 protects mice from high-fat diet-induced metabolic stress by limiting oxidative damage to the mitochondria from brown adipose tissue", *Exp. Mol. Med.*, vol. 52, p. 12276, 2020. [http://dx.doi.org/10.1038/s12276-020-0451-8]

[64] R.J. Mailloux, "The emerging importance of the α-keto acid dehydrogenase complexes in serving as intracellular and intercellular signaling platforms for the regulation of metabolism", *Redox Biol.,* vol. 72, p. 103155, 2024. [http://dx.doi.org/10.1016/j.redox.2024.103155] [PMID: 38615490]

[65] R. Kumar, A. Kumar, A. Kumar, and M. Singh, "HPTLC and GC-MS analyses of biologically active extracts and fractions from premna latifolia roxb leaves reveal anti-inflammatory and analgesic effects", In: *Sys. Rev. Pharm.*, vol. 13, no. 8, pp. 811-818, 2021.

[66] X. Hou, Y. Chen, X. Li, X. Gu, W. Dong, J. Shi, and S. Ji, "Protein succinylation: Regulating metabolism and beyond," Front. Nutr., vol. 11, 2024.

[67] R. Kumar, B. Kumar, A. Kumar, A. Kumar, and M. Singh, "GC-MS analysis of phytocomponents in the methanol extract of Premna latifolia Roxb", *Pharmacognosy Res.,* vol. 14, no. 1, pp. 19-23, 2021.

[http://dx.doi.org/10.5530/pres.14.1.4]

[68] M. Leshets, Y.B.H. Silas, N. Lehming, and O. Pines, "Fumarase: From the TCA cycle to DNA damage response and tumor suppression", *Front. Mol. Biosci.,* vol. 5, p. 68, 2018. [http://dx.doi.org/10.3389/fmolb.2018.00068] [PMID: 30090811]

[69] C.S. Fermaintt, and S.A. Wacker, "Malate dehydrogenase as a multi-purpose target for drug discovery", *Essays Biochem.,* vol. 68, no. 2, pp. 147-160, 2024. [http://dx.doi.org/10.1042/EBC20230081] [PMID: 38818725]

[70] P. Neupane, S. Bhuju, N. Thapa, and H.K. Bhattarai, "ATP synthase: Structure, function and inhibition", *Biomol. Concepts,* vol. 10, no. 1, pp. 1-10, 2019. [http://dx.doi.org/10.1515/bmc-2019-0001] [PMID: 30888962]

[71] F. Rajas, A. Gautier-Stein, and G. Mithieux, "Glucose-6 phosphate, a central Hub for liver carbohydrate metabolism", *Metabolites,* vol. 9, no. 12, p. 282, 2019. [http://dx.doi.org/10.3390/metabo9120282] [PMID: 31756997]

[72] H. Pandey, S. Srivastava, A. Kumar, R. Kumar, and Y. Bhusan Tripathi, "Pharmacokinetics study of puerarin absorption in blood after consumption of Pueraria tuberosa water extract (PTWE) by rats", *Explor. Anim. Med. Res.,* vol. 9, no. 2, pp. 188-196, 2019.

[73] V. Vi, and T.A. World, *Encyclopedia of Encyclopedia of.* vol. V. Anc. World, 1950, pp. 1-3.

[74] K. Shimizu, "Metabolic regulation of a bacterial cell system with emphasis on *Escherichia coli* metabolism," In: *ISRN*, W. Antonin, C. F. S. Bonafe, and T. B. Ng, Eds., 2013, pp. 1–47. [http://dx.doi.org/10.1155/2013/645983] [PMID: 25937963]

[75] M.C. Petersen, and G.I. Shulman, "Mechanisms of Insulin Action and Insulin Resistance", *Physiol. Rev.,* vol. 98, no. 4, pp. 2133-2223, 2018. [http://dx.doi.org/10.1152/physrev.00063.2017] [PMID: 30067154]

[76] P.L. Kastritis, and A.C. Gavin, "Enzymatic complexes across scales", *Essays Biochem.,* vol. 62, no. 4, pp. 501-514, 2018. [http://dx.doi.org/10.1042/EBC20180008] [PMID: 30315098]

[77] R.L. Smith, M.R. Soeters, R.C.I. Wüst, and R.H. Houtkooper, "Metabolic flexibility as an adaptation to energy resources and requirements in health and disease", *Endocr. Rev.,* vol. 39, no. 4, pp. 489-517, 2018. [http://dx.doi.org/10.1210/er.2017-00211] [PMID: 29697773]

[78] L. Bar-Peled, and N. Kory, "Principles and functions of metabolic compartmentalization", *Nat. Metab.,* vol. 4, no. 10, pp. 1232-1244, 2022. [http://dx.doi.org/10.1038/s42255-022-00645-2] [PMID: 36266543]

[79] Z. Tao, and Z. Cheng, "Hormonal regulation of metabolism—recent lessons learned from insulin and estrogen", *Clin. Sci. (Lond.),* vol. 137, no. 6, pp. 415-434, 2023. [http://dx.doi.org/10.1042/CS20210519] [PMID: 36942499]

[80] R. Kumar, S. Dutt, A. Goyal, A. Kumar, and B. Kumar, "Nanotechnology in gastrointestinal endoscopy," In: *Nanobiotechnology and Artificial Intelligence in Gastrointestinal Diseases*, 2024, pp. 5-1, 5-32. [http://dx.doi.org/10.1088/978-0-7503-6134-7ch5]

[81] P.V. Röder, B. Wu, Y. Liu, and W. Han, "Pancreatic regulation of glucose homeostasis", *Exp. Mol. Med.,* vol. 48, no. 3, p. e219, 2016. [http://dx.doi.org/10.1038/emm.2016.6] [PMID: 26964835]

[82] A. Hamid, R. Kumar, V. K. Chaturvedi, S. Dutt, G. Sulabh, V. Kumar, and D. P. Yadav, "Artificial intelligence (AI) based colonoscopy," In: *Nanobiotechnology and Artificial Intelligence in Gastrointestinal Diseases*, 2024, pp. 12-1, 12-23. [http://dx.doi.org/10.1088/978-0-7503-6134-7ch12]

[83] I. Rix, C. Nexøe-Larsen, N.C. Bergmann, A. Lund, and F.K. Knop, "Glucagon Physiology", In: *South Dartmouth, ,* MA. K. R. Feingold, B. Anawalt, M. R. Blackman, A. Boyce, G. Chrousos, E. Corpas, W. W. de Herder, K. Dhatariya, K. Dungan, J. Hofland, S. Kalra, G. Kaltsas, N. Kapoor, C. Koch, P. Kopp, M. Korbonits, C. S. Kovacs, W. Kuohung, B. Laferrère, M. Levy, E. A. McGee, R. McLachlan, M. New, J. Purnell, R. Sahay, A. S. Shah, F. Singer, M. A. Sperling, C. A. Stratakis, D. L. Trence, D. P. Wilson, Eds., , 2000.

[84] L. Thau, J. Gandhi, and S. Sharma, Physiology. Cortisol: Treasure Island, FL, 2024.

[85] A. Kumar, P. Singh, R. Kumar, and S. Dutt, "Novel drug delivery systems for inflammatory bowel disease," In: Nanobiotechnology and Artificial Intelligence in Gastrointestinal Diseases, 2024, pp. 4-1, 4-24.
[http://dx.doi.org/10.1088/978-0-7503-6134-7ch4]

[86] M. C. Petersen and G. I. Shulman, "Mechanisms of Insulin Action and Insulin Resistance.," *Physiol. Rev.*, vol. 98, no. 4, pp. 2133–2223, Oct. 2018,
[http://dx.doi.org/10.1152/physrev.00063.2017]

[87] A.Y. Yeung, and P. Tadi, Physiology, Obesity Neurohormonal Appetite And Satiety Control, *StatPearls*, StatPearls Publishing, 2023.

[88] A.L. Morris, and S.S. Mohiuddin, *Biochemistry.* Nutrients: Treasure Island, FL, 2024.

[89] E. Dilektasli, "Estimation of the influence of the environmental factors on the accumulation of phytochemicals and antioxidant capacity in the ivy leaves (*Hedera helix* L.)"., *J. Ethnopharmacol.,* vol. 27, no. 3, pp. 1-2, 2007.
[http://dx.doi.org/10.1016/0022-4804(91)90184-N]

[90] R. Kumar, Y. Singh, M. Piplani, P. Bhateja, R. Garg, and B. Kumar, "The chemistry and pharmacological study of Copaifera spp. oils", *Chem. Insid. Spices Herbs Res. Dev.,* no. vol. 38, pp. 79-116, 2024.
[http://dx.doi.org/10.2174/9789815196832124040006]

[91] B. Guha, and A. Ghosh, "A cytomorphological study of the endocrine pancreas of some Indian birds", *Gen. Comp. Endocrinol.,* vol. 34, no. 1, pp. 38-44, 1978.
[http://dx.doi.org/10.1016/0016-6480(78)90241-1] [PMID: 340340]

[92] M. Hatting, C.D.J. Tavares, K. Sharabi, A.K. Rines, and P. Puigserver, "Insulin regulation of gluconeogenesis", *Ann. N. Y. Acad. Sci.,* vol. 1411, no. 1, pp. 21-35, 2018.
[http://dx.doi.org/10.1111/nyas.13435] [PMID: 28868790]

[93] G. Wilcox, "Insulin and insulin resistance", *Clin. Biochem. Rev.,* vol. 26, no. 2, pp. 19-39, 2005.
[PMID: 16278749]

[94] C. Thakur, and F. Chen, "Connections between metabolism and epigenetics in cancers", *Semin. Cancer Biol.,* vol. 57, pp. 52-58, 2019.
[http://dx.doi.org/10.1016/j.semcancer.2019.06.006] [PMID: 31185282]

[95] A.G. Watts, S.E. Kanoski, G. Sanchez-Watts, and W. Langhans, "The physiological control of eating: signals, neurons, and networks", *Physiol. Rev.,* vol. 102, no. 2, pp. 689-813, 2022.
[http://dx.doi.org/10.1152/physrev.00028.2020] [PMID: 34486393]

[96] S. Hædersdal, A. Lund, F.K. Knop, and T. Vilsbøll, "The role of glucagon in the pathophysiology and treatment of type 2 diabetes", *Mayo Clin. Proc.,* vol. 93, no. 2, pp. 217-239, 2018.
[http://dx.doi.org/10.1016/j.mayocp.2017.12.003] [PMID: 29307553]

[97] A.J. Richard, U. White, C.M. Elks, and J.M. Stephens, "Adipose Tissue: Physiology to Metabolic Dysfunction", In: *South Dartmouth, MA*, 2000.K. R. Feingold, B. Anawalt, M. R. Blackman, A. Boyce, G. Chrousos, E. Corpas, W. W. de Herder, K. Dhatariya, K. Dungan, J. Hofland, S. Kalra, G. Kaltsas, N. Kapoor, C. Koch, P. Kopp, M. Korbonits, C. S. Kovacs, W. Kuohung, B. Laferrère, M. Levy, E. A. McGee, R. McLachlan, M. New, J. Purnell, R. Sahay, A. S. Shah, F. Singer, M. A. Sperling, C. A. Stratakis, D. L. Trence, and D. P. Wilson Eds, 2020,

[98] M. Bullón Vela, *Chapter 6 - Obesity and Nonalcoholic Fatty Liver Disease: Role of Oxidative Stress*, 2018, pp. 111-133.

[99] V.M. Gershuni, S.L. Yan, and V. Medici, "Nutritional Ketosis for Weight Management and Reversal of Metabolic Syndrome", *Curr. Nutr. Rep.,* vol. 7, no. 3, pp. 97-106, 2018. [http://dx.doi.org/10.1007/s13668-018-0235-0] [PMID: 30128963]

[100] J. Kailashiya, A. Mukherjee, and D. Dash, "Essentials of medical biochemistry", *Indian J. Med. Res.,* vol. 145, no. 4, pp. 576-577, 2017. [http://dx.doi.org/10.4103/0971-5916.213764]

[101] R. Jeannotte, "Metabolic pathways: nitrogen metabolism", In: *Encycl. Food Microbiol.*, 2014, pp. 544-560. [http://dx.doi.org/10.1016/B978-0-12-384730-0.00199-3]

[102] T. Sato, Y. Yoshida, A. Morita, N. Mori, and S. Miura, "Glycerol-3-phosphate dehydrogenase 1 deficiency induces compensatory amino acid metabolism during fasting in mice", *Metabolism,* vol. 65, no. 11, pp. 1646-1656, 2016. [http://dx.doi.org/10.1016/j.metabol.2016.08.005] [PMID: 27733253]

[103] C.E. Geisler, S. Ghimire, R.L. Bogan, and B.J. Renquist, "Role of ketone signaling in the hepatic response to fasting", *Am. J. Physiol. Gastrointest. Liver Physiol.,* vol. 316, no. 5, pp. G623-G631, 2019. [http://dx.doi.org/10.1152/ajpgi.00415.2017] [PMID: 30767679]

[104] J. Zhang, Y. Zheng, L. Martens, and A.F.H. Pfeiffer, "The regulation and secretion of glucagon in response to nutrient composition: Unraveling their intricate mechanisms", *Nutrients,* vol. 15, no. 18, p. 3913, 2023. [http://dx.doi.org/10.3390/nu15183913] [PMID: 37764697]

[105] S.A. Fernandes, and C. Demetriades, "The multifaceted role of nutrient sensing and mTORC1 signaling in physiology and aging", *Front. Aging,* vol. 2, p. 707372, 2021. [http://dx.doi.org/10.3389/fragi.2021.707372] [PMID: 35822019]

[106] C. Pignatti, S. D'Adamo, C. Stefanelli, F. Flamigni, and S. Cetrullo, "Nutrients and pathways that regulate health span and life span", *Geriatrics (Basel),* vol. 5, no. 4, p. 95, 2020. [http://dx.doi.org/10.3390/geriatrics5040095] [PMID: 33228041]

[107] R.W. Carthew, "Gene regulation and cellular metabolism: An essential partnership", *Trends Genet.,* vol. 37, no. 4, pp. 389-400, 2021. [http://dx.doi.org/10.1016/j.tig.2020.09.018] [PMID: 33092903]

[108] J.D. Mul, K.I. Stanford, M.F. Hirshman, and L.J. Goodyear, "Exercise and regulation of carbohydrate metabolism", *Prog. Mol. Biol. Transl. Sci.,* vol. 135, pp. 17-37, 2015. [http://dx.doi.org/10.1016/bs.pmbts.2015.07.020] [PMID: 26477909]

[109] R.A. DeBose-Boyd, and J. Ye, "SREBPs in lipid Metabolism, insulin signaling, and beyond", *Trends Biochem. Sci.,* vol. 43, no. 5, pp. 358-368, 2018. [http://dx.doi.org/10.1016/j.tibs.2018.01.005] [PMID: 29500098]

[110] S. Kersten, "Integrated physiology and systems biology of PPARα", *Mol. Metab.,* vol. 3, no. 4, pp. 354-371, 2014. [http://dx.doi.org/10.1016/j.molmet.2014.02.002] [PMID: 24944896]

[111] V. Haberle, and A. Stark, "Eukaryotic core promoters and the functional basis of transcription initiation", *Nat. Rev. Mol. Cell Biol.,* vol. 19, no. 10, pp. 621-637, 2018. [http://dx.doi.org/10.1038/s41580-018-0028-8] [PMID: 29946135]

[112] L. A. Pennacchio, W. Bickmore, A. Dean, M. A. Nobrega, and G. Bejerano, "Enhancers: five essential questions", *Nat. Rev. Genet.,* vol. 14, no. 4, pp. 288-295, 2013. [http://dx.doi.org/10.1038/nrg3458]

[113] V. Panwar, A. Singh, M. Bhatt, R.K. Tonk, S. Azizov, A.S. Raza, S. Sengupta, D. Kumar, and M. Garg, "Multifaceted role of mTOR (mammalian target of rapamycin) signaling pathway in human health and disease", *Signal Transduct. Target. Ther.,* vol. 8, no. 1, p. 375, 2023. [http://dx.doi.org/10.1038/s41392-023-01608-z] [PMID: 37779156]

[114] M.M. Mihaylova, and R.J. Shaw, "The AMPK signalling pathway coordinates cell growth, autophagy and metabolism", *Nat. Cell Biol.,* vol. 13, no. 9, pp. 1016-1023, 2011. [http://dx.doi.org/10.1038/ncb2329] [PMID: 21892142]

[115] J. Loscalzo, and D.E. Handy, "Epigenetic modifications: basic mechanisms and role in cardiovascular disease (2013 Grover Conference series)", *Pulm. Circ.,* vol. 4, no. 2, pp. 169-174, 2014. [http://dx.doi.org/10.1086/675979] [PMID: 25006435]

[116] L. Kovatsi, A. Vidaki, D. Fragou, and D. Syndercombe Court, "Epigenetic fingerprint," In: *Personalized Epigenetics*, 2015, pp. 221–243. [http://dx.doi.org/10.1016/B978-0-12-420135-4.00008-5]

[117] J.L. Miller, and P.A. Grant, "The role of DNA methylation and histone modifications in transcriptional regulation in humans", *Subcell. Biochem.,* vol. 61, pp. 289-317, 2013. [http://dx.doi.org/10.1007/978-94-007-4525-4_13] [PMID: 23150256]

[118] X. Li, G. Egervari, Y. Wang, S.L. Berger, and Z. Lu, "Regulation of chromatin and gene expression by metabolic enzymes and metabolites", *Nat. Rev. Mol. Cell Biol.,* vol. 19, no. 9, pp. 563-578, 2018. [http://dx.doi.org/10.1038/s41580-018-0029-7] [PMID: 29930302]

[119] E. Seto, and M. Yoshida, "Erasers of histone acetylation: the histone deacetylase enzymes", *Cold Spring Harb. Perspect. Biol.,* vol. 6, no. 4, p. a018713, 2014. [http://dx.doi.org/10.1101/cshperspect.a018713] [PMID: 24691964]

[120] K.J. Moore, K.J. Rayner, Y. Suárez, and C. Fernández-Hernando, "The role of microRNAs in cholesterol efflux and hepatic lipid metabolism", *Annu. Rev. Nutr.,* vol. 31, no. 1, pp. 49-63, 2011. [http://dx.doi.org/10.1146/annurev-nutr-081810-160756] [PMID: 21548778]

[121] A.M. Mahmoud, "An overview of epigenetics in obesity: The role of lifestyle and therapeutic interventions", *Int. J. Mol. Sci.,* vol. 23, no. 3, p. 1341, 2022. [http://dx.doi.org/10.3390/ijms23031341] [PMID: 35163268]

[122] J. Plaza-Diaz, D. Izquierdo, Á. Torres-Martos, A.T. Baig, C.M. Aguilera, and F.J. Ruiz-Ojeda, "Impact of physical activity and exercise on the epigenome in skeletal muscle and effects on systemic metabolism", *Biomedicines,* vol. 10, no. 1, p. 126, 2022. [http://dx.doi.org/10.3390/biomedicines10010126] [PMID: 35052805]

[123] C. Kandler, A. Zapko-Willmes, J. Richter, and R. Riemann, "Synergistic and dynamic genotypeenvironment interplays in the development of personality differences", The Handbook of Personality Dynamics and Processes, 1st ed. J. F. Rauthmann, Ed., Elsevier, 2021, pp. 155–181. [http://dx.doi.org/10.1016/B978-0-12-813995-0.00007-8]

[124] Y.L. Wu, Z.J. Lin, C.C. Li, X. Lin, S.K. Shan, B. Guo, M.H. Zheng, F. Li, L.Q. Yuan, and Z. Li, "Epigenetic regulation in metabolic diseases: mechanisms and advances in clinical study", *Signal Transduct. Target. Ther.,* vol. 8, no. 1, p. 98, 2023. [http://dx.doi.org/10.1038/s41392-023-01333-7] [PMID: 36864020]

[125] O. Ramos-Lopez, "Epigenetic Biomarkers of Metabolic Responses to Lifestyle Interventions", *Nutrients,* vol. 15, no. 19, p. 4251, 2023. [http://dx.doi.org/10.3390/nu15194251] [PMID: 37836535]

[126] Y. Cheng, C. He, M. Wang, X. Ma, F. Mo, S. Yang, J. Han, and X. Wei, "Targeting epigenetic regulators for cancer therapy: mechanisms and advances in clinical trials", *Signal Transduct. Target. Ther.,* vol. 4, no. 1, p. 62, 2019. [http://dx.doi.org/10.1038/s41392-019-0095-0] [PMID: 31871779]

[127] Y. Li, S.N. Saldanha, and T.O. Tollefsbol, "Impact of epigenetic dietary compounds on

transgenerational prevention of human diseases", *AAPS J.,* vol. 16, no. 1, pp. 27-36, 2014. [http://dx.doi.org/10.1208/s12248-013-9538-7] [PMID: 24114450]

[128] S.T. Keating, and A. El-Osta, "Epigenetics and Metabolism", *Circ. Res.,* vol. 116, no. 4, pp. 715-736, 2015.
[http://dx.doi.org/10.1161/CIRCRESAHA.116.303936] [PMID: 25677519]

CHAPTER 2

Understanding Incomplete Metabolism

Anjna Sharma[1], **Hitesh Chopra**[2,*], **Sunil Dutt**[3] and **Rajesh Kumar**[3]

[1] *Chitkara University School of Pharmacy, Chitkara University, Solan, Himachal Pradesh, India*

[2] *Department of Biosciences, Saveetha School of Engineering, Saveetha Institute of Medical and Technical Sciences, Chennai, Tamil Nadu, India*

[3] *Maharaja Agrasen School of Pharmacy, Maharaja Agrasen University, Solan, Himachal Pradesh, India*

Abstract: An incomplete metabolism is a state in which the body is unable to completely break down certain chemicals, which results in the buildup of hazardous metabolites. In addition to enzyme deficits and hereditary problems, this condition may also be brought on by other causes that interfere with the normal metabolic processes. There are many different systems in the body that may be affected by incomplete metabolism. The symptoms that might occur from incomplete metabolism can vary greatly depending on the exact metabolic disease. Here, we will talk about the significance of early intervention in preventing or minimizing long-term problems, as well as the role that genetic testing plays in determining the precise metabolic abnormality. The chapter will also emphasize the difficulties associated with regulating incomplete metabolism as well as the necessity of conducting further research in order to produce medications that are more successful.

Keywords: Alcohol metabolism, Incomplete metabolism, Maple syrup urine disease, Phenylketonuria.

INTRODUCTION

The body of an individual is subjected to a variety of external elements, including foodstuffs, poisons, dirt, airborne pollutants and sometimes some medicines [1]. All of these alien particles are referred to as xenobiotics. "Xeno" refers to something that is not native to our body and is therefore foreign [2]. The body has a system to expel these foreign particles once they enter it. This process is referred to as metabolism. Since medications are alien particles, they are also considered xenobiotics. Another name for the drug metabolism process is biotransformation [3]. Certain medications are metabolized to create hazardous byproducts which

* **Corresponding author Hitesh Chopra:** Department of Biosciences, Saveetha School of Engineering, Saveetha Institute of Medical and Technical Sciences, Chennai, Tamil Nadu, India; E-mail: chopraontheride@gmail.com

can cause cancer or mutations. The body's primary location of metabolism is the liver. Some medications undergo first-pass metabolism before entering the bloodstream; this is known as the "first-pass effect." The gastrointestinal tract, skin, kidneys, and other organs participate in drug metabolism. Drug metabolism occurs in two stages, known as phase I and phase II metabolic processes [4].

Hydroxylation, dealkylation, various oxidations, reduction, and hydrolysis are examples of phase-I processes. The main Phase-I enzymes are members of the cytochrome P450 family of enzymes (CYPs). The conjugation of several accessory groups, such as glutathione and glucuronic acid, to drug molecules occurs during Phase-II reactions. This process frequently results in improved water solubility, which aids in excretion [5].

COMPLETE *VS.* INCOMPLETE DRUG METABOLISM

The term "complete metabolism" refers to the processes involving the complete conversion of a drug into its metabolites to facilitate the excretion of a drug from the body easily by utilizing Phase I and Phase II reactions [6]. This involves the breakdown of metabolite into a water-soluble compound to be excreted through urine and bile. Whereas in incomplete metabolism drug is not fully converted to its metabolite, leaving the parent drug or some partially formed metabolites within the body. This can be due to saturation of enzymes, drug interactions, genetic factors like polymorphisms in metabolic enzymes and some formulation factors [7].

A drug's duration of action, the possibility of the drug or its metabolites building up, and the danger of adverse effects or toxicity can all be brought on by incomplete metabolism (Fig. **1**). To control these hazards, dose adjustments and monitoring could be required [8].

CHARACTERISTICS OF INCOMPLETE METABOLISM

Incomplete metabolism can have several implications for the effectiveness of a drug, side effects or toxic effects of a drug, as well as the overall response of a patient to a drug [9]. The key characteristics include the elevation of the pharmacological effect of a drug like losartan, clopidogrel & tamoxifen, which needs to be metabolized to an active form or prodrug. The incomplete metabolism or activation may lead to reduced pharmacological effect and extended duration of action and accumulation.

Incomplete metabolism refers to an incomplete breakdown of food, which leads to the generation of intermediate compounds that are toxic to human health. Normally, inside the human body, there are majority of processes that take place simultaneously to complete the process of metabolism [10]. These processes are often governed by bioenergetics and require proper oxygen supply, but no adequate oxygen to influence such processes may lead to incomplete metabolism. There are several factors that cause incomplete metabolism, including alcohol metabolism and Acetaldehyde accumulations, Phenylketonuria, maple syrup urine disease, and galactosemia [11, 12].

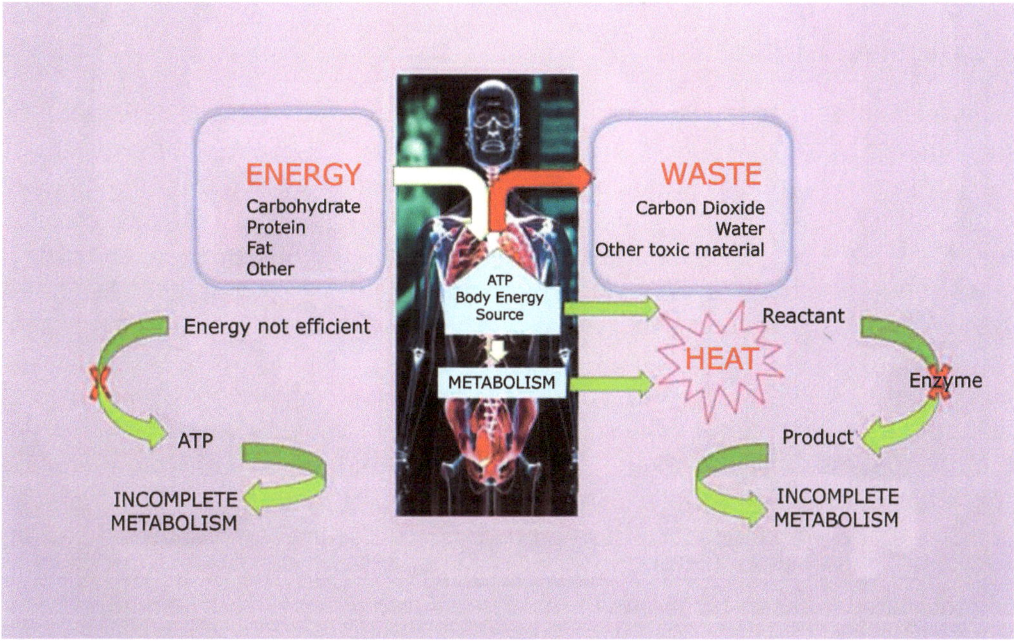

Fig. (1). Process of complete and incomplete metabolism.

Alcohol Metabolism and Acetaldehyde Accumulation

The liver is the major site for Alcohol metabolism, which occurs through the involvement of liver enzyme known as alcohol dehydrogenase (ADH), which converts ethanol to acetaldehyde. Further, the metabolism of acetaldehyde involves enzymatic reduction with the help of Aldehyde dehydrogenase (ALDH), which changes the acetaldehyde to acetate [13]. Excessive alcohol intake or ALDH deficiency causes incomplete alcohol metabolism leading to acetaldehyde accumulation. Additionally, incomplete metabolism of alcohol is commonly seen in individuals with variation in ALDH2 genes, causing less enzymatic activation [14, 15]. This triggers the toxic effects responsible for the symptoms like nausea, flushing and liver damage and also causes cancer [16]. The chronic accumulation however, leads to the generation of raw oxygen stress, inflammation and severe health conditions like hepatitis and cirrhosis. The acetaldehyde has carcinogenic properties that cause serious health hazards. The accumulation of acetaldehyde leads to DNA and protein adducts, damaging cellular structure and functions [17]. These also contribute to the development of alcoholic liver diseases. Moreover, these toxic bi-products of alcohol also disrupt mitochondrial functions by disrupting bioenergetic processes such as the electron transport chain [18]. The acetaldehyde accumulation not only impacts the cellular pathways, but the impact of these disruptions can be seen in the brain also. The impact targeting the brain arises due to alcohol-induced neurotoxicity [19].

Phenylketonuria (PKU)

Phenylketonuria is a genetic disorder that is often caused by a deficiency of phenylalanine hydroxylase, which is necessary for the conversion of phenylalanine to tyrosine [20]. This results in the accumulation of phenylalanine in the blood and brain. This triggers neurotoxicity and various clinical manifestations [21]. The incomplete metabolism of this amino acid disrupts normal brain functioning. Phenylalanine is important for protein synthesis but a high level of it triggers neurotoxicity [22]. Additionally, these high levels interfere with amino acid transport across the brain barrier, leading to a deficiency of catecholamines, which in turn leads to cognitive impairment, neurological symptoms and behavioral issues [23, 24]. Another major consequence of the incomplete metabolism of phenylalanine is the formation of phenylpyruvate and other phenyl ketones, leading to the development of neurotoxic characteristics [25, 26]. If phenylketonuria is left untreated, it can lead to hypomyelination, microcephaly and intellectual disability. The severity of PKU vary individual to individual and depends on the level of enzyme deficiency [27]. The management of PKU involves strict dietary control, which limits phenylalanine intake associated with neurotoxicity. Early detection of these incomplete metabolic processes at the newborn stage is crucial for the prompt initiation of a low-phenylalanine diet, which helps mitigate the adverse effects of metabolic dysfunction. In recent years, treatment involving enzyme replacement therapy and the use of supplements can enhance the AH activity, which helps to manage the phenylamine levels in PKU patients [28, 29].

Maple Syrup Urine Disease (MSUD)

MSUD is an inherited disorder that often occurs due to branched-chain alpha ketoacid dehydrogenase complex. This enzyme complex is important for the breakdown of branched amino acids like leucine, isoleucine and valine [30]. This condition results from genetic mutations affecting the encoding components of the branched-chain α-keto acid dehydrogenase complex, thereby impairing the catabolism of branched-chain amino acids (BCAAs) [31]. The incomplete or dysfunctional metabolism of these amino acids causes accumulation in the body and leads to severe neurological and physical symptoms. The pathology of MSUD involves BCKD complex, which cannot influence decarboxylation at branched chain alpha-keto acids, byproducts of BCAA metabolism [32, 33]. This often results in toxic intermediate accumulations in brain, leading to encephalopathy, a life-threatening condition which is characterized by swelling of brain and therefore neurological decline. This accumulation results in symptoms like maple syrup like odour in urine and sweat. The incomplete metabolism also leads to systemic disruptions beyond BCAAs and their products. These intermediate alterations in lipid metabolism reveal that the disorders affect in the worse way than the normal amino acid catabolism. The adolescent MSUD patients also showed improper metabolism, which leads to lipid homeostasis [34, 35].

Galactosemia

Galactosemia is a rare genetic disorder caused by a deficiency of enzyme (glactose-1-phosphhate uridylyl transferase (GALT)) responsible for galactose metabolism, a sugar primarily found in milk and dairy products including [36]. GALT plays a crucial role in Leloir pathway, which is a metabolic route that converts galactose into glucose. The incomplete metabolism of galactose often arises due to GALT deficiency or dysfunction, which also causes its accumulation that triggers toxic effects [37]. Galactosemia has significant consequences in newborns who completely rely on milk as a primary source of nutrition [38]. The galactose itself doesn't accumulate, but it accumulates in the form of galactose- 1-phosphate, which is toxic to tissues and highly perfused organs [39]. This toxicity can cause severe symptoms like jaundice, renal failure and hepatomegaly in newborns and in the long term, these can cause cataract, speech difficulties and ovarian insufficiency in the older population. The major problem of galactosemia is that it occurs due to incomplete metabolism, which leads to galactose buildup in tissues, causing harmful responses to cellular processes by disrupting them. These metabolic blocks also cause an increased level of galactitol, which is a sugar alcohol derived from galactose, targeting specifically the eye and causing cataracts [40]. Even after dietary changes that involve strict avoidance of these milk sugars, patients still experience long-term complications. These complications show that the toxic effects of galactose accumulations are not fully comprehended by dietary restrictions alone. The exact molecular mechanism of galactose and its influence in galactosemia is not well understood yet, but it likely involves multiple pathways, which may include oxidative stress, incomplete protein glycosylation and bioenergetic processes at the cellular level [41, 42]. This disorder is being treated by gene therapy and enzyme replacement therapy and involves biomarker inhibition that bypasses metabolic block and mitigates the effect of this disorder [43].

PHYSIOLOGICAL EFFECTS OF INCOMPLETE METABOLISM

Incomplete metabolism causes metabolic pathway disruptions by which they are unable to process the substrates, including fatty acids, glucose, and amino acids, and also leads to intermediate metabolite accumulations which trigger physiological consequences [44]. These disruptions also effect various tissues and organs, contributing to being the host of many diseases and metabolic disorders. The influence of these disorders, in turn, leads to physiological effects that are significant in conditions like inborn error of metabolism, diabetes and mitochondrial disturbances [45].

Accumulation of Intermediate Metabolites

The most serious consequence of incomplete metabolism is the buildup of intermediate metabolites, which are normally converted into final products and eliminated from the body [46]. These intermediate products are toxic to cells, leading to cellular damage and dysfunction. This inability to expel the intermediate product from the body leads to the accumulation of intermediate products which disrupt various cellular processes (Fig. **2**). In some cases, such as in diabetes, the excess of glucose, if not metabolized properly, can cause the formation of advanced glycation end-products (AGEs), that are harmful and can damage lipids, protein and DNA [47]. These AGEs contribute toward the development of neuropathy, retinopathy and nephropathy. Additionally, the intermediate build-up causes feedback inhibitions, where an excessive build-up of these substances inhibits enzymatic activity in the pathway and also exacerbates the metabolic dysfunctions [48].

Energy Production Inefficiencies

Energy production is a critical metabolic function that involves various bioenergetic pathways like glycolysis, citric acid cycle and oxidative phosphorylation. All of these pathways are involved in the formation and generation of ATP. The incomplete metabolism impairs these pathways and leads to inefficient energy production. Additionally, mitochondrial disorder, mutations in mitochondrial DNA or nuclear genes that specifically encode for mitochondrial proteins can impair the *etc.* functions and reduce ATP formation [49]. This energy deterioration leads to fatigue, muscle weakness and organ dysfunctions. In the diabetes, insulin resistance or deficiency leads to incomplete oxidation of glucose, which forces cells to rely on fatty oxidation. This shift leads to ketone bodies accumulation, resulting in ketoacidosis [50]. Energy production is also evident in cancer cells, which completely rely on aerobic glycolysis instead of phosphorylation, causing accumulation of lactate, which further contributes to tumor acidic microenvironment [51].

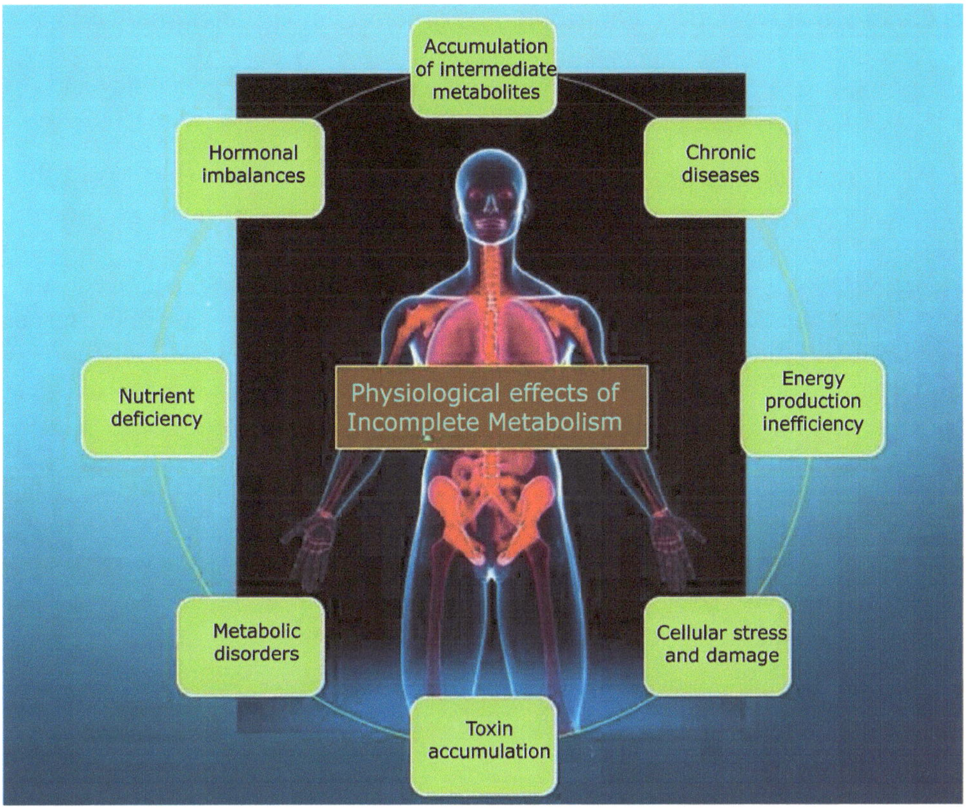

Fig. (2). Illustrating physiological effects of incomplete metabolism in the human body.

Cellular Stress and Damage

The incomplete metabolism leads to increased cellular stress, such as oxidative stress, which arises when there is an imbalance between Reactive Oxygen Species generation and their detoxifications. The ROS also generated as a byproduct of cellular metabolism during mitochondrial oxidative phosphorylation [52]. However, if metabolic pathways are incomplete, it will lead to overproduction of ROS, which leads to oxidative damage to protein, lipid and DNA. For instance, in fatty acid oxidation, incomplete metabolism leads to fatty acid accumulation, which leads to lipid peroxidation. This development of oxidative stress leads to damage to the cellular membranes, impacts enzymatic functions and triggers apoptosis. In neurodegenerative diseases, incomplete metabolism of neurotransmitters and energy production failure may lead to neuronal death and progressive cognitive decline [53]. Additionally, the accumulation of toxic intermediates can activate inflammatory pathways that contribute to chronic inflammation. These are particularly evident in metabolic syndrome, in which incomplete metabolism of glucose leads to activation of pro-inflammatory cell mediators such as cytokines, contributing to atherosclerosis, insulin resistance and cardiovascular complications [54, 55].

INCOMPLETE METABOLISM DIAGNOSIS AND DETECTION

Early detection and treatment of inborn errors of metabolism (IEMs) rely on prompt neonatal screening and genetic profiling. Delay in diagnosis might lead to sudden metabolic decompensation, gradual neurological damage, or even death [56]. DBS testing is used to screen newborns for IEMs in the majority of industrialized countries, as well as in many developing ones. In 1980s, targeted metabolomics using tandem mass spectrometry (MS/MS) was launched as a reliable method for newborn screening. This technology was chosen because it has the ability to screen a variety of IEMs were previously not screened. Furthermore, it only requires a single test conducted on a dried blood spot (DBS). Newborn screening programs, initially introduced in public health care, have since grown to encompass over 50 hereditary metabolic diseases. These illnesses include amino acidopathies, organic acidemias, fatty acid oxidation disorders, and lysosomal storage disorders [57]. Although newborn screening is highly valuable in clinical settings and plays a crucial role in public health care, it only provides a limited overview of a small number of metabolic changes. Additionally, conventional newborn screening programs may not be able to detect some IEMs.

The laboratory method for screening IEMs may be described in four steps: (I) a general metabolic screening test; (II) specialized assays to measure metabolites; (III) investigations on enzymes; and (IV) DNA analysis. At present, MS/MS methods are often employed for newborn screening of inborn errors of metabolism (IEMs) because of their promptness, sensitivity, and specificity. These techniques need small sample volumes and allow for efficient processing of a large number of samples [58].

In addition, the sole utilization of tailored metabolomic tests is confined to certain sets of substances for chosen IEM illnesses. The issue with focused techniques is the occurrence of incorrect negative and positive diagnoses. This problem might be resolved by expanding the spectrum of analytics used in metabolic screening. The selection of particular studies for an individual patient is now determined based on their clinical symptomatology. Nevertheless, this approach is very dependent on the accuracy and comprehensiveness of the clinical data sent to the metabolic laboratory [59]. Consequently, there is a potential for false-negative results if a metabolic test is not conducted due to an inadequate description of the patient's symptoms. Diagnostic profiling in the metabolic laboratory is essential for accurately diagnosing individual patients with IEMs that are not covered by newborn screening.

In addition, IEMs might exhibit a variety of symptoms and metabolites that overlap with each other. The identification of new metabolic abnormalities is likely hindered by focusing exclusively on a narrow range of established metabolic pathways [60, 61].

Newborn screening procedures enhance the identification of IEM; however they should not be solely relied upon. False-positive and false-negative screening tests can occur due to early screening, insufficient exposure to protein or carbohydrates, medicines, or transfusions [62, 63]. Furthermore, it is worth noting that the screening test results may not be accessible during the initial days of an individual's existence, a critical period when some IEMs may manifest. In addition, newborn screening programs do not provide screening for all IEMs. For instance, the urea cycle abnormalities known as ornithine transcarboxylase deficit and carbamoyl phosphate synthetase deficiency cannot be identified with the newborn screening tools now available [64]. Enhanced detection relies on a strong level of suspicion. The American College of Medical Genetics provides ACTion (ACT) sheets that include guidelines for evaluating aberrant results from newborn screening [65]. The state newborn screening program frequently provides these guidelines and other useful information, which may be accessed through their websites.

MANAGEMENT AND TREATMENT STRATEGIES

Dietary Strategies

The objective of dietary treatment is to achieve the normalization of branched-chain amino acids (BCAAs), with a specific focus on leucine, by limiting the consumption of BCAAs without negatively affecting growth and cognitive development [66]. Long-term adherence to dietary treatment is necessary. There are several baby formulae [such as BCAD 1] and meals for all ages (such Nutricia products and MSUD Express) that may be purchased without BCAAs or with lower amounts of BCAAs. Continuous dietary assistance is essential. Patients with maple syrup urine illness determine their leucine consumption based on individual measurements of plasma branched-chain amino acids (BCAAs). Regularly monitoring plasma levels of branched-chain amino acids (BCAAs) at suitable intervals during the first 6-12 months of life is recommneded. Alongside nutritional treatment, it is advisable to provide thiamine [10-20 mg/d] for a duration of 4 weeks in order to assess the individual's thiamine response [67].

Due to the absence of universally recognized standards, the approach to managing PKU differs among nations. Nevertheless, it is widely acknowledged that dietary control and/or pharmacologic therapy are crucial [68]. The primary component of the diet is limiting phenylalanine consumption and providing other important amino acids, vitamins, minerals, and energy through the use of medical foods and low-protein meals [69]. It is imperative to abstain from consuming aspartame. Phenylalanine is a key constituent of aspartame and is present in several artificially sweetened foods and beverages, as well as certain vitamins and medications. A 12-ounce can of diet drink sweetened with aspartame contains around 105 milligrams of phenylalanine, which is equivalent to 25-50% of the typical daily dose. The majority of neonates with PKU need a daily dietary intake of phenylalanine ranging from 40 mg/kg/d to 60 mg/kg/d in order to ensure normal growth and development. Continuing to breastfeed is generally feasible and should only be discontinued if explicitly advised by a local health authority or treatment center. As the pace of growth decreases, the amount of phenylalanine needed decreases as well [70]. Most older children and adults may manage a daily intake of 200 mg to 400 mg of phenylalanine. In order to avoid a lack of the necessary amino acid phenylalanine, it is crucial to supply the body with a sufficient amount of this natural compound. The diet necessitates the complete removal of all high-protein items, including meat, dairy, nuts, and legumes. Starchy foods such as bread, potatoes, maize, and beans need to be limited as well (a single slice of bread or a modest serving of fries has around 120-150 mg of phenylalanine). Medical foods are necessary to complement essential amino acids, vitamins, and minerals. The majority of these substances are ingested in the form of a powdered substance that is dissolved in a liquid, such as formula. Supplements such as capsules, amino acid bars, and amino acids incorporated into meals are becoming accessible (Fig. **3**).

Therapy Involving Enzyme Replacement

It is commonly employed to substitute a lacking or inadequate enzyme in individuals with a hereditary enzyme deficiency condition. Usually, the enzyme is altered to increase its duration of action, enhance its effectiveness, make it more resistant to breakdown, or direct it to a particular organ, tissue, or cell type. The initial efficacious enzyme replacement therapy used the utilization of plasma-derived pure human alpha-1-antitrypsin [A1AT] to treat A1AT deficiency. A1AT deficiency is linked to the development of emphysema at a young age due to the absence of leukocyte elastase inhibitor, resulting in gradual damage to the lungs. Preliminary investigations indicated that augmentation treatment may be advantageous by increasing the levels of A1AT in the bloodstream by the infusion of the enzyme extracted from human serum. Subsequent research demonstrated the therapeutic effectiveness of this treatment, especially in individuals with mild to moderate lung impairment. Additionally, it was found to be safe since it did not result in viral hepatitis, despite being derived from human plasma [64 - 66].

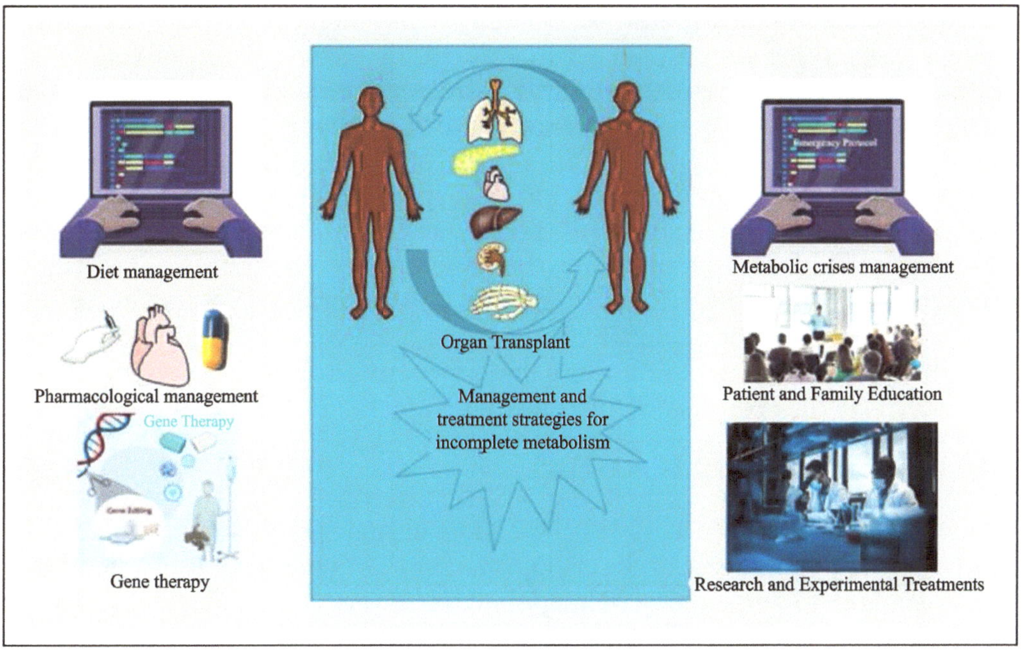

Fig. (3). Management and treatment strategies for incomplete metabolism.

Another effective kind of enzyme replacement treatment was developed for Gaucher disease, a genetic condition characterized by a lack of lysosomal acid β-glucocerebrosidase enzyme. Following that, modified versions of glucocerebrosidase have been created and are currently considered the established treatment for type 1 Gaucher illness.

Enzyme replacement treatment operates on the premise that the patient is administered the deficient enzyme by frequent intravenous infusion. The enzyme idursulfase, used for treating Hunter syndrome, is produced using recombinant gene technology in a continuous human cell line. Clinical trials were begun in people with Hunter syndrome based on the promising outcomes shown in animal tests [63]. Following many months of therapy, a notable drop in the excretion of urine glycosaminoglycan, as well as a reduction in the size of the liver and spleen, were noted. Idursulfase was approved in 2006 as an enzyme replacement medication for those suffering from Hunter syndrome. The medication is given through a vein at a dosage of 0.5 milligrams per kilogram of body weight once a week [71, 72]. Findings from the Hunter Outcome Survey (HOS), a worldwide registry for hunters, indicate that the significant advantages shown in clinical trials, such as enhanced performance in the 6-minute walk test, improved lung function, and expanded shoulder range of motion, were maintained over time. Enzyme replacement treatment, albeit effective, has several limits and cannot address all elements of the illnesses equally. This medication does not elicit a response from some organs, such as cartilage, bone, and heart valve [73]. In addition, some signs of Hunter syndrome are irreversible even with prolonged therapy. When an enzyme is administered intravenously, it is unable to cross the blood-brain barrier, so it does not affect the functioning of the central nervous system [74]. The administration of idursulfase is often well-tolerated, and the occurrence of side effects, such as fever, headache, cough, respiratory tract infections, and diarrhea, aligns with the anticipated consequences observed in untreated individuals with Hunter syndrome. The majority of adverse events due to infusion are of mild or moderate intensity and often occur between weeks 4 and 12 of therapy. Overall, idursulfase has a risk-benefit profile that is comparable to that observed in other forms of replacement therapy [3, 70, 75].

PUBLIC HEALTH IMPLICATIONS

With an assessed prevalence of 3.4:1000 and a carrier identification of 6.8:1000, the ENS [Expanded Newborn Screening] database has the benefit of identifying over 50 IEM, endocrinopathies, hemoglobinopathies, and other illnesses. This is a significant advantage as it demonstrates how important it is to build a screening program that could be administered to every infant in Mexico's whole country. When other findings from the past are taken into consideration, this conclusion outweighs the expectations. An investigation that was carried out in the Mexican state of Nuevo Leon revealed that the incidence of IEM was 1 in 5000, taking into account just 30 disorders that were screened by mass spectrophotometer in a sample size of 42,264 newborns. In contrast, our comparison findings revealed that the incidence was 1 in 2500. When linked to the rate of 1 in 295 that was found in this study, other nations, such as the United States, have reported an incidence rate that is as high as 1 in 500. This is in contrast to the over 50 IEM, endocrinopathies, hemoglobinopathies, and other illnesses shown in the study. On the other hand, when particularly looking for G6PD, the incidence that was observed was roughly one in four hundred, which is lower than the incidence rate that was identified by Lin et al., who stated that it was one in ninety.

Depending on the number of illnesses that are tested for and the technology that is utilized for their detection, the prevalence of a variety of disorders, including IEM, endocrinopathies, hemoglobinopathies, and others, may experience variations. The following are some other variables that should be taken into consideration: variances in genotypes between populations and races, consanguinity, differences in cutoff values between laboratories, and varying sample volumes.

When investigating the disorders that were screened and the procedures that were utilized, a group of researchers from Mexico concluded that there was a significant amount of variation among the medical facilities. It is impossible to make a comparison between the data, and determining the true illness incidence is nothing more than a question of guesswork. During the second-tier testing, additional mutation detection was performed for cystic fibrosis, metabolic syndrome, as well as amino acid and acylcarnitine profiles. A closer calculation of the actual incidence of these disorders in the Mexican hospital may be obtained by the utilization of this protocol, which allows for the identification of a greater number of diseases than previous newborn screening programs.

The G6PD was the illness that was found the most commonly, which is consistent with the worldwide incidence records that were generated in the past. In addition, this research identified carriers, mostly for hemoglobinopathies, which is advantageous for the families since they were able to obtain genetic counseling as a result of this discovery. Additionally, the significant detection of TNT was also found, which accounted for 14 additional cases throughout the investigation. In the short term, this condition is considered to be harmless; nevertheless, there is data that suggests that, in the long run, it may produce certain learning impairments. Once more, the significance of early detection and the accurate collection of all data linked to the status of the baby is emphasized. These are the factors that enable one to correctly identify patients in order to offer sufficient treatment for these newborns. There are still flaws in the sample collection that require immediate attention and correction. These inaccuracies include the procedure, shipment, and storage of the samples, as well as the accurate reporting of an infant's age, food, medicines, transfusions, ethnicity, and other relevant information. However, it is well knowledge that the objective of screening is to dramatically cut down on the percentage of false-negative findings, which ultimately leads to an increase in the number of false-positive results. Regarding the follow-up, once a good result is identified, the education of the parents becomes a priority. This allows for the implementation of an appropriate diet and treatment alternatives, which helps to avoid extra stress and prevents lasting damage or even death. As a result, afflicted individuals will be able to receive treatment in a timely manner. Preventive actions were done in a timely way for all of the G6PD instances that were presented in the cases that were presented. Genetic counseling was offered to each and every family that had a newborn who had been diagnosed.

It is, therefore, necessary for patients who have inborn errors of metabolism to have follow-up care on a regular basis in order to effectively manage their illness. In order to establish and determine the optimal level of metabolic management for certain patients, monitor them as frequently as once per week. It is possible that some patients who are stable will need to be monitored every three to four months. Patients who have disorders that are more stable are visited every six months, nine months, or annually. The medical care of the child's disease falls within the purview of the metabolic geneticist or genetic epidemiologist. In addition to doing a medication review, the nursing staff is responsible for providing the initial clinical examination, which includes the child's vital signs and growth metrics information. Education about the proper administration of medicine is one of the ways in which our clinical pharmacist offers help to our families. This is because many metabolic problems may be managed by making changes to the patient's diet. When it comes to gaining access to resources and specialized services, the social worker assists families that require further assistance.

CONCLUSION

The development of gene therapy, which is centered on the insertion of genes and editing of genes, has occurred concurrently with the advancement of technology in the field of engineering nucleic acids, nucleases, and viruses. Important early milestones have given rise to a great deal of optimism regarding inherited metabolic disorders (IMDs) that, according to the current standard of treatment, offer little to no therapeutic benefit. These complex biotechnologies, which include gene addition mediated by adeno-associated viral vectors (AAV) and integrating vectors that rely upon lentiviral and CRISPR-Cas9-driven gene-editing platforms, are currently the foundation upon which licensed therapeutic solutions for monogenic illnesses are built. In a great number of uncommon IMDs, the use of gene therapy has been investigated. This has made it possible to develop a wide range of interesting new medicines for genetic and metabolic illnesses. Even though first-in-man clinical trials are growing, there are still many obstacles to overcome in this fast-developing sector. These obstacles include the creation of safer and more effective vectors, the development of technologies that are more easily available, and the creation of new regulatory paradigms that will speed up approvals. Today, a one-size-fits-all method remains difficult for most illnesses given that even within a rare IEM patient group, phenotypic variability, unpredictable disease course and uncertainty surrounding the natural history can further complicate the risk-benefit balance for clinical trials.

REFERENCES

[1] B. Fredricsson, K. Carlström, B. Kjessler, J. Lindstedt, L. Plöen, M. Ritzén, and B. de la Torre, "Incomplete androgen insensitivity: Asymmetry in morphology and steroid profile and metabolism of the gonads. An analysis of a case.," *Acta Endocrinol (Copenh)*, vol. 110, no. 4, pp. 564–571, 1985.

[2] T.G. Hornby, C.R. Kinnaird, C.L. Holleran, M.R. Rafferty, K.S. Rodriguez, and J.B. Cain, "Kinematic, muscular, and metabolic responses during exoskeletal-, elliptical-, or therapist-assisted stepping in people with incomplete spinal cord injury," *Phys. Ther. Rehabil. J.*, vol. 92, no. 10, pp.

1278–1291, 2012.
Available from: https://academic.oup.com/ptj/article/92/10/1278/2735127

[3] Luonan Chen, XSun Zhang, Lingyun Wu, and Yong Wang, *7th Int. Conf. Syst. Biol. (ISB),* 2013

[4] H. Dong, Y. Sun, L. Nie, A. Cui, P. Zhao, W.K. Leung, and Q. Wang, "Metabolic memory: mechanisms and diseases", *Signal Transduct. Target. Ther.,* vol. 9, no. 1, p. 38, 2024. [http://dx.doi.org/10.1038/s41392-024-01755-x] [PMID: 38413567]

[5] A.L. McCall, J.S. Dhindsa, A.M. Bailey, L.A. Pucci, L.M. Strickland, and M.K. ElMallah, "Glycogen accumulation in smooth muscle of a Pompe disease mouse model", *J. Smooth Muscle Res.,* vol. 57, no. 0, pp. 8-18, 2021.
[http://dx.doi.org/10.1540/jsmr.57.8] [PMID: 33883348]

[6] E. Holmes, J. Wist, R. Masuda, S. Lodge, P. Nitschke, T. Kimhofer, R.L. Loo, S. Begum, B. Boughton, R. Yang, A.C. Morillon, S.T. Chin, D. Hall, M. Ryan, S.H. Bong, M. Gay, D.W. Edgar, J.C. Lindon, T. Richards, B.B. Yeap, S. Pettersson, M. Spraul, H. Schaefer, N.G. Lawler, N. Gray, L. Whiley, and J.K. Nicholson, "Incomplete Systemic Recovery and Metabolic Phenoreversion in Post-Acute-Phase Nonhospitalized COVID-19 Patients: Implications for Assessment of Post-Acute COVID-19 Syndrome", *J. Proteome Res.,* vol. 20, no. 6, pp. 3315-3329, 2021. [http://dx.doi.org/10.1021/acs.jproteome.1c00224] [PMID: 34009992]

[7] V.J. Clemente-Suárez, A. Martín-Rodríguez, L. Redondo-Flórez, C. López-Mora, R. Yáñez-Sepúlveda, and J.F. Tornero-Aguilera, "New Insights and Potential Therapeutic Interventions in Metabolic Diseases", *Int. J. Mol. Sci.,* vol. 24, no. 13, p. 10672, 2023.
[http://dx.doi.org/10.3390/ijms241310672] [PMID: 37445852]

[8] S.W. Chang, M.L. Cheng, M.S. Shiao, C.T. Yeh, C.H. Wang, C.M. Fan, C.T. Chiu, and M.L. Chang, "Recovery of lipid metabolic alterations in hepatitis C patients after viral clearance: Incomplete restoration with accelerated ω-oxidation", *J. Clin. Lipidol.,* vol. 12, no. 3, pp. 756-766, 2018.
[http://dx.doi.org/10.1016/j.jacl.2018.02.011] [PMID: 29574072]

[9] M. Adília, B. Chócha, and P. Monteiro, "A model validation pipeline for healthy tissue genome-scale metabolic models." Dec. 22, 2021. Accessed: Oct. 28, 2025. [Online]. Available: https://hdl.handle.net/1822/82451

[10] G. Zampieri, S. Vijayakumar, E. Yaneske, and C. Angione, "Machine and deep learning meet genome-scale metabolic modeling", *PLOS Comput. Biol.,* vol. 15, no. 7, p. e1007084, 2019.
[http://dx.doi.org/10.1371/journal.pcbi.1007084] [PMID: 31295267]

[11] J. H. Veerkamp, H. T. B. van Moerkerk, J. F. C. Glatz, and V. W. M. van Hinsbergh, "Incomplete palmitate oxidation in cell-free systems of rat and human muscles," Biochimica et Biophysica Acta (BBA) - Lipids and Lipid Metabolism, vol. 753, no. 3, pp. 399–410, 1983.

[12] M. Mirabelli, E. Chiefari, B. Arcidiacono, D.M. Corigliano, F.S. Brunetti, V. Maggisano, D. Russo, D.P. Foti, and A. Brunetti, "Mediterranean diet nutrients to turn the tide against insulin resistance and related diseases", *Nutrients,* vol. 12, no. 4, p. 1066, 2020.
[http://dx.doi.org/10.3390/nu12041066] [PMID: 32290535]

[13] P. Blackburn, J. Gass, F. Pinto e Vairo, K. Farnham, H. Atwal, S. Macklin, E. Klee, and P. Atwal, "Maple syrup urine disease: mechanisms and management", *Appl. Clin. Genet.,* vol. 10, pp. 57-66, 2017.
[http://dx.doi.org/10.2147/TACG.S125962] [PMID: 28919799]

[14] C.D.M. van Karnebeek, M. Shevell, J. Zschocke, J.B. Moeschler, and S. Stockler, "The metabolic evaluation of the child with an intellectual developmental disorder: Diagnostic algorithm for identification of treatable causes and new digital resource", *Mol. Genet. Metab.,* vol. 111, no. 4, pp. 428-438, 2014.
[http://dx.doi.org/10.1016/j.ymgme.2014.01.011] [PMID: 24518794]

[15] M. Deponte, "The incomplete glutathione puzzle: Just guessing at numbers and figures?", *Antioxid. Redox Signal.,* vol. 27, no. 15, pp. 1130-1161, 2017.

[http://dx.doi.org/10.1089/ars.2017.7123] [PMID: 28540740]

[16] G. Sun, W. Huang, and L. Wang, "Molecular findings among patients for whole exome sequencing and mitochondrial genome assessment", Clinica Chimica Acta, vol. 561, 2024", [http://dx.doi.org/10.21203/rs.3.rs-3309204/v1]

[17] A.S. Vickram, S. Manikandan, T. Richard, S.V. Lakshmi, and H. Chopra, "Targeted gene therapy: Promises and challenges in disease management", *J. BioX Res.,* vol. 7, p. 0007, 2024. [http://dx.doi.org/10.34133/jbioxresearch.0007]

[18] A. Sharma, A. Magotra, S.K. Rath, P. Wazir, U. Nandi, S. Koul, P.L. Sangwan, A.P. Gupta, and G. Singh, "In-vitro and in-vivo pharmacokinetics of IS01957, p-coumaric acid derivative using a validated LC–ESI–MS/MS method in mice plasma", *J. Pharm. Investig.,* vol. 48, no. 5, pp. 565-574, 2018.
[http://dx.doi.org/10.1007/s40005-017-0350-8]

[19] M. Adeva-Andany, M. López-Ojén, R. Funcasta-Calderón, E. Ameneiros-Rodríguez, C. Donapetry-García, M. Vila-Altesor, and J. Rodríguez-Seijas, "Comprehensive review on lactate metabolism in human health", *Mitochondrion,* vol. 17, pp. 76-100, 2014.
[http://dx.doi.org/10.1016/j.mito.2014.05.007] [PMID: 24929216]

[20] T.R. Koves, J.R. Ussher, R.C. Noland, D. Slentz, M. Mosedale, O. Ilkayeva, J. Bain, R. Stevens, J.R.B. Dyck, C.B. Newgard, G.D. Lopaschuk, and D.M. Muoio, "Mitochondrial overload and incomplete fatty acid oxidation contribute to skeletal muscle insulin resistance", *Cell Metab.,* vol. 7, no. 1, pp. 45-56, 2008.
[http://dx.doi.org/10.1016/j.cmet.2007.10.013] [PMID: 18177724]

[21] B.C. Carter, and B.P. Bean, "Sodium entry during action potentials of mammalian neurons: incomplete inactivation and reduced metabolic efficiency in fast-spiking neurons", *Neuron,* vol. 64, no. 6, pp. 898-909, 2009.
[http://dx.doi.org/10.1016/j.neuron.2009.12.011] [PMID: 20064395]

[22] N. Longo, M. Frigeni, and M. Pasquali, "Carnitine transport and fatty acid oxidation", *Biochim. Biophys. Acta Mol. Cell Res.,* vol. 1863, no. 10, pp. 2422-2435, 2016.
[http://dx.doi.org/10.1016/j.bbamcr.2016.01.023] [PMID: 26828774]

[23] Sasaki, T. Ito, and Y. Sasaki, *The complete genome sequence of Lactobacillus bulgaricus reveals extensive and ongoing reductive evolution.,* 1993. Available from: http://www.pnas.orgcgidoi10.1073pnas.0603024103

[24] N. Wiernsperger, A.L. Sylvia, and F.F. Jobsis, “Incomplete transient ischemia: A non-destructive evaluation of in vivo cerebral metabolism and hemodynamics in rat brain,” AHA Scientific Journal, vol. 12, 1981.
[http://dx.doi.org/10.1161/01.STR.12.6.864]

[25] A. Sharma, A. Davis, and P.S. Shekhawat, "Hypoglycemia in the preterm neonate: etiopathogenesis, diagnosis, management and long-term outcomes", *Transl. Pediatr.,* vol. 6, no. 4, pp. 335-348, 2017.
[http://dx.doi.org/10.21037/tp.2017.10.06] [PMID: 29184814]

[26] P. A. Steen, J. D. Michenfelder, and J. H. Milde, “Incomplete transient ischemia: A non-destructive evaluation of in vivo cerebral metabolism and hemodynamics in rat brain,” *AHA Scientific Journal*, vol. 12, 1981.
[http://dx.doi.org/10.1002/ana.410060503]

[27] C.H. Schilling, M.W. Covert, I. Famili, G.M. Church, J.S. Edwards, and B.O. Palsson, "Genome-scale metabolic model of Helicobacter pylori 26695", *J. Bacteriol.,* vol. 184, no. 16, pp. 4582-4593, 2002.
[http://dx.doi.org/10.1128/JB.184.16.4582-4593.2002] [PMID: 12142428]

[28] *Appraisal of inhalation method/juge*. Available from: http://ahajournals.org

[29] S. Lee, S.A. Kim, J. Hong, Y. Kim, G. Hong, S. Baik, K. Choi, M.K. Lee, and K.R. Lee, "Identification of genetic variants related to metabolic syndrome by next-generation sequencing",

Diabetol. Metab. Syndr., vol. 14, no. 1, p. 119, 2022.
[http://dx.doi.org/10.1186/s13098-022-00893-y] [PMID: 35999587]

[30] M. Rosenthal, D. L. Martel, and J. C. LaManna, “Effects of incomplete and complete ischemia on mitochondrial functioning measured in intact cerebral cortex of cats,” *Exp. Neurol.*, vol. 52, no. 3, pp. 433–446, 1976.

[31] M. Pourfarzam, and F. Zadhoush, "Newborn Screening for Inherited Metabolic Disorders", *News Views,* 2013.

[32] J.F. Israel, D.D. Campbell, J.H. Kahn, and T.G. Hornby, "Metabolic costs and muscle activity patterns during robotic- and therapist-assisted treadmill walking in individuals with incomplete spinal cord injury", *Phys. Ther.,* vol. 86, no. 11, pp. 1466-1478, 2006.
[http://dx.doi.org/10.2522/ptj.20050266] [PMID: 17079746]

[33] S.H. Adams, C.L. Hoppel, K.H. Lok, L. Zhao, S.W. Wong, P.E. Minkler, D.H. Hwang, J.W. Newman, and W.T. Garvey, "Plasma acylcarnitine profiles suggest incomplete long-chain fatty acid β-oxidation and altered tricarboxylic acid cycle activity in type 2 diabetic African-American women", *J. Nutr.,* vol. 139, no. 6, pp. 1073-1081, 2009.
[http://dx.doi.org/10.3945/jn.108.103754] [PMID: 19369366]

[34] N. Raben, T. Fukuda, A.L. Gilbert, D. de Jong, B.L. Thurberg, R.J. Mattaliano, P. Meikle, J.J. Hopwood, K. Nagashima, K. Nagaraju, and P.H. Plotz, "Replacing acid α-glucosidase in Pompe disease: recombinant and transgenic enzymes are equipotent, but neither completely clears glycogen from type II muscle fibers", *Mol. Ther.,* vol. 11, no. 1, pp. 48-56, 2005.
[http://dx.doi.org/10.1016/j.ymthe.2004.09.017] [PMID: 15585405]

[35] “Changes in brain energy metabolism, neurotransmitters, and choline during and after incomplete cerebral ischemia in spontaneously hypertensive rats,” Neurochem. Res., vol. 20, no. 1, pp. 23–30, 1995.

[36] M. Rieger, O. Kappeli, and A. Fiechter, "The role of limited respiration in the incomplete oxidation of glucose by Saccharomyces cerevisiae," J. Gen. Microbiol., vol. 129, pp. 653–661, 1983.

[37] L.H. Koens, J.J. de Vries, F. Vansenne, T.J. de Koning, and M.A.J. Tijssen, "How to detect late-onset inborn errors of metabolism in patients with movement disorders – A modern diagnostic approach", *Parkinsonism Relat. Disord.,* vol. 85, pp. 124-132, 2021.
[http://dx.doi.org/10.1016/j.parkreldis.2021.02.029] [PMID: 33745796]

[38] T. Fullston, E.M.C.O. Teague, N.O. Palmer, M.J. DeBlasio, M. Mitchell, M. Corbett, C.G. Print, J.A. Owens, and M. Lane, "Paternal obesity initiates metabolic disturbances in two generations of mice with incomplete penetrance to the F 2 generation and alters the transcriptional profile of testis and sperm microRNA content", *FASEB J.,* vol. 27, no. 10, pp. 4226-4243, 2013.
[http://dx.doi.org/10.1096/fj.12-224048] [PMID: 23845863]

[39] H. Tsukamoto, and X.P. Xi, "Incomplete compensation of enhanced hepatic oxygen consumption in rats with alcoholic centrilobular liver necrosis", *Hepatology,* vol. 9, no. 2, pp. 302-306, 1989.
[http://dx.doi.org/10.1002/hep.1840090223] [PMID: 2912830]

[40] A.T. Bertrand, L. Renou, A. Papadopoulos, M. Beuvin, E. Lacène, C. Massart, C. Ottolenghi, V. Decostre, S. Maron, S. Schlossarek, M.E. Cattin, L. Carrier, M. Malissen, T. Arimura, and G. Bonne, "DelK32-lamin A/C has abnormal location and induces incomplete tissue maturation and severe metabolic defects leading to premature death", *Hum. Mol. Genet.,* vol. 21, no. 5, pp. 1037-1048, 2012.
[http://dx.doi.org/10.1093/hmg/ddr534] [PMID: 22090424]

[41] A. Kozioł-Kozakowska, D. Januś, A. Stępniewska, E. Szczudlik, A. Stochel-Gaudyn, and M. Wójcik, "Beyond the metabolic syndrome: Non-obvious complications of obesity in children", *Children (Basel),* vol. 10, no. 12, p. 1905, 2023.
[http://dx.doi.org/10.3390/children10121905] [PMID: 38136107]

[42] A. Sharma, A. Gour, S. Bhatt, S.K. Rath, T.A. Malik, A. Dogra, P.L. Sangwan, S. Koul, S.T. Abdullah, G. Singh, and U. Nandi, "Effect of IS01957, a para-coumaric acid derivative on

pharmacokinetic modulation of diclofenac through oral route for augmented efficacy", *Drug Dev. Res.,* vol. 80, no. 7, pp. 948-957, 2019.
[http://dx.doi.org/10.1002/ddr.21574] [PMID: 31318064]

[43] *Negl6n P. Department of Surgery.*

[44] M.D. Brand, "Uncoupling to survive? The role of mitochondrial inefficiency in ageing", *Exp. Gerontol.,* vol. 35, no. 6-7, pp. 811-820, 2000.
[http://dx.doi.org/10.1016/S0531-5565(00)00135-2] [PMID: 11053672]

[45] A.M. Fenuta, and A.L. Hicks, "Metabolic demand and muscle activation during different forms of bodyweight supported locomotion in men with incomplete SCI", *BioMed Res. Int.,* vol. 2014, pp. 1-10, 2014.
[http://dx.doi.org/10.1155/2014/632765] [PMID: 24971340]

[46] L. Velázquez-López, G. Santiago-Díaz, J. Nava-Hernández, A.V. Muñoz-Torres, P. Medina-Bravo, and M. Torres-Tamayo, "Mediterranean-style diet reduces metabolic syndrome components in obese children and adolescents with obesity", *BMC Pediatr.,* vol. 14, no. 1, p. 175, 2014.
[http://dx.doi.org/10.1186/1471-2431-14-175] [PMID: 24997634]

[47] F. Kneepkens, R.J. Vonk, and J. Fernandes, "Incomplete intestinal absorption of fructose," Arch. Dis. Child., vol. 59, no. 8, pp. 735–738, 1984.

[48] S. Ciężki, E. Odyjewska, A. Bossowski, and B. Głowińska-Olszewska, "Not Only Metabolic Complications of Childhood Obesity", *Nutrients,* vol. 16, no. 4, p. 539, 2024.
[http://dx.doi.org/10.3390/nu16040539] [PMID: 38398863]

[49] R.A. Swanson, K. Farrell, and B.A. Stein, "Astrocyte energetics, function, and death under conditions of incomplete ischemia: A mechanism of glial death in the penumbra," Glia. , vol. 21, no. 1, pp. 142–153, 1997.

[50] O. Riddle, G.C. Smith, and C.S. Moran, "Effects of complete and incomplete hypophysectomy on basal metabolism of pigeons," In: *A Biographical Memoir*, G. W. Corner, Ed, vol. 18, 1934.

[51] P.J. Osther, A.B. Hansen, K. Engel, and P. Kildeberg, "Pathophysiology of incomplete renal tubular acidosis in recurrent renal stone formers: Evidence of disturbed calcium, bone and citrate metabolism," *Urol. Res.*, vol. 21, pp. 169–173, 1993.

[52] G Chechik, G Heitz, G Elidan, P Abbeel, and D. Koller, "Max-margin classification of incomplete data," In: *Advances in Neural Information Processing Systems 19: Proceedings of the 2006 Conference*, B. Schölkopf, J. Platt, and T. Hofmann, Eds. Cambridge, MA: MIT Press, 2007.
[http://dx.doi.org/10.7551/mitpress/7503.003.0034]

[53] J. Kressler, M.S. Nash, P.A. Burns, and E.C. Field-Fote, "Metabolic responses to 4 different body weight-supported locomotor training approaches in persons with incomplete spinal cord injury", *Arch. Phys. Med. Rehabil.,* vol. 94, no. 8, pp. 1436-1442, 2013.
[http://dx.doi.org/10.1016/j.apmr.2013.02.018] [PMID: 23473703]

[54] J.A. Bell, M.A. Reed, L.A. Consitt, O.J. Martin, K.R. Haynie, M.W. Hulver, D.M. Muoio, and G.L. Dohm, "Lipid partitioning, incomplete fatty acid oxidation, and insulin signal transduction in primary human muscle cells: effects of severe obesity, fatty acid incubation, and fatty acid translocase/CD36 overexpression", *J. Clin. Endocrinol. Metab.,* vol. 95, no. 7, pp. 3400-3410, 2010.
[http://dx.doi.org/10.1210/jc.2009-1596] [PMID: 20427507]

[55] C. Kazunga, and M.D. Aitken, *Products from the Incomplete Metabolism of Pyrene by Polycyclic Aromatic Hydrocarbon-Degrading Bacteria,* vol. 66, 2000.
Available from: https://journals.asm.org/journal/aem

[56] H.R. Zielke, C.L. Zielke, P.J. Baab, and R.M. Collins, "Large neutral amino acids auto exchange when infused by microdialysis into the rat brain: implication for maple syrup urine disease and phenylketonuria," *Neurochem. Int.*, vol. 40, no. 4, pp. 347–354, 2002.

[57] J. Kressler, T. Wymer, and A. Domingo, "Respiratory, cardiovascular and metabolic responses during

different modes of overground bionic ambulation in persons with motor-incomplete spinal cord injury: A case series", *J. Rehabil. Med.,* vol. 50, no. 2, pp. 173-180, 2018. [http://dx.doi.org/10.2340/16501977-2281] [PMID: 29068039]

[58] S. Illsinger, and A.M. Das, "Impact of selected inborn errors of metabolism on prenatal and neonatal development", *IUBMB Life,* vol. 62, no. 6, pp. 403-413, 2010. [http://dx.doi.org/10.1002/iub.336] [PMID: 20503433]

[59] For Personal Use, *Only Reproduce with Permission from Elsevier Ltd Rapid Review Enzyme-Replacement Therapy.* Available from: http://neurology.thelancet.com

[60] A. Gaya Sicilia, M. Pilar Sánchez-Jiménez, A. Peirotén, A. Landete, and J. Maria Medina, "Incomplete metabolism of phytoestrogens by gut microbiota from children under the age of three," *Int. J. Food Sci. Nutr.*, vol. 69, no. 3, pp. 334–343, 2018. Available from: http://hdl.handle.net/10261/289261

[61] E. Carbonneau, A.M. de Passillé, J. Rushen, B.G. Talbot, and P. Lacasse, "The effect of incomplete milking or nursing on milk production, blood metabolites, and immune functions of dairy cows", *J. Dairy Sci.,* vol. 95, no. 11, pp. 6503-6512, 2012. [http://dx.doi.org/10.3168/jds.2012-5643] [PMID: 22981581]

[62] S. Silvestri, P. Orlando, T. Armeni, L. Padella, F. Brugè, G. Seddaiu, G.P. Littarru, and L. Tiano, "Coenzyme Q10 and α-lipoic acid: antioxidant and pro-oxidant effects in plasma and peripheral blood lymphocytes of supplemented subjects", *J. Clin. Biochem. Nutr.,* vol. 57, no. 1, pp. 21-26, 2015. [http://dx.doi.org/10.3164/jcbn.14-130] [PMID: 26236096]

[63] M. Agana, J. Frueh, M. Kamboj, D.R. Patel, and S. Kanungo, "Common metabolic disorder (inborn errors of metabolism) concerns in primary care practice", *Ann. Transl. Med.,* vol. 6, no. 24, pp. 469-469, 2018. [http://dx.doi.org/10.21037/atm.2018.12.34] [PMID: 30740400]

[64] S. Berthoumieux, M. Brilli, H. de Jong, D. Kahn, and E. Cinquemani, "Identification of metabolic network models from incomplete high-throughput datasets", *Bioinformatics,* vol. 27, no. 13, pp. i186-i195, 2011. [http://dx.doi.org/10.1093/bioinformatics/btr225] [PMID: 21685069]

[65] K.J. Barry, and N.R. Wainwright, "Biosynthetic induction of a secondary metabolite by a marine bacterium under nutritional stress: Potential role of the incomplete oxidation of an organic acid," *Biol. Bull.*, vol. 193, no. 2, pp. 274–275, 1997. Available from: http://www.journals.uchicago.edu/t-and-c

[66] E.L. MacLeod, and D.M. Ney, "Nutritional management of phenylketonuria", *Ann. Nestle [Eng.],* vol. 68, no. 2, pp. 58-69, 2010. [http://dx.doi.org/10.1159/000312813] [PMID: 22475869]

[67] D.M. Frausto, C.B. Forsyth, A. Keshavarzian, and R.M. Voigt, "Dietary Regulation of Gut-Brain Axis in Alzheimer's Disease: Importance of Microbiota Metabolites", *Front. Neurosci.,* vol. 15, p. 736814, 2021. [http://dx.doi.org/10.3389/fnins.2021.736814] [PMID: 34867153]

[68] Y. Wu, Y. Wang, A. Hu, X. Shu, W. Huang, J. Liu, B. Wang, R. Zhang, M. Yue, and C. Yang, "Lactobacillus plantarum-derived postbiotics prevent Salmonella-induced neurological dysfunctions by modulating gut-brain axis in mice," *Front. Nutr.*, vol. 9, p. 946096, 2022. doi: 10.3389/fnut.2022.946096. PMID: 35967771; PMCID: PMC9365972.

[69] S. Yannicelli, and A. Ryan, "Improvements in behaviour and physical manifestations in previously untreated adults with phenylketonuria using a phenylalanine□restricted diet: a national survey", *J. Inherit. Metab. Dis.,* vol. 18, no. 2, pp. 131-134, 1995. [http://dx.doi.org/10.1007/BF00711747] [PMID: 7564227]

[70] Z. Zhang, Q. Peng, D. Huo, S. Jiang, C. Ma, H. Chang, K. Chen, C. Li, Y. Pan, and J. Zhang,

"Melatonin Regulates the Neurotransmitter Secretion Disorder Induced by Caffeine Through the Microbiota-Gut-Brain Axis in Zebrafish (Danio rerio)", *Front. Cell Dev. Biol.,* vol. 9, p. 678190, 2021. [http://dx.doi.org/10.3389/fcell.2021.678190] [PMID: 34095150]

[71] VK Gauttam, K Munjal, H Chopra, A Ahmad, MK Rana, and MA Kamal, "A mechanistic review on therapeutic potential of medicinal plants and their pharmacologically active molecules for targeting metabolic syndrome", *Curr. Pharm. Des.,* vol. 30, no. 1, pp. 10-30, 2024. [http://dx.doi.org/10.2174/0113816128274446231220113957]

[72] A. Taraskina, O. Ignatyeva, D. Lisovaya, M. Ivanov, L. Ivanova, V. Golovicheva, G. Baydakova, D. Silachev, V. Popkov, T. Ivanets, D. Kashtanova, V. Yudin, V. Makarov, I. Abramov, M. Lukashina, V. Rakova, A. Zagainova, D. Zorov, E. Plotnikov, G. Sukhikh, and S. Yudin, "Effects of Traumatic Brain Injury on the Gut Microbiota Composition and Serum Amino Acid Profile in Rats", *Cells,* vol. 11, no. 9, p. 1409, 2022. [http://dx.doi.org/10.3390/cells11091409] [PMID: 35563713]

[73] K. Thapa, K. Shivam, H. Khan, A. Kaur, K. Dua, S. Singh, and T.G. Singh, "Emerging targets for modulation of immune response and inflammation in stroke", *Neurochem. Res.,* vol. 48, no. 6, pp. 1663-1690, 2023. [http://dx.doi.org/10.1007/s11064-023-03875-2] [PMID: 36763312]

[74] N. Kanojia, K. Thapa, G. Kaur, A. Sharma, V. Puri, and N. Verma, "Update on Therapeutic potential of emerging nanoformulations of phytocompounds in Alzheimer's and Parkinson's disease", *J. Drug Deliv. Sci. Technol.,* vol. 79, p. 104074, 2023. [http://dx.doi.org/10.1016/j.jddst.2022.104074]

[75] B.D. Needham, R. Kaddurah-Daouk, and S.K. Mazmanian, "Gut microbial molecules in behavioural and neurodegenerative conditions", *Nat. Rev. Neurosci.,* vol. 21, no. 12, pp. 717-731, 2020. [http://dx.doi.org/10.1038/s41583-020-00381-0] [PMID: 33067567]

CHAPTER 3

Metabolic Pathways and Intermediates

Jyoti[1], **Saurav Kumar**[2], **Yogendra Singh**[1,*] and **Sunil Dutt**[3]

[1] *School of Pharmaceutical Sciences, MVN University, Palwal, Haryana, India*

[2] *Department of Pharmaceutical Sciences, Gurugram University, Gurugram, Haryana, India*

[3] *Maharaja Agrasen School of Pharmacy, Maharaja Agrasen University, Solan, Himachal Pradesh, India*

Abstract: Metabolism is the chemical process that keeps living organisms alive by providing cells with energy to maintain normal cell functions and metabolic building blocks to produce new cellular components. Metabolism is, therefore, central to all life and comprises myriad biochemical pathways organised by myriad intermediates to form networks that convert nutrient substrates to metabolic products. This chapter describes the metabolic pathways and metabolic intermediates present in all cells, the enzymes that catalyse these reactions, the mechanistic details behind the biochemistry, and the roles that they play in obtaining energy, biosynthesis and cell regulation. It explores systems-level control of metabolic flux to adapt cell functions to environmental changes and discusses the serious consequences that result when control breaks down and so-called metabolic 'disinsiders' take over – as in metabolic disorders such as diabetes and obesity. It also discusses novel technologies and research paradigms driving progress in the world of metabolism, including new approaches to data analysis and drug discovery, and will define the future of this field in the coming decades. In this chapter, we describe the process of primary metabolism: how cells obtain energy and build new biological molecules such as proteins, lipids and nucleic acids.

Keywords: Biochemistry, Cell function, Metabolic pathways, Metabolism.

INTRODUCTION

The word "metabolism," derived from the Greek word meaning "change, " refers to the sum of all chemical processes that are targeted toward enabling life to take place in the cells of living organisms. A combination word originates from the Greek roots "meta," meaning "over," and "ballein," meaning "to throw. The totality of the essential chemical processes occurring inside a cell is referred to as metabolism. The last three decades of the sixteenth century represented the start

* **Corresponding author Yogendra Singh**: School of Pharmaceutical Sciences, MVN University, Palwal, Haryana, India; E-mail: yogesigh1978@gmail.com

of the first recorded and more detailed studies on metabolism. This age represented the time when instrumentation started to be coupled with direct observation so that quantification, and therefore validation, was introduced to the sciences, in this instance, within biological systems [1]. Metabolic balance can no longer be achieved under extreme external parameters or when internal control is contaminated by genetic or epigenetic alterations. This leads, in response to aging, to the evolution of disorders such as metabolic syndrome or cancer and, ultimately, death [2]. The chemical apparatus that allows cells to convert some chemical feedstocks into chemical energy (ATP, NADPH) and into building blocks (small organic molecules, activated coenzymes) used to synthesize biomass is generally referred to as metabolism. In practice, all the metabolic reactions are steady-state processes (almost) catalysed by enzymes. Non-enzymatic cellular responses are rare, as shown in Fig. (**1**).

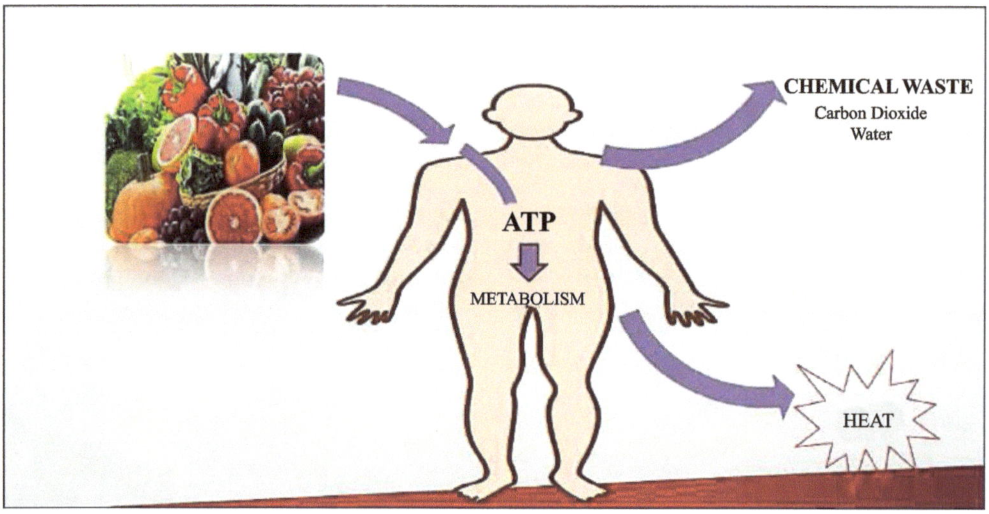

Fig. (1). Metabolic reactions.

Proteins catalyze metabolic processes, and biochemists have spent decades studying the *in vitro* kinetic characteristics of these enzymes. Through these initiatives, it became clear that acute modulation of enzyme activity is possible and essential to maintaining metabolic homeostasis. Allosteric interactions with the metabolites themselves or post-translational changes brought on by intracellular signal transduction pathways are two possible mechanisms by which regulation is mediated.

More recently, cell biologists have become interested in the regulation of enzymes after noticing that variations in growth circumstances frequently cause diffusely dispersed enzymes to condense into one or more distinct foci that are readily observed by light microscopy. The most appropriate way to characterize this transition from a soluble to a condensed state is as a phase separation.

Anabolism uses energy to produce biomolecular polymers and macromolecules. When they are reduced to smaller molecules during metabolism, energy is released. Living things always need energy to maintain cellular and whole-body functions. Mechanical work (contraction and movement of cells), active transfer of ions and substrates (K^+, Mg^{2+}, and Ca^{2+}, for example, in heart contraction), and the manufacture of complex macromolecules (like glycogen) all require energy [3, 4].

METABOLIC PATHWAY

Chemical products go through a succession of metabolic intermediates to become the substrate molecule or molecules in metabolic pathways and cycles, which ultimately result in a final product or products. Large molecules, for instance, are broken down into glucose by one metabolic pathway for carbohydrates. Chemical transformation occurs in all substrates in reactions that are part of metabolic cycles (if the reaction moieties are conserved) or pathways (if the reactions are arranged linearly). Substance deterioration that ultimately results in the creation of CO_2 is referred to as substrate oxidation. However, there are a number of reasons why an oxidative pathway like this could be disrupted. For example, during the oxidation of glucose, low oxygen levels cause lactate to be produced instead of CO_2 and water. Certain substrates, but not all of them, can also be transformed into the fundamental components of biosynthetic reactions [5, 6].

A metabolic pathway refers to a series of interconnected processes in which chemical events occur within a cell. The products, reactants, or intermediates of an enzymatic process whose modification entails a sequence of chemical reactions mediated by enzymes are referred to as metabolites [7]. The product of one enzyme usually becomes the substrate for the next enzyme in most metabolic pathways. The by-products are expelled out of the cell as they are considered waste. Depending on their location in eukaryotic cells and the significance they may have towards that compartment of a cell, the various metabolic pathways perform different functions. The electron transport chain and oxidative phosphorylation, for example, occur in the mitochondrial membrane, whereas fatty acid synthesis, glycolysis, and the pentose phosphate pathway are all located in the cytoplasm of a cell [8].

The pathways are needed by the organism for homeostasis, the flow of metabolites through a pathway is decided by the cellular requirement and substrate availability. The end product of a particular metabolic pathway may be stored for future use, utilised immediately, or may serve as the starting material in another metabolic pathway. In general, the breakdown of food and its conversion into energy are referred to as metabolism. Anabolism, or the accumulation of compounds, and catabolism, or the breakdown of substances, are the two components of metabolism [9, 10].

Catabolism

The process that breaks down proteins, lipids, or tissues into smaller cells (or fatty acids) is known as catabolism. This happens as a result of food digestion (wheat breaking down into glucose, for example, a catabolic pathway is a sequence of events that results in the net release of energy as a high-energy phosphate bond formed with the energy carriers guanosine triphosphate (GTP) and adenosine triphosphate (ATP), respectively, from guanosine diphosphate (GDP) and adenosine diphosphate (ADP) [11, 12].

Thus, the entire reaction is thermodynamically favourable, as the overall free energy of the final products is much less. An exergonic process is one in which energy-rich biomolecules are degraded into chemical energy in the form of ATP, GTP, NADH, NADPH, $FADH_2$, *etc.* Very often, ammonia, water, and carbon-d--oxide are produced as the final products in case the pathway is degradative or catabolic. Only when major precursors of the anabolic pathway are coupled with an endergonic anabolism reaction is the cell in a position to synthesize new macromolecules [13, 14].

When complex molecules are broken down into smaller ones, energy is released through the action of catabolic processes. The adenosine triphosphate (ATP) that is produced from this energy is frequently utilized to power different biological functions. Examples include the production of ATP and NADH by the oxidation of glucose to pyruvate. Acetyl-CoA is converted to carbon dioxide in the citric acid cycle. This produces ATP, NADH, and $FADH_2$. The oxidation of fatty acids is known as β-oxidation. The generated acetyl-CoA molecules subsequently go into the citric acid cycle (Fig. **2**). The breakdown of proteins into amino acids is known as proteolysis. After that, they can be burned or further catabolized [3].

Anabolism

The process of anabolism is the joining of smaller cells (nutrients, cells, or amino acids) to form larger structures (*e.g.*, your body attempting to repair a cut by forming new tissue and structures surrounding the wound). Energy investment is thus required for the building of macromolecules, such as polypeptides, nucleic acids, proteins, polysaccharides, and lipids through anabolic processes. Since it is the only anabolic process with a positive Gibbs free energy ($+\Delta G$), it is non-spontaneous [15, 16].

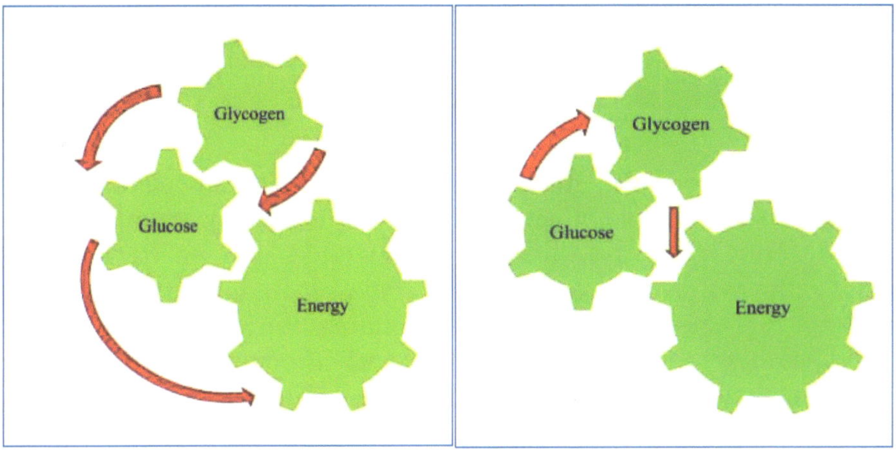

Fig. (2). Illustrating catabolism and anabolism.

As a result, chemical energy must be supplied through coupling with an exergonic process. Coupling the reaction of a catabolic pathway lowers the total activation energy of an anabolic pathway and enables the reaction to proceed. An endergonic response is non-spontaneous otherwise.

These are biosynthetic pathways requiring energy input to synthesize complex molecules from simpler ones. These pathways play a significant role in the synthesis of cellular components for cell growth and repair [17, 18].

Gluconeogenesis, Fatty Acid Synthesis, Protein Synthesis, Amphibolic Pathways

Gluconeogenesis is the production of glucose from precursors other than carbohydrates and involves lactate and amino acids [19].

Fatty acid synthesis is the building up of fatty acids from acetyl-CoA and malonyl-CoA, eventually getting incorporated into making lipids [20].

Protein synthesis involves the translation process, which results in the synthesis of proteins through the combination of amino acids.

Nucleotide Biosynthesis is a process associated with the generation of nucleotides, through which DNA and RNA are synthesized [21, 22].

Metabolic pathways that exhibit, according to the metabolic demand inside the cell, both catabolic and anabolic directions. These kinds of pathways are called amphibolic pathways. For example, citric acid cycle, which is largely catabolic, also supplies transitions for biosynthetic processes, including the production of nucleotide precursors and amino acids [23, 24].

Significance of Metabolic Pathways

The importance of metabolic pathways has recently increased due to their critical role in cellular survival and function. Energy-producing metabolic processes: Through metabolic pathways, foods are converted into energy for cellular operations. Pathways produce essential molecules, including lipids, nucleotides, and amino acids. They are also utilized as building blocks for development and repair. Additionally, they are employed in hazardous compounds that are broken down *via* routes and cell waste is recycled. Metabolic intermediates influence gene expression and cellular decisions by acting as signalling molecules (Fig. **3**).

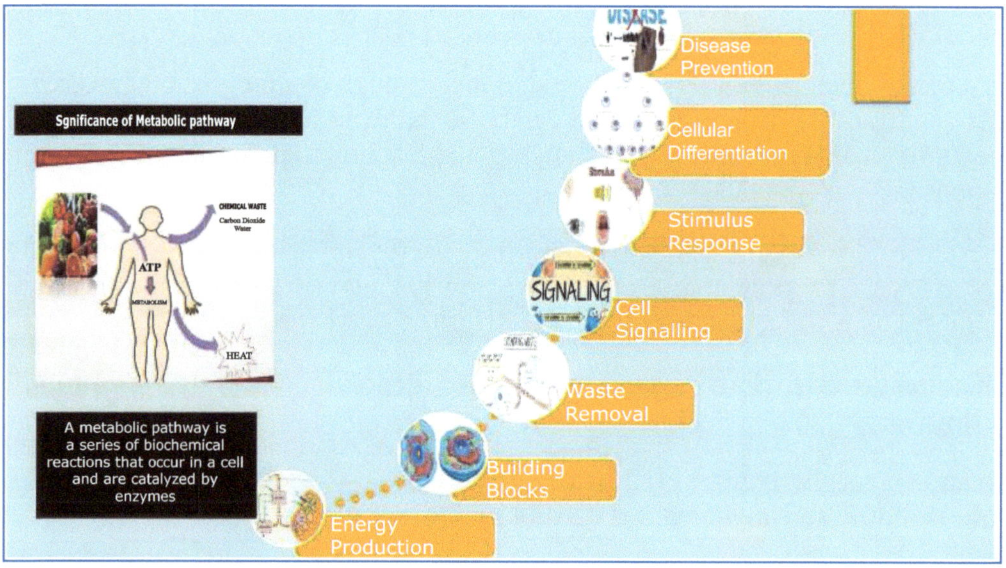

Fig. (3). Significance of metabolic pathway.

Through regulation of pH, electrolyte levels, and nutrient levels, routes maintain homeostasis and balance inside cells.

Metabolic pathways adapt to changes in the environment, stress, or developmental signals. Routes influence cell fate decisions by fostering tissue development and specialization as well as cell differentiation.

With appropriate metabolic function, diseases, including diabetes, cancer, and neurological disorders, can be prevented. The metabolic pathways are essential for energy generation, molecular synthesis, waste disposal, signaling, balance, adaptation, cellular differentiation, and disease prevention in both general organism health and cellular function [25, 26].

Cellular Metabolism

The cell mainly produces ATP through the processes of two pathways: glycolysis and oxidative phosphorylation. Even though less energetic, during the early development and proliferation phase of a tumor, aerobic glycolysis represents the major metabolic pathway. Once upon a time, this used to be considered one of the major reasons for cancer cell multiplication since there was a suspicion of mitochondrial abnormalities [27]. Further investigation of this high glycolytic activity, however, uncovered that Warburg metabolism is indeed an integral part of the pathophysiology of cancer. More recent investigations have shown that even under non-hypoxic conditions, aerobic glycolysis is a requirement for normal cell proliferation and differentiation. Why glycolysis remains essential for cell growth and differentiation is still unknown.However, many investigations suggest a few possible functions. As glycolysis is a key node in these many anabolic processes, an increase in aerobic glycolysis reflects a demand for increased biomass, including phospholipids, nucleotides, and amino acids. There are two shunts in glycolysis that make these biomolecules. The first of the glycolytic shunts is the pentose phosphate pathway [28]. It is made up of an oxidative branch that degrades the enzyme glucose-6-phosphate into pentose sugars and NADPH. The former is a key reducing agent in many biosynthetic pathways, while pentose sugars are required to synthesize nucleotides. Fructose-6-phosphate and glyceraldehyde-3-phosphate are transformed into ribose- 5-phosphate and erythrose-4-phosphate, respectively, in the non-oxidative arm of the PPP. Both PPP branches participate in the synthesis of nucleotides and amino acids, especially non-essential amino acids [29].

Metabolite transport between pathways depends on a series of reactions that are kept in a dynamic equilibrium. For example, it is well known that the availability of NAD^+ during the metabolism of carbohydrates has a significant impact on the glycolytic flux. As a result, a higher rate of NAD^+ regeneration is required for this pathway to function more quickly. This is often accomplished by the process known as lactic fermentation, which uses pyruvate as the last electron acceptor to produce lactate. This kind of procedure occurs because the pace of aerobic respiration cannot maintain the rate at which NAD^+ is replenished to meet the high demand for this intermediate in glycolysis, which is necessary to fulfill ATP demands. Despite the fact that this process mostly takes place in anaerobic environments, a substantial body of recent research indicates that oxygen is merely one of many participants in this cross-talk.

The scientific community has made thorough information about the metabolites, cofactors, and enzymes involved in each metabolic process for various physiological conditions, stimuli, and insults publicly available due to the fundamental role that metabolism plays in cellular function. When examining this data at the level of cellular functionality, it is important to keep in mind the dynamic interactions among various metabolic nodes that control metabolism and cellular activity.

Hence, knowing the complex relationships between each metabolic route and its primary regulatory sites is essential to integrating metabolism toward a deeper comprehension of physiology and pathophysiology [30, 31].

Network Metabolism

The context of a cell's entire metabolism was explained by its metabolic pathways. The cellular metabolic reaction network is a system of enzyme activities and transport mechanisms that regulate the concentrations of various metabolites by replenishing or reducing them. This network enables the analysis of both the metabolic system's capabilities and its structural organization. The intricate network of chemical and physical reactions necessary for the upkeep of life is known as metabolism [32, 33].

The architecture and topology of the metabolic network are defined by the set of reactions and, consequently, by the pathways contained in each metabolic network. The complete set of reactions is controlled by complex biological control schemes, which enable the metabolic network of the cell to achieve its objectives. Metabolism is the very diverse array of enzymatic reactions and transport processes capable of converting organic compounds into the several molecules needed to support cellular life [34].

To fully exploit the production potential of metabolic networks, cells must regulate the organization of the relevant molecules and pathways that govern these capabilities.In metabolic networks, the enzyme or gene product constitutes the functional unit by acting in one type of chemical reaction or assisting one type of moving process. Metabolism regulation ranges from the regulation of reaction in many levels, including enzymatic and activation/inhibition, all the way to regulating the function of genes undertaking transcription [35, 36].

The cells can control their metabolic pathways by physically manipulating the individual functional units comprising the metabolic network. Thus, trying to understand the regulating mechanisms that the cell uses in operating its metabolic networks will allow analysis and understanding of metabolic pathways. Due to the complexity of metabolic networks, the primary challenge in graphically representing metabolic pathways is determining the overall model that includes the correct components—namely, the metabolites.

The major challenge in the representation of metabolic pathways is that there can be many ways around, thus being able to link the substrates to the intended metabolites, given the large number of ways a pathway can go. Moreover, identification of the most efficient pathway might be cumbersome or require an excessively high number of constructs to be built and screened in the absence of a well-defined objective coupled with a ranking system that is automated [37 - 40].

Several *in silico* methods have been developed, keeping in mind the properties of the intermediate metabolites, enzymes, metabolic flux, and the host cell. Such bio-computational tools are used to generate the metabolic route and allow visualization by the researcher. Therefore, the computational approaches assist in faster analysis of the metabolic network [41]. The process entailed putting all the routes that could result in the creation of a particular compound in easy-to-understand detail. Thus, finding the ideal variations of potential chemicals, such as enzymes or small molecules, could further lead to an upgrading process efficiency. In addition to providing tools for analysing metabolic routes to emphasize the most pertinent information and simultaneously create a network representation that can ease the visual analysis, computational resources aid in the retrieval of information about metabolic pathways [42].

METABOLIC INTERMEDIATES AND THEIR ROLES

Essential substances called metabolic intermediates are created during the many metabolic processes in living things. They serve as bridges between substrates and products during a cell's metabolism. These substances are important players in the chain of biological reactions known as metabolic pathways, but they are neither the starting molecules (substrates) nor the finished products. These intermediates are essential to several activities, including the pentose phosphate pathway, the citric acid cycle (Krebs cycle), and glycolysis. Developing an understanding of these intermediates is crucial to understanding how cells produce necessary chemicals, obtain energy, and preserve homeostasis [43].

Metabolic intermediates are studied because their measurement provides insight into the way organisms take nutrients and convert them into energy and cellular building blocks, a study that is basic to the discipline of biochemistry and molecular biology. These metabolites can reveal details about the metabolic state of individual cells, tissues, or the entire body. They can also function as biomarkers for a wide range of disorders. For example, the build-up of specific intermediates could be a sign of metabolic disorders such as metabolic syndrome or inherited metabolic abnormalities.

Furthermore, research in metabolic intermediates is driven by advances in high-throughput technologies like mass spectrometry and NMR spectroscopy. These technologies are relevant to the study of disease because changes in the levels of these molecules can alter the progression of disease in such cases as diabetes and cancer. Such technologies now allow the performance of really extensive profiling of metabolites, hence giving a deeper understanding of metabolic pathways and their intermediates [44].

KEY ROLES IN METABOLIC PATHWAYS:

Glycolysis, Citric Acid Cycle (Krebs Cycle), and Oxidative Phosphorylation

The metabolic pathways that are linked and necessary for the synthesis of cellular energy include glycolysis, the Krebs cycle, and oxidative phosphorylation. During glycolysis, two pyruvate molecules are produced from glucose in the cytoplasm, leaving two ATP and two NADH molecules as by-products. Providing energy and building blocks for subsequent routes, this anaerobic mechanism is the first stage of glucose metabolism. When pyruvate from glycolysis enters the mitochondria, acetyl-CoA is produced from it, and it enters the citric acid cycle. The citric acid cycle, or Krebs cycle, is housed in the mitochondrial matrix. Here, acetyl-CoA joins forces with oxaloacetate to create citrate, which then goes through a sequence of events that result in the production of CO_2, NADH, $FADH_2$, and GTP (or ATP). The subsequent step, oxidative phosphorylation, depends on these high-energy electron carriers ($FADH_2$ and NADH) (Fig. **4**) [28, 43].

Electrons from $FADH_2$ and NADH must travel through the electron transport chain to reach the inner mitochondrial membrane in order to participate in oxidative phosphorylation. This results in a proton gradient across the membrane, which drives ATP synthase to make ATP as a result. When oxygen acts as the final acceptor of electrons, water is formed as a product of this process. The major source of ATP in aerobic organisms, oxidative phosphorylation efficiently produces energy for cellular functions [23].

Three key metabolic pathways—glycolysis, oxidative phosphorylation, and the Krebs cycle—combine to power cellular energy generation. First, the process occurs in the cytoplasm and is known as glycolysis. This mechanism splits a single glucose molecule into two pyruvate molecules. This process results in a net gain of two ATP and two NADH molecules. After entering the mitochondria, the produced pyruvate is changed into Acetyl- CoA, on the other hand, it exits *via* the citric acid cycle. This cycle begins with acetyl-CoA attached to oxaloacetate, which is then oxidized systematically into high-energy electron carriers (NADH and $FADH_2$), one molecule of GTP (or ATP) and two molecules of CO_2. While oxidative phosphorylation is occurring on the inner mitochondrial membrane, these electron carriers provide their electrons to the transport chain. When electrons flow through the chain and pump protons across the membrane, a gradient of protons is created. The manufacture of ATP is started by this gradient-driven ATP synthase, which pumps protons back into the mitochondrial matrix. Water is generated by the reaction of protons and electrons with oxygen, the final electron acceptor. Because oxidative phosphorylation is the most efficient energy production process in the cell, the majority of ATP produced during aerobic respiration is due to this process (Fig. **5**). Since so many pathways converge on efficiently converting glucose into usable energy, maintaining cellular function and overall metabolic health depends on this [43, 45, 46].

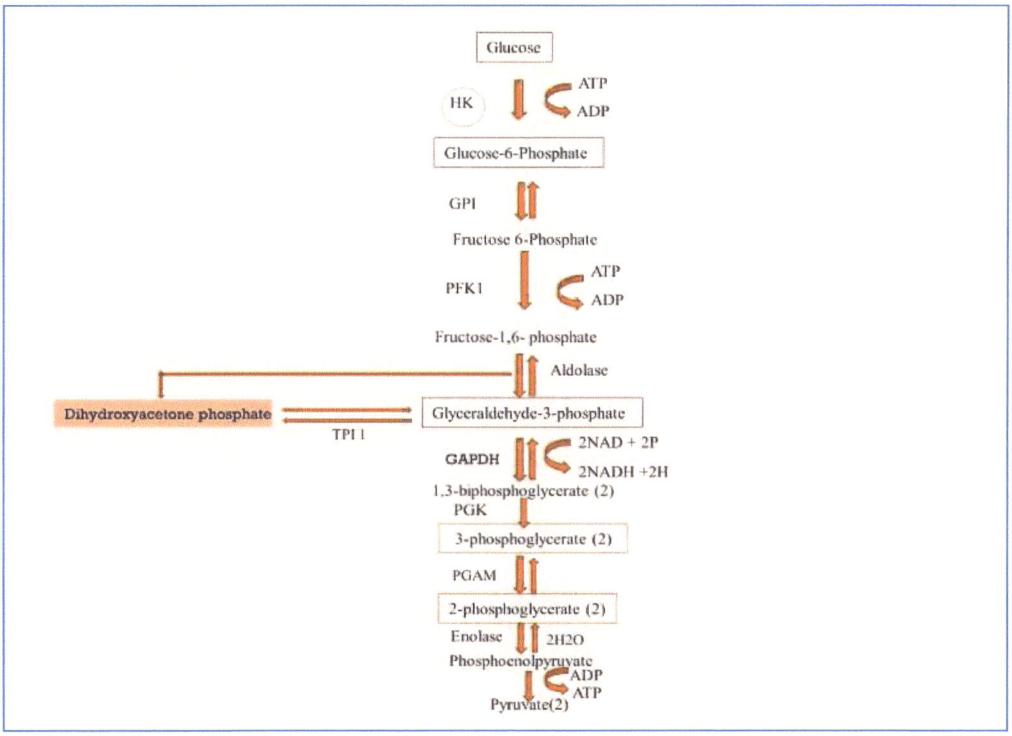

Fig. (4). Flow diagram of the metabolic pathway of glycolysis, citric acid cycle (Krebs cycle) and oxidative Phosphorylation.

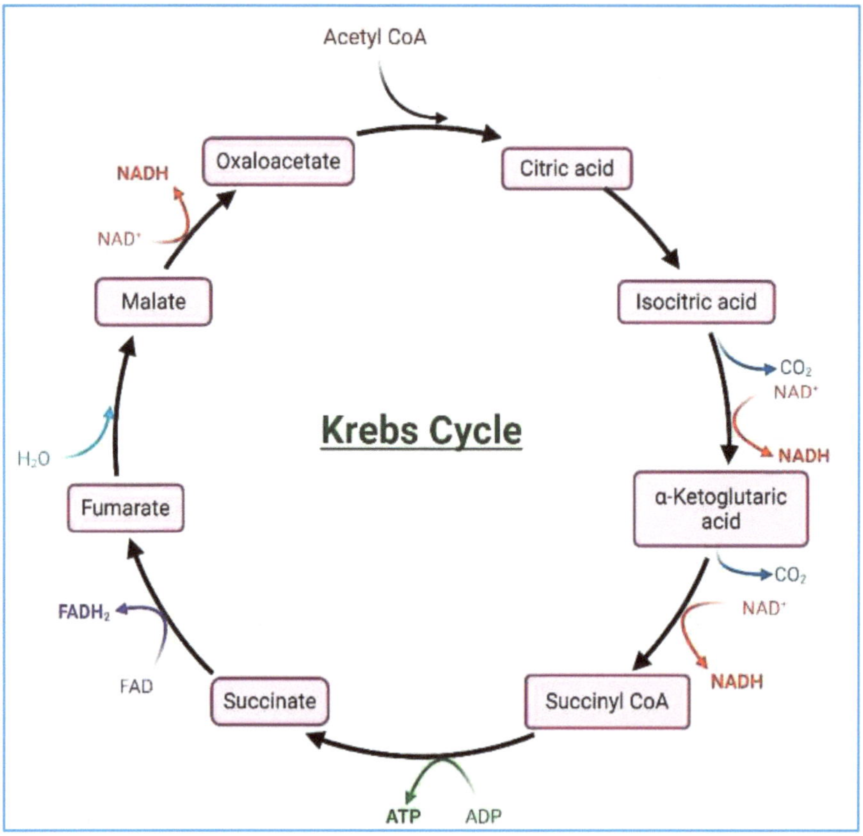

Fig. (5). Kreb's cycle.

Pentose Phosphate Pathway

The pentose phosphate pathway, also known as the hexose monophosphate shunt, runs parallel to glycolysis. It has two stages: the oxidative stage and the non-oxidative stage, with the cytoplasm hosting the majority of the processes. Producing ribose-5-phosphate, which is a building block for the synthesis of nucleotides and nucleic acids, and NADPH are the two main objectives of the PPP, which is needed in various anabolic reactions, including fatty acid and nucleotide synthesis.

During the oxidative phase, glucose-6-phosphate dehydrogenase catalyzes the conversion of glucose-6-phosphate to 6-phosphogluconolactone, producing NADPH in the process. This is a very critical step because NADPH is one of the major reducing agents not only for maintaining the cellular redox state but also for most biosynthesis. Hydrolysis of 6-phosphogluconolactone to 6-phosphogluconate by hydrolytic activity produces yet another NADPH and releases a molecule of CO_2 to ribulose-5-phosphate.

In the non-oxidative phase, carbohydrates are interconverted to generate ribose-5phosphate for nucleotide biosynthesis or regenerate glucose6-phosphate for continued NADPH production. *Ribose-5phosphate* can be isomerized and epimerized to xylulose-5-phosphate or ribose-5-phosphate. These sugars can then undergo a series of transketolase and transaldolase reactions that shuffle carbons to form the intermediates glyceraldehyde-3-phosphate and fructose-6-phosphate that can also be used as glycogen or starch precursors. The PPP is of critical importance for the anabolic pathway, and its role is also engaged in the regeneration of NADPH, which protects cells from oxidative stress. It has the possibility to be flexible in the way cells modulate the product of the pathway and, by such means, allows cells to adapt to metabolic requirements. The pathway is absolutely vital for cells with high metabolic demands, such as those proliferating cells that grow rapidly and the biosynthetic cells producing lipids (Fig. **6**) [8, 17, 47].

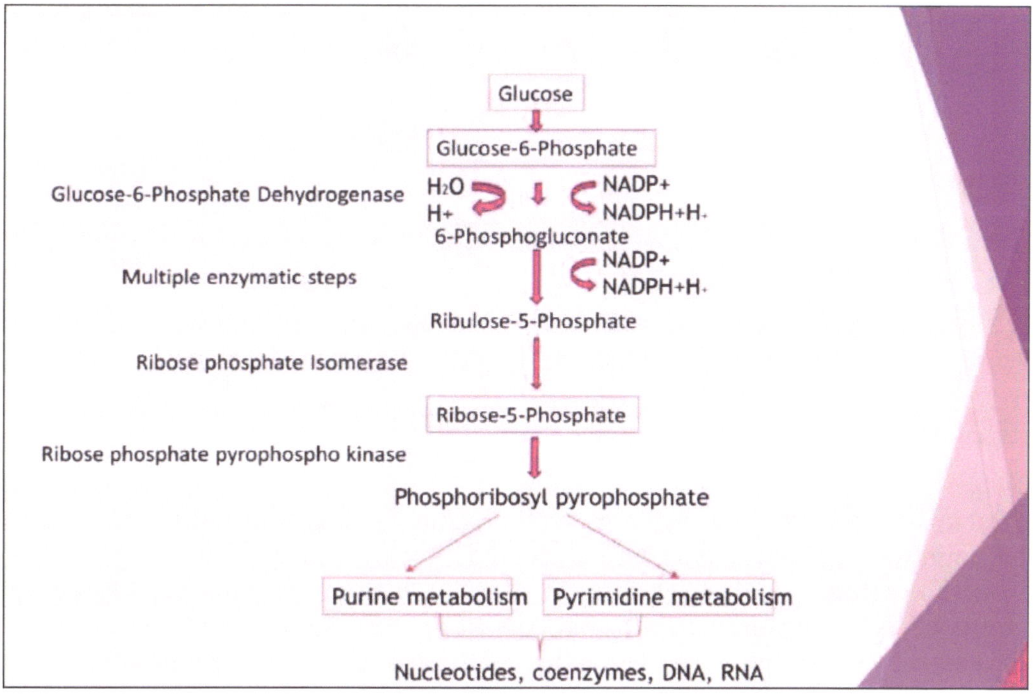

Fig. (6). Flow diagram of pentose phosphate pathway.

Fatty Acid and Amino Acid Metabolism

Key metabolic processes involved in the degradation of fatty acids and amino acids are important to cellular activity, as these metabolic processes provide energy, structural elements, as well as signaling molecules. Lipolysis is a process where triglycerides degrade into glycerol and free fatty acids; hence, it is the initial stage of the breakdown of fatty acids. Once within the mitochondria, these free fatty acids undergo beta-oxidation, which is the removal of two carbon units from the fatty acid chain in order to produce acetyl-CoA one after the other. Acetyl- CoA, once it gets into the Krebs cycle, produces ATP, NADH, and FADH2, which in turn will be used for oxidative phosphorylation to produce more ATP. Long-chain fatty acids, which are necessary for membrane synthesis and energy storage, can also be synthesized in the cytoplasm by means of the fatty acid synthase complex starting with acetyl-CoA, carboxylation to form malonyl-CoA and then extending (Fig. 7) [48, 49].

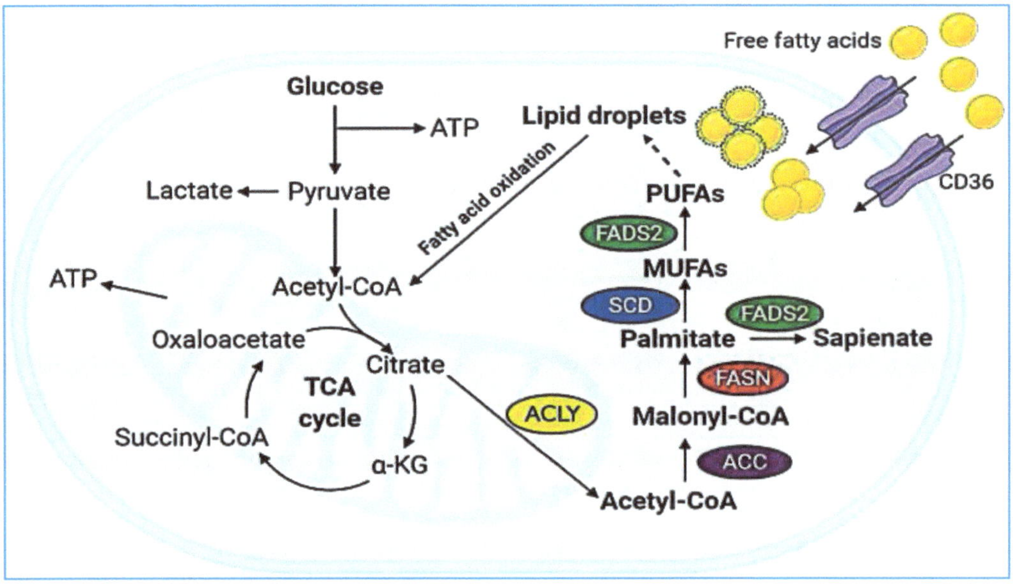

Fig. (7). Flow diagram of Fatty acid metabolism.

Protein synthesis, neurotransmitter synthesis, and energy metabolism all depend on amino acids. The synthesis and degradation of amino acids are processes involved in amino acid metabolism [50]. When the amino group is taken out of amino acids during the catabolic process, a carbon skeleton and ammonia are created. The generation of ATP is aided by the conversion of carbon skeletons to a variety of intermediates, such as pyruvate, acetyl-CoA, or alpha-ketoglutarate, entering the citric acid cycle. Ammonia is transformed into urea *via* the liver's urea cycle, which the kidneys then eliminate. A small number of amino acids are also required as building blocks for the production of neurotransmitters, nucleotides, and other bioactive compounds. In order to provide a supply for protein synthesis and other processes needing these macromolecules, amino acids can also be produced from either the intermediates of the citric acid cycle or glycolysis [46, 51].

INTEGRATION AND REGULATION OF METABOLIC PATHWAYS

The basic mechanisms that guarantee the effective and coordinated operation of the numerous biochemical reactions within cells are the integration and regulation of metabolic pathways. To meet the needs of cells and organisms for energy, building blocks for macromolecules, and homeostasis maintenance, these processes entail the meticulous balance of several metabolic pathways [52].

Integration of Metabolic Pathways

Integration describes how many metabolic routes are integrated, with the products of one pathway acting as substrates for those of another, resulting in a web of biochemical events. For example, the pathways of glycolysis, citric acid cycle, and oxidative phosphorylation are interconnected in the production of energy. Pyruvate, the by-product of glycolysis, enters the citric acid cycle, and oxidative phosphorylation is fuelled by the high-energy electron carriers (NADH and $FADH_2$) produced in the citric acid cycle. In a similar vein, acetyl-CoA, a by-product of fatty acid metabolism, can likewise join the citric acid cycle, connecting the metabolism of carbohydrates and lipids. Further tying protein metabolism with these key pathways is the conversion of amino acids, which are produced from the breakdown of proteins, into intermediates of the citric acid cycle or glycolysis [53].

Regulation of Metabolic Pathways: Regulation prevents the unnecessary or excessive accumulation of intermediates or end products by ensuring that metabolic pathways function at rates that correspond to the needs of the cell [54]. Various strategies are employed to modulate metabolic pathways, such as:

Allosteric Regulation, Feedback Inhibition, Hormonal Regulation, Covalent Modification, Gene Expression

Allosteric Regulation: Allosteric sites, which are locations other than the active site where molecules bind, are often used to regulate the activity of enzymes in metabolic processes. For instance, phosphofructokinase-1 (PFK-1), an enzyme necessary for glycolysis and an indicator of sufficient energy in the cell, is allosterically inhibited by high concentrations of ATP.

Feedback inhibition is the process by which the end result of a pathway inhibits the synthesis of an earlier stage, hence preventing the overproduction of that product [55]. For example, high citrate levels can obstruct PFK-1 and reduce glycolytic flow when the citric acid cycle is saturated.

In **hormonal regulation**, there are two key hormones that regulate metabolic processes: insulin and glucagon. Insulin promotes the storage of glucose as glycogen and enhances lipid synthesis, whereas glucagon increases glucose production and drives the breakdown of glycogen during fasting.

Covalent modification leads addition or removal of phosphate groups, which makes an enzyme active or inactive, and is known as phosphorylation and dephosphorylation. This type of control allows enzyme activity to be altered rapidly and reversibly in response to cellular inputs [56].

Gene expression suggests long-term control of metabolic pathways can result from variations in gene expression, which can boost or decrease the production of metabolism-related enzymes in response to cellular demands [19].

METABOLIC INTERMEDIATES

Molecules known as metabolic intermediates are created and used throughout certain stages of metabolic pathways. These pathways are a sequence of chemical events that take place inside a cell and result in the creation or disintegration of molecules, making it easier for nutrients to be converted into energy and other essential substances. These compounds function as essential precursors or metabolites of physiologically important molecules but are neither the initial substrates nor the end products of metabolic pathways. By serving as "middlemen" between the initial substrate (or reactant) and the finished product *via* a series of enzymatically catalyzed reactions, metabolic intermediates facilitate these processes (Fig. **8**).

Metabolic intermediates are important for cellular function, especially in the regulation of metabolic processes and the creation of treatment solutions, even if they may not seem important at first [57].

Types of Metabolic Intermediates

Based on the particular metabolic processes they participate in, metabolic intermediates are categorized. The primary kinds of metabolic intermediates are listed below.

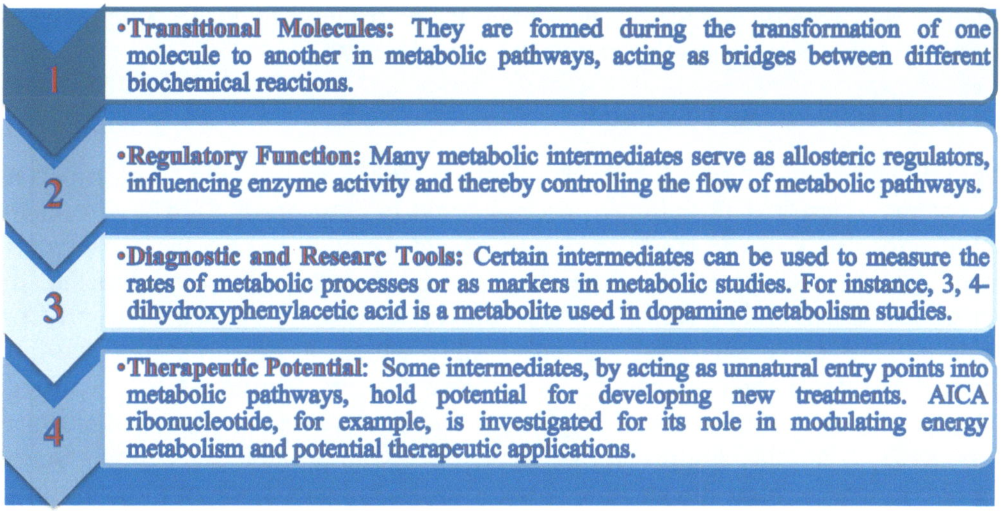

Fig. (8). Development of therapeutic strategies in the metabolic pathways.

Carbohydrate Metabolism Intermediates

Carbohydrate metabolism is an intricate web of metabolic reactions that transform carbs into energy and essential intermediates, essential for the health of cells and organisms. The first step in metabolism is glucose, which is phosphorylated to produce glucose-6-phosphate (G6P), a key intermediate that can go through the pentose phosphate pathway (PPP) or glycolysis. G6P is transformed into fructose-6-phosphate (F6P) and fructose-1,6-bisphosphate (F1,6BP) during glycolysis. F1,6BP splits into two molecules with three carbons each: glyceraldehyde- 3-phosphate (G3P) and dihydroxyacetone phosphate (DHAP). As glycolysis proceeds, G3P produces 1,3-bisphosphoglycerate (1,3BPG), a high-energy intermediate that phosphate-groups ADP to generate ATP and 3-phosphoglycerate (3PG).

As glycolysis proceeds, 3PG is changed into 2-phosphoglycerate (2PG) and phosphoenolpyruvate (PEP). PEP then goes through one last substrate-level phosphorylation to produce pyruvate and extra ATP. A flexible intermediate, pyruvate can enter the citric acid cycle as acetyl-CoA in an aerobic environment or change into lactate in an anaerobic one. In addition, acetyl-CoA is essential for the synthesis of fatty acids and is produced by the citric acid cycle, which forms citrate when it mixes with oxaloacetate (OAA). Intermediates such as α-ketoglutarate, succinate, and malate are produced along with ATP, NADH, and $FADH_2$, which power the electron transport chain as the cycle proceeds. These intermediaries play a crucial role in preserving energy homeostasis, metabolic balance, and general health because they are also engaged in gluconeogenesis, amino acid synthesis, and other biochemical pathways [17, 58].

Lipid Metabolism Intermediates

Lipid metabolism comprises hydrolysis and synthesis of lipids to form important intermediates critical in cellular processes and energy production. There is the initial hydrolysis of the triglycerides to glycerol and free fatty acids. Glycerol enters glycolysis, while the free fatty acids undergo beta-oxidation in the mitochondria and end up forming acetyl-CoA. A crucial intermediary, acetyl-CoA feeds into the citric acid cycle to make ATP, which is used to synthesise fatty acids, cholesterol, and ketone bodies during fasting. Malonyl-CoA, a crucial intermediary that comes from acetyl-CoA and is required for the production of fatty acids, is another significant product. Phosphatidic acid and diacylglycerol are two more essential intermediates required for the synthesis of phospholipids and triglycerides. These chemicals are critical for preserving the integrity of the membrane, energy homeostasis, and general cellular health [17, 59, 60].

Amino Acid Metabolism Intermediates and Nucleotide Metabolism Intermediates

Intermediates produced by amino acid metabolism are essential for biosynthesis, cellular control, and the manufacture of energy. Amino acids lose their amino groups by the process of deamination, resulting in keto acids such as oxaloacetate and alpha-ketoglutarate, which are used to produce ATP in the citric acid cycle. The conversion of pyruvate and other intermediates produced from amino acids into acetyl-CoA connects protein catabolism and energy metabolism. Their versatility in cellular processes is further demonstrated by the fact that these intermediates are precursors for the synthesis of lipids, glucose, and other amino acids [61].

Nucleotide metabolism is the synthesis and breakdown of nucleotides, including important intermediates such as uridine monophosphate (UMP) in pyrimidine metabolism and inosine monophosphate (IMP) in purine metabolism. Whereas UMP produces CTP and TMP, which are necessary for the synthesis of DNA and RNA, IMP acts as a precursor for ATP and GTP. Furthermore, as they connect nucleotide metabolism to the pentose phosphate pathway, intermediates such as PRPP (phosphoribosyl pyrophosphate) are essential to the production of both pyrimidines and purines. These intermediaries are essential for cellular function and the maintenance of genomic integrity since they are needed for cell reproduction, repair, and signalling [62].

Citric acid Cycle (TCA Cycle) Intermediates

The citric acid cycle, often known as the TCA cycle, is a vital metabolic mechanism that produces high-energy electron carriers like NADH and $FADH_2$ and oxidizes acetyl-CoA into carbon dioxide. Acetyl-CoA and oxaloacetate combine to create citrate at the start of the cycle. After that, citrate is changed into isocitrate, which is then oxidized to alpha-ketoglutarate, which is then produced as NADH. Further oxidation and decarboxylation of alpha-ketoglutarate results in succinyl-CoA, which is then converted into NADH in a subsequent step. When succinate is produced from succinyl-CoA, ATP or GTP is produced. $FADH_2$ is then created when succinate is oxidized to fumarate. After fumarate is hydrated to malate, it is oxidized to renew oxaloacetate and produce one more NADH molecule. These intermediaries support the TCA cycle's uninterrupted flow and act as starting points for a number of biosynthetic processes, like the process of making amino acids and nucleotides. The TCA cycle intermediates are essential for linking the metabolism of proteins, lipids, and carbohydrates, ensuring efficient energy production and providing the building blocks for essential biomolecules [43, 63].

ROLE OF INTERMEDIATES IN METABOLIC PATHWAYS

Essential to the complex web of biochemical processes that keep life alive are metabolic intermediates. They are vital connectors that help substrates in metabolic pathways become converted into finished products. From controlling enzyme activity to guaranteeing the appropriate flow and integration of metabolic processes, these intermediates serve a variety of functions (Fig. **9**).

Facilitation of Biochemical Reactions

Allosteric Regulation of Enzyme Activity

Integration and Coordination of Metabolic Pathways

Storage and Mobilization of Energy

Precursor for Biosynthetic Pathways

Role in Detoxification and Waste Elimination

Signalling Molecules

Clinical Relevance and Diagnostic Markers

Fig. (9). Key role of metabolic intermediates.

Glucose-6-Phosphate as a Metabolic Intermediate

Glucose-6-phosphate (G6P) is a critical metabolic intermediate involved in several cellular processes. Gluconeogenesis, glycolysis, the pentose phosphate pathway (PPP), and glycogen metabolism are just a few of the metabolic processes it links. It is a crucial node in the metabolism of carbohydrates. The capacity of G6P to transport carbon generated from glucose into these many metabolic pathways, contingent on cellular requirements, energy status, and nutrient availability, is what gives it significance [64].

The first intermediate in the process of **glycolysis**, which converts glucose into pyruvate and subsequently into ATP and NADH, is G6P. Hexokinase or glucokinase, two liver enzymes, catalyze the phosphorylation of glucose to generate G6P, which traps glucose inside the cell and commits it to additional metabolism. Phosphoglucose isomerase subsequently isomerizes G6P to fructose-6-phosphate (F6P). This is the most critical phase because it prepares G6P for further phosphorylation and cleavage into two three-carbon molecules that would eventually produce the energy-yielding compound ATP [47, 65]. In contrast to glycolysis, it is a crucial substrate for the process known as gluconeogenesis, which produces glucose from non-carbohydrate sources such as lactate, amino acids, and glycerol. Glucose-6-phosphatase catalyzes the final step of gluconeogenesis, a process where G6P is dephosphorylated to form free glucose. In order to maintain blood glucose levels, particularly during periods of fasting or intense activity, this glucose is released into the bloodstream [64]. Another important metabolic pathway branching from G6P is the hexose monophosphate shunt or pentose phosphate pathway. This process involves glucose-6-phosphate dehydrogenase, or G6PD, which oxidizes G6P to form 6-phosphoglucono-δ-lactone and NADPH. The PPP produces NADPH utilized in biosynthetic reactions such as that of fatty acids and nucleotides, as well as in the maintenance of cellular redox state by recycling reduced glutathione. G6P is also an important chemical for anabolic processes and cellular resistance to oxidative stress since the PPP also yields ribose-5-phosphate, which acts as a precursor for the synthesis of nucleotides and nucleic acid because ribose-5-phosphate, a precursor to nucleotide and nucleic acid synthesis, is also produced by the PPP [47, 66]. G6P is essential for glycogen metabolism, which includes synthesis and glycogenolysis. G6P is transformed into glucose-1-phosphate (G1P) during glycogen formation by phosphoglucomutase. G1P is then activated to produce UDP-glucose, which glycogen synthase then polymerizes into glycogen. G1P is created during glycogenolysis and is subsequently changed back into G6P by phosphoglucomutase. While G6P is mostly focused on glycolysis in muscle cells to meet the immediate energy demands of muscular contraction, it can also be dephosphorylated in liver cells to release free glucose into the bloodstream [67].

G6P regulates other enzymes allosterically as well. Glucose synthesis is enhanced when glucose is plentiful due to the potent allosteric activator of glycogen synthase, G6P. As an alternative, it decreases the breakdown of glycogen by inhibiting glycogen phosphorylase. Glycogen synthesis and breakdown are balanced based on the cell's glucose supply and energy requirements, thanks to this dual regulatory function [68].

Pyruvate as a Metabolic Intermediate

As part of the primary process of glucose metabolism, glycolysis, pyruvate is a crucial metabolic intermediary. Pyruvate is the byproduct of glycolysis and is a crucial link in cellular metabolism, connecting the TCA cycle and other metabolic pathways to glycolysis. A cell's energy requirements, the availability of oxygen, and the organism's particular metabolic requirements all affect pyruvate's destiny [43].

MECHANISMS OF PYRUVATE IN VARIOUS BIOLOGICAL ACTIVITIES

Link to the Citric Acid Cycle (TCA Cycle)

Pyruvate plays key roles in several metabolic pathways, including gluconeogenesis, amino acid metabolism, lactate production and the Cori cycle, cellular signaling, and protection against oxidative stress; additionally, acetyl-CoA, derived from pyruvate, serves as a crucial metabolic intermediate.

In conversion to acetyl-CoA the pyruvate dehydrogenase complex (PDC) is mostly responsible for converting pyruvate into acetyl-CoA. The creation of NADH, $FADH_2$, and GTP/ATP, which are necessary for ATP synthesis in the mitochondria, is achieved by acetyl-CoA as it proceeds through the TCA cycle.

In anaplerotic reactions, pyruvate carboxylase, an anaplerotic enzyme that restores TCA cycle intermediates, can carboxylate pyruvate to oxaloacetate [69, 70].

Role in Glucose Synthesis: The gluconeogenic pathway, which generates glucose from non-carbohydrate sources, depends on pyruvate as a necessary substrate in the liver and kidney. Glucose is finally formed from pyruvate by the conversion of oxaloacetate by pyruvate carboxylase and phosphoenolpyruvate by PEP carboxykinase.

Regulation of Blood Sugar Levels: When carbohydrate intake is low, this mechanism is essential for sustaining blood glucose levels during fasting or vigorous activity [71, 72].

Precursor to Alanine: Alanine aminotransferase (ALT), which connects the metabolism of carbohydrates and amino acids, converts pyruvate to alanine. Because this reaction is reversible, alanine can help with gluconeogenesis by acting as a transporter of nitrogen.

Interconnection with Nitrogen Metabolism: Pyruvate can be turned back into pyruvate in the liver *via* the aspartate-from-pyruvate alanine cycle, which releases nitrogen in the form of urea [73, 74].

Pyruvate is converted to lactate during anaerobic glycolysis. Intense activity limits the amount of oxygen available to muscle cells, therefore this process is very crucial.

In the Cori Cycle, the muscle-produced lactate is carried to the liver, where gluconeogenesis transforms it back into glucose. After that, the muscles can absorb this glucose once more to create the Cori cycle, which is necessary to keep the energy balance during extended exercise [75, 76].

Both pyruvate itself and its byproducts, such as lactate and acetyl-CoA, can function as signaling molecules and affect several metabolic processes. For instance, pyruvate levels can influence pyruvate dehydrogenase activity, which in turn controls the ratio of glycolysis to oxidative phosphorylation and regulates metabolic pathways.

Impact on Gene Expression: It has been demonstrated that pyruvate and its metabolites affect gene expression by modifying gamma coactivator 1-alpha (PGC-1α) and hypoxia-inducible factor (HIF), two transcription factors that are involved in energy metabolism and the cell's response to hypoxia, are active [77, 78].

According to reports, pyruvate possesses antioxidant qualities that shield (ROS) and protect cells from oxidative damage. Raising the amounts of glutathione, a vital intracellular antioxidant, can help strengthen the cellular antioxidant defence mechanism.

Neuroprotective Effects: Pyruvate has demonstrated possible neuroprotective benefits in neurological tissues by lowering oxidative stress and supporting metabolism during ischemia [79].

Acetyl-CoA, also known as acetyl coenzyme A, is a critical metabolic intermediate to which proteins, lipids, and carbohydrates are metabolized. For this reason, it plays a crucial role in the synthesis of cholesterol and other isoprenoids, the TCA cycle, and the synthesis and oxidation of fatty acids. Acetyl-CoA serves as an acetyl carrier; it assists in transferring two-carbon units across metabolic pathways [63, 80].

Alpha-Ketoglutarate as a Metabolic Intermediate

In the tricarboxylic acid cycle (TCA cycle, sometimes referred to as the Krebs or citric acid cycle), alpha-ketoglutarate (α-KG), sometimes called 2-oxoglutarate, is an essential metabolic step. It is essential for many metabolic pathways, amino acid metabolism, and the synthesis of cellular energy. The regulation of metabolic flux in both anabolic and catabolic processes depends on alpha-ketoglutarate, a crucial component in the movement of carbon and nitrogen throughout the cell [81, 82].

We will focus on the roles of alpha-ketoglutarate in various biological processes, including the TCA cycle, amino acid metabolism, redox balance and oxidative stress, nitrogen metabolism, involvement in biosynthetic pathways, its impact on aging and longevity, its role in cancer metabolism and neurodegenerative diseases, as well as the function of oxaloacetate as a metabolic intermediate.

Energy Production: The α-KGDH catalyzes isocitrate to produce α-KG, which then undergoes further conversion through oxidative decarboxylation into succinyl-CoA. This reaction is coupled with the reduction of NAD^+ into NADH, which then, in turn, will assist the *ETC* in going through oxidative phosphorylation to generate ATP. Generation of α-KG ensures the energy balance in the cell.

The TCA cycle's regulation indicates that reactions are highly controlled, and one of the cycle's main regulatory sites is α-KGDH. It reacts to cellular energy levels and is feedback suppressed by succinyl-CoA and NADH, two of its products. High ATP concentrations also inhibit the enzyme [81].

Transamination Reactions: Alpha-ketoglutarate is also necessary for transamination reactions. In this reaction, when α-KG receives an amino group from an amino acid, glutamate is formed along with the appropriate alpha-keto acid. This pathway is important in deaminating amino acids and removing excess nitrogen from the body.

Glutamate and Glutamine Synthesis: Glutamate, formed from α-KG, can be subsequently transformed into glutamine, an amino acid involved in nitrogen transport, protein synthesis, and the regulation of acid-base balance. Glutamate and glutamine also enable the production of various other amino acids, along with neurotransmitters like GABA [83, 84].

Antioxidant Properties: Alpha-ketoglutarate is able to scavenge for responsive oxygen species and, therefore, can reduce oxidative stress, showing antioxidant capabilities. It acts as a substrate for enzymes that control the levels of hypoxia-inducible factors, for instance, prolyl hydroxylases. α-KG may modulate cellular responses to both oxidative stress and hypoxia due to its regulation of HIF activity.

Energy metabolism and mitochondrial function depend on α-KG, a crucial TCA cycle intermediate. Deficient α-KG metabolism can result in mitochondrial malfunction, which exacerbates oxidative damage and the generation of more ROS [85, 86].

As a byproduct of producing glutamate and glutamine, alpha-ketoglutarate has an indirect role in the urea cycle. In animals, urea is the main excretory form of nitrogen, and these amino acids aid in its production. The body's nitrogen homeostasis is preserved by α-KG's participation in nitrogen metabolism.

Amino Acid Catabolism: When there is a breakdown of proteins or an excess of amino acids consumed, α-KG helps with the breakdown of amino acids by taking up amino groups, turning them into glutamate, and then producing urea and ammonia [87].

Biosynthesis of Non-Essential Amino Acids: Several non-essential amino acids, including proline, arginine, glutamate, and glutamine, are biosynthesized from alpha-ketoglutarate. These amino acids are crucial for cell signaling, protein synthesis, and the production of other biomolecules.

Production of Neurotransmitters: As α-KG plays a part in the synthesis of glutamate, which is generated from glutamate by the enzyme glutamate decarboxylase, it indirectly supports the creation of neurotransmitters such as GABA. Neuronal excitability is regulated by GABA, a significant inhibitory neurotransmitter in the CNS [88, 89].

Research on model animals like *C. elegans* and mice has shown that giving α-KG supplements can extend longevity and delay the onset of age-related disorders, which is the role of lifespan extension. Some of the proposed mechanisms include improved mitochondrial function, reduced oxidative stress, and metabolic and epigenetic processes associated with aging. A caloric restriction mimetic has been proposed that alpha-ketoglutarate replicates the benefits of caloric restriction, a well-known strategy that extends life. Similar to calorie restriction, α-KG supplementation may enhance healthspan and longevity by impacting metabolic and signaling pathways [90, 91].

Modified Metabolism of α-KG in cancerous cells atypical cells often have altered α-KG metabolism to promote endurance and rapid cell division. For example, changes in the isocitrate dehydrogenase (IDH) enzyme might lead to the production of 2-hydroxyglutarate (2-HG), an oncometabolite, instead of α-KG. 2. HG may inhibit α-KG-dependent dioxygenases, which might lead to aberrant gene expression and promote the growth of cancer. Potential therapeutic drugs for the treatment of cancer include alpha-ketoglutarate and its analogs. The metabolic and epigenetic alterations linked to the advancement of cancer might potentially be undone by returning α-KG levels and activity to baseline [92].

Neuroprotective Effects: Research has demonstrated that alpha-ketoglutarate exhibits neuroprotective properties in models of neurodegenerative illnesses like Parkinson's and Alzheimer's. Its function in energy metabolism, antioxidant defense, and neurotransmitter synthesis regulation is suggested to mediate these effects [93].

Impact on Mitochondrial Dysfunction: Since α-KG plays a crucial role in mitochondrial function, any disruptions in its metabolism may lead to mitochondrial dysfunction, which is a characteristic of numerous neurodegenerative disorders. In some cases, α-KG supplements or methods to increase its production might be therapeutically beneficial [94].

In numerous important metabolic pathways, such as the tricarboxylic acid cycle, often known as the Krebs or citric acid cycle, and gluconeogenesis, oxaloacetate (OAA) is an essential metabolic intermediary. As a dicarboxylic acid containing four carbons, OAA is essential for the generation of energy, amino acid biosynthesis, and preservation of cellular metabolic equilibrium. It is an essential modulator of cellular metabolism and a hub for the integration of anabolic and catabolic pathways due to its many metabolic processes.

ROLES OF OXALOACETATE IN VARIOUS BIOLOGICAL ACTIVITIES

Role in the TCA Cycle

The following sections discuss oxaloacetate's roles in gluconeogenesis, amino acid biosynthesis and metabolism, the malate-aspartate shuttle, the glyoxylate cycle, redox homeostasis and oxidative stress, aging and lifespan regulation, neuroprotection, and cancer metabolism.

In aerobic organisms, the key metabolic hub for the creation of energy is the TCA cycle. The cycle starts when acetyl-CoA and oxaloacetate combine to form citrate. The electron transport chain utilizes high-energy electron carriers (NADH and $FADH_2$) produced by the TCA cycle to generate ATP, and the continuation of this process depends on the ongoing supply of these carriers. TCA cycle rate can be controlled by OAA availability. Slowing down the cycle can have an effect on ATP generation if OAA levels are low. Increasing OAA, on the other hand, can activate the cycle and increase energy production to fulfill cellular requirements [95].

An essential step in the gluconeogenic process is oxaloacetate, which the enzyme PEPCK transforms into PEP. When lactate, glycerol, and amino acids are used as non-carbohydrate precursors to synthesize glucose, this process is essential. Blood glucose levels must be maintained during fasting or vigorous physical exercise, and this process is mostly carried out by the liver and kidneys.

OAA is essential for preserving blood glucose homeostasis because of its role in gluconeogenesis. The brain and red blood cells, two tissues that are highly dependent on glucose, can always get it from gluconeogenesis, whether fasting or eating a low-carb diet [96].

Oxaloacetate is transaminated by aspartate aminotransferase to create the non-essential amino acid aspartate. Apart from its involvement in the urea cycle, aspartate gives nucleotides and other amino acids a nitrogen atom during their production.

A number of amino acids, including asparagine, are catabolized by OAA through deamination mechanisms, which direct the carbon skeletons of these amino acids into the TCA cycle for energy production [97].

The malate-aspartate shuttle is one of the most significant routes for transferring reducing equivalents (NADH) from the cytosol into the mitochondria. Cytosolic malate dehydrogenase in this shuttle converts OAA to malate, which is then carried out into the mitochondria. Malate is converted back to OAA inside the mitochondria, which replenishes NADH for the electron transport chain.

Maintaining the balance of NAD^+ and NADH in the cytosol is essential for glycolysis and other metabolic processes that require NAD^+. Oxaloacetate (OAA) contributes to this balance by participating in mitochondrial transport mechanisms that help shuttle reducing equivalents [96].

Oxaloacetate is involved in the glyoxylate cycle, an anabolic process that enables organisms to transform fatty acids into carbohydrates, in bacteria, fungi, and plants. This cycle allows the net synthesis of glucose from acetyl-CoA obtained from fatty acids by avoiding the TCA cycle's decarboxylation phases. When plant germination occurs, the glyoxylate cycle plays a crucial role in helping the plants turn stored lipids into sugars that will fuel growth. *in situ* ations where carbohydrates are scarce, bacteria can survive thanks to the glyoxylate cycle [71].

Oxaloacetate has been reported to have antioxidant properties that enable it to scavenge reactive oxygen species (ROS) and reduce oxidative stress in cells. OAA helps protect its components from oxidative damage by controlling the redox state of the cell.

The synthesis of reducing equivalents (NADPH) and other molecules that prevent oxidative damage can be prioritized during oxidative stress by changing the metabolic flux *via* the TCA cycle and other pathways involving OAA [98].

Oxaloacetate's involvement in the TCA cycle is essential for preserving mitochondrial efficiency, as mitochondrial failure is a defining feature of aging. Improved mitochondrial activity has been linked in a number of model species to longer lifespans and fewer age-related illnesses.

Research indicates that supplementing with oxaloacetate could replicate certain advantages of calorie restriction, like improved mitochondrial performance, decreased oxidative stress, and longer life span. It is thought that its function in regulating cellular metabolism and energy balance mediates these effects [99]. When it comes to neurodegenerative illnesses like Parkinson's and Alzheimer's, oxaloacetate has been investigated as a possible neuroprotective treatment. It may have protective benefits on neuronal cells because of its involvement in energy metabolism, antioxidant defense, and neurotransmitter production.

Oxaloacetate is converted to aspartate, which is then used in the synthesis of neurotransmitters such as glutamate and GABA. Therefore, it plays an essential role in maintaining neurotransmitter balance and supporting healthy brain function [100].

The metabolic pathways of cancer cells are frequently changed to facilitate their fast growth and multiplication. These alterations are linked to oxaloacetate metabolism because they can be redirected to biosynthetic pathways that create macromolecules required for cell division.

It is being investigated as a possible cancer treatment approach to target the mctabolic pathways involving oxaloacctatc, such as thc TCA cyclc and gluconeogenesis. Treatments that target OAA metabolism may prevent tumor growth and survival by impairing the metabolic flexibility of cancer cells. An adaptable and vital metabolic intermediate, oxaloacetate plays important functions in the synthesis of amino acids, gluconeogenesis, energy production, and redox equilibrium. It is crucial for preserving cellular homeostasis and adjusting to diverse physiological situations, as evidenced by its involvement in a number of metabolic processes. Moreover, oxaloacetate is a major study target for creating treatment plans against age-related illnesses, neurological conditions, and cancer due to its roles in aging, neuroprotection, and cancer metabolism. OAA's various biological activities and possible medical uses will probably become clearer as its roles in cellular metabolism are investigated further [101, 102].

CONCLUSION

Metabolism, the complex network of chemical processes that sustain cellular and organismal health, involves pathways that convert substrates to end products through enzymatic reactions. These pathways are categorized as anabolic (building complex molecules) and catabolic (breaking down molecules), essential for energy production, biosynthesis, and waste elimination. Key intermediates, like pyruvate and acetyl-CoA, play crucial roles in maintaining metabolic balance and signaling. Advances in technologies such as metabolomics and CRISPR-Cas9 have enhanced our understanding of these pathways and their roles in health and disease. Incomplete metabolism, where nutrients are not fully broken down, leads to the accumulation of toxic intermediates and depletion of vital metabolites, contributing to conditions like diabetes, obesity, cardiovascular diseases, and cancer. Dysregulation of energy metabolism, particularly in mitochondria, is central to many diseases, highlighting the importance of metabolic flexibility. Understanding these processes is vital for developing targeted therapies and improving health outcomes.

REFERENCES

[1] G. Di Conza, and P.C. Ho, "Metabolic adaptation of macrophages in chronic diseases", *Cancer Lett.*, vol. 414, pp. 250-256, 2018. [http://dx.doi.org/10.1016/j.canlet.2017.11.023] [PMID: 29175462]

[2] G. Fahed, L. Aoun, M. Bou Zerdan, S. Allam, M. Bou Zerdan, Y. Bouferraa, and H.I. Assi, "Metabolic syndrome: Updates on pathophysiology and management in 2021", *Int. J. Mol. Sci.*, vol. 23, no. 2, p. 786, 2022. [http://dx.doi.org/10.3390/ijms23020786] [PMID: 35054972]

[3] M. S. Patel and R. A. Harris, "Metabolic Regulation," In: *Encyclopedia of Cell Biology*, 2nd ed., R. A. Bradshaw, G. W. Hart, and P. D. Stahl, Eds., vol. 1. Elsevier, 2023, pp. 353–365. [http://dx.doi.org/10.1016/B978-0-12-821618-7.00147-4]

[4] A. Singh, L.B. Prasad, K. Shiv, R. Kumar, and S. Garai, "Synthesis, characterization, and in vitro antibacterial and cytotoxic study of Co(II), Ni(II), Cu(II), and Zn(II) complexes of N- (4-methoxybenzyl) N-(phenylethyl) dithiocarbamate ligand", *J. Mol. Struct.*, vol. 1288, no. May, p. 135835, 2023. [http://dx.doi.org/10.1016/j.molstruc.2023.135835]

[5] M.S. Saghiv, and M.S. Sagiv, "Metabolism", In: *Basic Exercise Physiology: Clinical and Laboratory Perspectives.* Springer International Publishing: Cham, 2020, pp. 33-147. [http://dx.doi.org/10.1007/978-3-030-48806-2_2]

[6] A. Singh, R. Kumar, K. Shiv, S.K. Pandey, M.K. Bharty, R.J. Butcher, and L.B. Prasad, "Synthesis, crystal structure and screening for anticonvulsant and antianxiety activities of three new Ni(II), Cu(II), and Zn(II) dithiocarbamate complexes", *J. Mol. Struct.,* vol. 1298, no. P2, p. 137052, 2024. [http://dx.doi.org/10.1016/j.molstruc.2023.137052]

[7] R. Jeannotte, "Metabolic pathways: nitrogen metabolism", In: *Encyclopedia of Food Microbiology.* M. L. Tortorello and C. A. Batt, Eds., Elsevier, 2014, pp. 544–560. [http://dx.doi.org/10.1016/B978-0-12-384730-0.00199-3]

[8] T. TeSlaa, M. Ralser, J. Fan, and J.D. Rabinowitz, "The pentose phosphate pathway in health and disease", *Nat. Metab.,* vol. 5, no. 8, pp. 1275-1289, 2023. [http://dx.doi.org/10.1038/s42255-023-00863-2] [PMID: 37612403]

[9] K.A. Kripps, P.R. Baker II, J.A. Thomas, H.E. Skillman, L. Bernstein, S. Gaughan, C. Burns, C.R. Coughlin II, S.E. McCandless, A.A. Larson, A. Kochar, C.F. Stillman, E.M. Wymore, E.G. Hendricks, M. Woontner, and J.L.K. Van Hove, "REVIEW: Practical strategies to maintain anabolism by intravenous nutritional management in children with inborn metabolic diseases", *Mol. Genet. Metab.,* vol. 133, no. 3, pp. 231-241, 2021. [http://dx.doi.org/10.1016/j.ymgme.2021.04.007] [PMID: 33985889]

[10] A. Singh, R. Kumar, R. Patel, Trishna, R. N. Gautam, M. K. Bharty, and L. B. Prasad, "Dithiocarbamate-based novel anti-histaminic agents: synthesis, characterization, crystal structure and thermal study", In: Dalt. Trans, vol. 33, 2024. [http://dx.doi.org/10.1039/D4DT01777C]

[11] R. van Dijk, U. Beuers, and P.J. Bosma, "Gene replacement therapy for genetic hepatocellular jaundice", *Clin. Rev. Allergy Immunol.,* vol. 48, no. 2-3, pp. 243-253, 2015. [http://dx.doi.org/10.1007/s12016-014-8454-7] [PMID: 25315738]

[12] "COVID-19: a challenge towards the sustainability of health in platform era", In: *NeuroQuantology* vol. 20. , 2022, no. 11, pp. 2503-2519. [http://dx.doi.org/10.47750/pnr.2022.13]

[13] A. Judge, and M.S. Dodd, "Metabolism", *Essays Biochem.,* vol. 64, no. 4, pp. 607-647, 2020. [http://dx.doi.org/10.1042/EBC20190041] [PMID: 32830223]

[14] R. Kumar, S. Dutt, A. Goyal, A. Kumar, and B. Kumar, "Nanotechnology in gastrointestinal endoscopy," In: *Nanobiotechnology and Artificial Intelligence in Gastrointestinal Diseases*, V. K. Chaturvedi, A. K. Singh, J. Singh, and D. P. Yadav, Eds., 2024, pp. 5-1–5-32. [http://dx.doi.org/10.1088/978-0-7503-6134-7ch5]

[15] J.J. McCarthy, and K.A. Esser, "Anabolic and catabolic pathways regulating skeletal muscle mass", *Curr. Opin. Clin. Nutr. Metab. Care,* vol. 13, no. 3, pp. 230-235, 2010. [http://dx.doi.org/10.1097/MCO.0b013e32833781b5] [PMID: 20154608]

[16] A. Hamid, R. Kumar, V. K. Chaturvedi, S. Dutt, G. Sulabh, V. Kumar, and D. P. Yadav, "Artificial intelligence (AI) based colonoscopy," In: *Nanobiotechnology and Artificial Intelligence in Gastrointestinal Diseases*, V. K. Chaturvedi, A. K. Singh, J. Singh, and D. P. Yadav, Eds., 2024, pp. 12-1–12-23. [http://dx.doi.org/10.1088/978-0-7503-6134-7ch12]

[17] F. Rajas, A. Gautier-Stein, and G. Mithieux, "Glucose-6 Phosphate, a Central Hub for Liver Carbohydrate Metabolism", *Metabolites,* vol. 9, no. 12, p. 282, 2019. [http://dx.doi.org/10.3390/metabo9120282] [PMID: 31756997]

[18] A. Kumar, P. Singh, R. Kumar, and S. Dutt, "Novel drug delivery systems for inflammatory bowel disease," In: *Nanobiotechnology and Artificial Intelligence in Gastrointestinal Diseases*, V. K. Chaturvedi, A. K. Singh, J. Singh, and D. P. Yadav, Eds., 2024, pp. 4-1–4-24. [http://dx.doi.org/10.1088/978-0-7503-6134-7ch4]

[19] M. Hatting, C.D.J. Tavares, K. Sharabi, A.K. Rines, and P. Puigserver, "Insulin regulation of

gluconeogenesis", *Ann. N. Y. Acad. Sci.,* vol. 1411, no. 1, pp. 21-35, 2018. [http://dx.doi.org/10.1111/nyas.13435] [PMID: 28868790]

[20] S. Kersten, "Integrated physiology and systems biology of PPARα", *Mol. Metab.,* vol. 3, no. 4, pp. 354-371, 2014. [http://dx.doi.org/10.1016/j.molmet.2014.02.002] [PMID: 24944896]

[21] T. Sato, Y. Yoshida, A. Morita, N. Mori, and S. Miura, "Glycerol-3-phosphate dehydrogenase 1 deficiency induces compensatory amino acid metabolism during fasting in mice", *Metabolism,* vol. 65, no. 11, pp. 1646-1656, 2016. [http://dx.doi.org/10.1016/j.metabol.2016.08.005] [PMID: 27733253]

[22] A. Kumar, B. Kumar, R. Kumar, A. Kumar, M. Singh, V. Tiwari, A. Trigunayat, P. Paul, and P. Singh, "Acute and subacute toxicity study of ethanolic extract of Calotropis procera (Aiton) Dryand flower in Swiss albino mice", *Phytomed. Plus,* vol. 2, no. 2, p. 100224, 2022. [http://dx.doi.org/10.1016/j.phyplu.2022.100224]

[23] J. Pelley, "Citric Acid Cycle, Electron Transport Chain, and Oxidative Phosphorylation," In: *Elsevier's Integrated Biochemistry*, 2007, pp. 55–63. [http://dx.doi.org/10.1016/B978-0-323-03410-4.50013-4]

[24] R. Dhiman, "Novel LC-MS/MS bioanalytical method validation for gabapentin in human plasma in accordance with current regulatory guidances", *Int J Pharm Chem Sci,* vol. 4, 2015no. 4, pp. 462-468. Available from: http://www.ijpcsonline.com/files/08-10-15/07-1035.pdf

[25] A. Judge, and M.S. Dodd, *Metabolism,* vol. 0, no. August, pp. 1-41, 2020.

[26] "Preclinical acute and repeated dose toxicity of Pueraria tuberosa (PTWE) on Charles Foster rats", *Int J Pharm Sci Res.* Available from: https://ijpsr.com/bft-article/preclinical-acute-and-repeated-dose-toxicity-of--ueraria-tuberosa-ptwe-on-charles-foster-rats/

[27] S.T. Keating, and A. El-Osta, "Epigenetics and Metabolism", *Circ. Res.,* vol. 116, no. 4, pp. 715-736, 2015. [http://dx.doi.org/10.1161/CIRCRESAHA.116.303936] [PMID: 25677519]

[28] R. Chaudhry, and M. Varacallo, "Biochemistry, Glycolysis", *StatPearls,* 2023.

[29] E. Gümüş, and H. Özen, "Glycogen storage diseases: An update", *World J. Gastroenterol.,* vol. 29, no. 25, pp. 3932-3963, 2023. [http://dx.doi.org/10.3748/wjg.v29.i25.3932] [PMID: 37476587]

[30] R.J. DeBerardinis, and C.B. Thompson, "Cellular metabolism and disease: what do metabolic outliers teach us?", *Cell,* vol. 148, no. 6, pp. 1132-1144, 2012. [http://dx.doi.org/10.1016/j.cell.2012.02.032] [PMID: 22424225]

[31] R. Dhiman, D. Kumar, B. Kumar, and B.L. Pandey, "Quantitative determination of atorvastatin, ortho-hydroxy atorvastatin, para-hydroxy atorvastatin in human plasma using rosuvastatin as internal standard by LC-MS / MS", *Int. J. Pharm. Chem. Sci.,* vol. 4, no. 4, pp. 487-500, 2016.

[32] T.R. Singh, A. Shukla, B. Taoufik, A. Moussa, and B. Vannier, "Disease Biomarker Discovery", In: *Encyclopedia of Bioinformatics and Computational Biology.,* S. Ranganathan, M. Gribskov, K. Nakai, C. Schönbach, Eds., Academic Press: Oxford, 2019, pp. 476-488. [http://dx.doi.org/10.1016/B978-0-12-809633-8.20137-4]

[33] M. Singh, B. Kumar, A. K Singh, C. Shekhar Azad, M. K Yadav, A. Kumar, R. Kumar, A. Kumar, and P. Chaudhary, "Pharmacokinetic screening to Estimate the drug likeliness characteristics of selected Herbal Anticancer Drugs", *Res. J. Pharm. Technol.,* vol. 16, no. 7, pp. 3422-3426, 2023. [http://dx.doi.org/10.52711/0974-360X.2023.00566]

[34] C. Rodríguez-Caso, and R. Montañez, "Topology of Metabolic Reaction Networks", In: *Encyclopedia of Systems Biology.,* W. Dubitzky, O. Wolkenhauer, K-H. Cho, H. Yokota, Eds., Springer New York: New York, 2013, pp. 2185-2188.

[http://dx.doi.org/10.1007/978-1-4419-9863-7_1175]

[35] A. Gutteridge, M. Kanehisa, and S. Goto, "Regulation of metabolic networks by small molecule metabolites", *BMC Bioinformatics,* vol. 8, no. 1, p. 88, 2007. [http://dx.doi.org/10.1186/1471-2105-8-88] [PMID: 17352833]

[36] B. Kumar, R. Kumar, H. Patel, and A. Kumar, "Development and validation of an RP-HPLC method for simultaneous estimation of amlodipine besylate and indapamide in bulk and tablet dosage form", *Indo Am. J. Pharm. Sci.,* vol. 4, no. 11, pp. 4278-4287, 2017.

[37] A.V. Pandit, S. Srinivasan, and R. Mahadevan, "Redesigning metabolism based on orthogonality principles", *Nat. Commun.,* vol. 8, no. 1, p. 15188, 2017. [http://dx.doi.org/10.1038/ncomms15188] [PMID: 28555623]

[38] "B. Kumar, R. Kumar, A. Kumar, and R. Patel, "Simultaneous RP-HPLC method development and validation of fenofibrate and rosuvastatin calcium in bulk and tablet dosage form,"", *Indo Am. J. Pharm. Sci.,* vol. 4, no. 11, pp. 4288-4297, 2017.

[39] K. Ajay, N. Manmath Kumar, K. Brijesh, K. Ashutosh, K. Rajesh, K. Vikas, and S. Ambrish Kumar, "Toxicity (acute and subacute) assessment and in-vivo antiurolithiatic activity of ethanolic extract of Caesalpinia bonducella seed in albino Wistar rat", *J. Appl. Pharm. Sci.,* vol. 12, no. 2, pp. 187-197, 2022. [http://dx.doi.org/10.7324/JAPS.2022.120220]

[40] R. Kumar, B. Kumar, and A. Kumar, *Stability indicating method development and validation of abiraterone in human plasma by using LC-MS/MS.* vol. 3. , 2017, no. 5, pp. 12-26.

[41] W.B. Copeland, B.A. Bartley, D. Chandran, M. Galdzicki, K.H. Kim, S.C. Sleight, C.D. Maranas, and H.M. Sauro, "Computational tools for metabolic engineering", *Metab. Eng.,* vol. 14, no. 3, pp. 270-280, 2012. [http://dx.doi.org/10.1016/j.ymben.2012.03.001] [PMID: 22629572]

[42] S.C. Booth, A.M. Weljie, and R.J. Turner, "Computational tools for the secondary analysis of metabolomics experiments", *Comput. Struct. Biotechnol. J.,* vol. 4, no. 5, p. e201301003, 2013. [http://dx.doi.org/10.5936/csbj.201301003] [PMID: 24688685]

[43] R.J. Mailloux, "The emerging importance of the α-keto acid dehydrogenase complexes in serving as intracellular and intercellular signaling platforms for the regulation of metabolism", *Redox Biol.,* vol. 72, p. 103155, 2024. [http://dx.doi.org/10.1016/j.redox.2024.103155] [PMID: 38615490]

[44] S. Veerappa, and J. McClure, "Intermediary metabolism", *Anaesth. Intensive Care Med.,* vol. 21, no. 3, pp. 162-167, 2020. [http://dx.doi.org/10.1016/j.mpaic.2020.01.003]

[45] J.J. Zimmerman, A. von Saint André-von Arnim, and J. McLaughlin, "Cellular Respiration", In: *Pediatric Critical Care.,* B.P. Fuhrman, J.J. Zimmerman, Eds., 4th ed. Mosby: Saint Louis, 2011, pp. 1058-1072. [http://dx.doi.org/10.1016/B978-0-323-07307-3.10074-6]

[46] A. Salminen, A. Kauppinen, M. Hiltunen, and K. Kaarniranta, "Krebs cycle intermediates regulate DNA and histone methylation: Epigenetic impact on the aging process", *Ageing Res. Rev.,* vol. 16, pp. 45-65, 2014. [http://dx.doi.org/10.1016/j.arr.2014.05.004] [PMID: 24910305]

[47] "A. Stincone, A. Prigione, T. Cramer, M. M. C. Wamelink, K. Campbell, E. Cheung, V. Olin-Sandoval, N.-M. Grüning, A. Krüger, M. T. Alam, M. A. Keller, M. Breitenbach, K. M. Brindle, J. D. Rabinowitz, and M. Ralser, "The return of metabolism: biochemistry and physiology of the pentose phosphate pathway,"", *Biol. Rev. Camb. Philos. Soc.,* vol. 90, no. 3, pp. 927-963, 2015. [http://dx.doi.org/10.1111/brv.12140.The] [PMID: 25243985]

[48] E. Dilektasli, "Estimation of the influence of the environmental factors on the accumulation of

phytochemicals and antioxidant capacity in the ivy leaves (Hedera helix L.)", *J. Ethnopharmacol.,* vol. 27, no. 3, pp. 1-2, 2007.
[http://dx.doi.org/10.1016/0022-4804(91)90184-N]

[49] I. Velikyan, "Radionuclides for Imaging and Therapy in Oncology", In: *Cancer Theranostics.,* X. Chen, S. Wong, Eds., Academic Press: Oxford, 2014, pp. 285-325.
[http://dx.doi.org/10.1016/B978-0-12-407722-5.00017-7]

[50] N.S. Chandel, "Amino Acid Metabolism", *Cold Spring Harb. Perspect. Biol.,* vol. 13, no. 4, p. a040584, 2021.
[http://dx.doi.org/10.1101/cshperspect.a040584] [PMID: 33795250]

[51] I. Rix, C. Nexøe-Larsen, N.C. Bergmann, A. Lund, and F.K. Knop, "Glucagon Physiology", In: *South Dartmouth, MA,* K. R. Feingold, B. Anawalt, M. R. Blackman, A. Boyce, G. Chrousos, E. Corpas, W. W. de Herder, K. Dhatariya, K. Dungan, J. Hofland, S. Kalra, G. Kaltsas, N. Kapoor, C. Koch, P. Kopp, M. Korbonits, C. S. Kovacs, W. Kuohung, B. Laferrère, M. Levy, E. A. McGee, R. McLachlan, M. New, J. Purnell, R. Sahay, A. S. Shah, F. Singer, M. A. Sperling, C. A. Stratakis, D. L. Trence, D. P. Wilson, Eds., , 2000.

[52] J. Zhang, Y. Zheng, L. Martens, and A.F.H. Pfeiffer, "The regulation and secretion of glucagon in response to nutrient composition: Unraveling their intricate mechanisms", *Nutrients,* vol. 15, no. 18, p. 3913, 2023.
[http://dx.doi.org/10.3390/nu15183913] [PMID: 37764697]

[53] T. McKee, and J.R. McKee, "Integration of Metabolism", *Science Trove,* Aug. 31, 2020.
[http://dx.doi.org/10.1093/hesc/9780190847685.003.0016]

[54] L. Bar-Peled, and N. Kory, "Principles and functions of metabolic compartmentalization", *Nat. Metab.,* vol. 4, no. 10, pp. 1232-1244, 2022.
[http://dx.doi.org/10.1038/s42255-022-00645-2] [PMID: 36266543]

[55] T. Ross, and D.S. Nichols, "Ecology of bacteria and fungi in foods: influence of temperature", In: *Encyclopedia of Food Microbiology (Second Edition),* C. A. Batt and M. Lou Tortorello, Eds. Oxford: Academic Press, 2014, pp. 602-609.
[http://dx.doi.org/10.1016/B978-0-12-384730-0.00087-2]

[56] Z. Tao, and Z. Cheng, "Hormonal regulation of metabolism—recent lessons learned from insulin and estrogen", *Clin. Sci. (Lond.),* vol. 137, no. 6, pp. 415-434, 2023.
[http://dx.doi.org/10.1042/CS20210519] [PMID: 36942499]

[57] R. Haas, D. Cucchi, J. Smith, V. Pucino, C.E. Macdougall, and C. Mauro, "Intermediates of metabolism: From bystanders to signalling molecules", *Trends Biochem. Sci.,* vol. 41, no. 5, pp. 460-471, 2016.
[http://dx.doi.org/10.1016/j.tibs.2016.02.003] [PMID: 26935843]

[58] H.A. Krebs, "Intermediate metabolism of carbohydrates", *Nature,* vol. 138, no. 3485, pp. 288-289, 1936.
[http://dx.doi.org/10.1038/138288b0]

[59] L. Badimon, and G. Chiva-Blanch, "Lipid metabolism in dyslipidemia and familial hypercholesterolemia", In: *The Molecular Nutrition of Fats.,* V.B. Patel, Ed., Academic Press, 2019, pp. 307-322.
[http://dx.doi.org/10.1016/B978-0-12-811297-7.00024-X]

[60] D. Gyamfi, E. Ofori Awuah, and S. Owusu, "Lipid metabolism: An overview", In: *The Molecular Nutrition of Fats.,* V.B. Patel, Ed., Academic Press, 2019, pp. 17-32.
[http://dx.doi.org/10.1016/B978-0-12-811297-7.00002-0]

[61] Z.N. Ling, Y.F. Jiang, J.N. Ru, J.H. Lu, B. Ding, and J. Wu, "Amino acid metabolism in health and disease", *Signal Transduct. Target. Ther.,* vol. 8, no. 1, p. 345, 2023.
[http://dx.doi.org/10.1038/s41392-023-01569-3] [PMID: 37699892]

[62] E. Stashi, B. York, and B.W. O'Malley, "Steroid receptor coactivators: servants and masters for control of systems metabolism", *Trends Endocrinol. Metab.,* vol. 25, no. 7, pp. 337-347, 2014. [http://dx.doi.org/10.1016/j.tem.2014.05.004] [PMID: 24953190]

[63] I. Choi, H. Son, and J.H. Baek, "Tricarboxylic acid (TCA) cycle intermediates: Regulators of immune responses", *Life (Basel),* vol. 11, no. 1, p. 69, 2021. [http://dx.doi.org/10.3390/life11010069] [PMID: 33477822]

[64] L.R. Engelking, "Gluconeogenesis," In: *Textbook of Veterinary Physiological Chemistry*, 3rd ed., 2015, pp. 225–230. [http://dx.doi.org/10.1016/B978-0-12-391909-0.50037-2]

[65] A.M. Mahmoud, "An overview of epigenetics in obesity: The role of lifestyle and therapeutic interventions", *Int. J. Mol. Sci.,* vol. 23, no. 3, p. 1341, 2022. [http://dx.doi.org/10.3390/ijms23031341] [PMID: 35163268]

[66] Z.E. Stine, B.J. Altman, A.L. Hsieh, A.M. Gouw, and C.V. Dang, "Deregulation of the cellular energetics of cancer cells", In: *Pathobiology of Human Disease.,* L.M. McManus, R.N. Mitchell, Eds., Academic Press: San Diego, 2014, pp. 444-455. [http://dx.doi.org/10.1016/B978-0-12-386456-7.01912-2]

[67] M.M. Adeva-Andany, M. González-Lucán, C. Donapetry-García, C. Fernández-Fernández, and E. Ameneiros-Rodríguez, "Glycogen metabolism in humans", *BBA Clin.,* vol. 5, pp. 85-100, 2016. [http://dx.doi.org/10.1016/j.bbacli.2016.02.001] [PMID: 27051594]

[68] B. Kobe, and B.E. Kemp, "Principles of kinase regulation", In: *Handbook of Cell Signaling.,* R.A. Bradshaw, E.A. Dennis, Eds., 2nd ed. Academic Press: San Diego, 2010, pp. 559-563. [http://dx.doi.org/10.1016/B978-0-12-374145-5.00074-7]

[69] W. Yuan, Y. Du, K. Yu, S. Xu, M. Liu, S. Wang, Y. Yang, Y. Zhang, and J. Sun, "The production of pyruvate in biological technology: A critical review", *Microorganisms,* vol. 10, no. 12, p. 2454, 2022. [http://dx.doi.org/10.3390/microorganisms10122454] [PMID: 36557706]

[70] R. Kumar, B. Kumar, and A. Kumar, "Pharmacokinetic profile of antitubercular drugs and their clinical applications in therapeutic drug monitoring", *Asian J Pharm,* vol. 3, 2017no. 5, pp. 145-154. Available from: http://ajpp.in/uploaded/p71.pdf

[71] S.Y. Chew, and L.T.L. Than, "Glucose metabolism and use of alternative carbon sources in medically-important fungi", In: *Encyclopedia of Mycology.,* Ó. Zaragoza, A. Casadevall, Eds., Elsevier: Oxford, 2021, pp. 220-229. [http://dx.doi.org/10.1016/B978-0-12-819990-9.00068-8]

[72] A. Kumar, "Assessment of pain-ameliorative potential of ethanolic extract of Calotropis procera and its active components characterization by HPTLC and GC-MS analysis", [http://dx.doi.org/10.31858/0975-8453.13.6.659-664]

[73] Z.N. Ling, Y.F. Jiang, J.N. Ru, J.H. Lu, B. Ding, and J. Wu, "Amino acid metabolism in health and disease", *Signal Transduct. Target. Ther.,* vol. 8, no. 1, p. 345, 2023. [http://dx.doi.org/10.1038/s41392-023-01569-3] [PMID: 37699892]

[74] R.D. Vijaya Durga, "Determination of S-amlodipine in human plasma using tizanidine as internal standard by LC/MS/MS method", *J. Bioanal. Biomed.,* vol. 7, no. 3, pp. 91-96, 2015. [http://dx.doi.org/10.4172/1948-593X.1000130]

[75] M. Dashty, "A quick look at biochemistry: Carbohydrate metabolism", *Clin. Biochem.,* vol. 46, no. 15, pp. 1339-1352, 2013. [http://dx.doi.org/10.1016/j.clinbiochem.2013.04.027] [PMID: 23680095]

[76] A. Kumar, M.K. Nandi, B. Kumar, A. Kumar, R. Kumar, A.K. Singh, and M. Singh, "Antiurolithiasis, antioxidant, anti-inflammatory, analgesic, and diuretic activity of ethanolic extract of seeds of Caesalpinia bonducella", *Int. J. Pharm. Investig.,* vol. 11, no. 3, pp. 306-311, 2021. [http://dx.doi.org/10.5530/ijpi.2021.3.54]

[77] X. Li, Y. Yang, B. Zhang, X. Lin, X. Fu, Y. An, Y. Zou, J.X. Wang, Z. Wang, and T. Yu, "Lactate metabolism in human health and disease", *Signal Transduct. Target. Ther.,* vol. 7, no. 1, p. 305, 2022. [http://dx.doi.org/10.1038/s41392-022-01151-3] [PMID: 36050306]

[78] R. Dhiman, V. Durga, K. Jayasankar, and S. Raj, "Solid phase extraction: Potential carbocisteine bioanalysis using carbocisteine 13c3 stable isotope: LC-MS/MS technique with pharmacokinetics on healthy human subjects", *Pharm. Lett.,* vol. 7, no. 8, pp. 126-134, 2015.

[79] X. Wang, E. Perez, R. Liu, L.J. Yan, R.T. Mallet, and S.H. Yang, "Pyruvate protects mitochondria from oxidative stress in human neuroblastoma SK-N-SH cells", *Brain Res.,* vol. 1132, no. 1, pp. 1-9, 2007. [http://dx.doi.org/10.1016/j.brainres.2006.11.032] [PMID: 17174285]

[80] R. Dhiman, V. Durga, and A. J. Rio, "Bioanalytical method for lornoxicam determination in human plasma by using piroxicam as internal standard by LC-MS/MS", *J Chem Pharm Res,* vol. 7, no. 8, pp. 54-60, 2015.

[81] N. Wu, M. Yang, U. Gaur, H. Xu, Y. Yao, and D. Li, "Alpha-Ketoglutarate: Physiological Functions and Applications", *Biomol. Ther. (Seoul),* vol. 24, no. 1, pp. 1-8, 2016. [http://dx.doi.org/10.4062/biomolther.2015.078] [PMID: 26759695]

[82] R. Dhiman, B. Kumar, D. Kumar, and A. Kumar, "Liquid-liquid extraction method for ziprasidone (ZRS) bioanalysis by using ZRS-D 8 (stable isotope) as internal standard", *J Chem Pharm Res,* vol. 8, no. 3, pp. 722-727, 2016.

[83] E. Riedel, M. Nündel, and H. Hampl, "alpha-Ketoglutarate application in hemodialysis patients improves amino acid metabolism", *Nephron,* vol. 74, no. 2, pp. 261-265, 1996. [http://dx.doi.org/10.1159/000189319] [PMID: 8893139]

[84] R. Kumar, B. Kumar, A. Kumar, A. Kumar, and M. Singh, "HPTLC and GC-MS analyses of biologically active extracts and fractions from Premna latifolia Roxb leaves", *J Neurol Disord,* 2021.

[85] S. Liu, L. He, and K. Yao, "The antioxidative function of alpha-ketoglutarate and its applications", *BioMed Res. Int.,* vol. 2018, pp. 1-6, 2018. [http://dx.doi.org/10.1155/2018/3408467] [PMID: 29750149]

[86] R. Kumar, B. Kumar, A. Kumar, A. Kumar, and M. Singh, "GC-MS analysis of phytocomponents in the methanol extract of Premna latifolia Roxb", *Pharmacognosy Res.,* vol. 14, no. 1, pp. 19-23, 2021. [http://dx.doi.org/10.5530/pres.14.1.4]

[87] C.D. Doucette, D.J. Schwab, N.S. Wingreen, and J.D. Rabinowitz, "α-ketoglutarate coordinates carbon and nitrogen utilization via enzyme I inhibition", *Nat. Chem. Biol.,* vol. 7, no. 12, pp. 894-901, 2011. [http://dx.doi.org/10.1038/nchembio.685] [PMID: 22002719]

[88] M.M. Bayliak, and V.I. Lushchak, "Pleiotropic effects of alpha-ketoglutarate as a potential anti-ageing agent", *Ageing Res. Rev.,* vol. 66, p. 101237, 2021. [http://dx.doi.org/10.1016/j.arr.2020.101237] [PMID: 33340716]

[89] H. Pandey, S. Srivastava, A. Kumar, R. Kumar, and Y. Bhusan Tripathi, "Pharmacokinetics study of puerarin absorption in blood after consumption of Pueraria tuberosa Water Extract (PTWE) By Rats", *Explor. Anim. Med. Res.,* vol. 9, no. 2, pp. 188-196, 2019.

[90] S.H. Naeini, L. Mavaddatiyan, Z.R. Kalkhoran, S. Taherkhani, and M. Talkhabi, "Alpha-ketoglutarate as a potent regulator for lifespan and healthspan: Evidences and perspectives", *Exp. Gerontol.,* vol. 175, p. 112154, 2023. [http://dx.doi.org/10.1016/j.exger.2023.112154] [PMID: 36934991]

[91] R. Kumar *et al.*, "HPTLC and GC-MS Analyses of Biologically Active Extracts and fractions from Premna latifolia Roxb Leaves reveal Anti-inflammatory and Analgesic Effects," vol. 12, no. 10, pp. 811–818, 2021.

[92] H. Abla, M. Sollazzo, G. Gasparre, L. Iommarini, and A.M. Porcelli, "The multifaceted contribution of

α-ketoglutarate to tumor progression: An opportunity to exploit?", *Semin. Cell Dev. Biol.,* vol. 98, pp. 26-33, 2020.
[http://dx.doi.org/10.1016/j.semcdb.2019.05.031] [PMID: 31175937]

[93] G.E. Hansen, and G.E. Gibson, "The α-ketoglutarate dehydrogenase complex as a hub of plasticity in neurodegeneration and regeneration", *Int. J. Mol. Sci.,* vol. 23, no. 20, p. 12403, 2022.
[http://dx.doi.org/10.3390/ijms232012403] [PMID: 36293260]

[94] D. Cheng, M. Zhang, Y. Zheng, M. Wang, Y. Gao, X. Wang, X. Liu, W. Lv, X. Zeng, K.N. Belosludtsev, J. Su, L. Zhao, and J. Liu, "α-Ketoglutarate prevents hyperlipidemia-induced fatty liver mitochondrial dysfunction and oxidative stress by activating the AMPK-pgc-1α/Nrf2 pathway", *Redox Biol.,* vol. 74, p. 103230, 2024.
[http://dx.doi.org/10.1016/j.redox.2024.103230] [PMID: 38875959]

[95] S. Kovac, A.Y. Abramov, and M.C. Walker, "Energy depletion in seizures: Anaplerosis as a strategy for future therapies", *Neuropharmacology,* vol. 69, pp. 96-104, 2013.
[http://dx.doi.org/10.1016/j.neuropharm.2012.05.012] [PMID: 22659085]

[96] M. Holeček, "Roles of malate and aspartate in gluconeogenesis in various physiological and pathological states", *Metabolism,* vol. 145, p. 155614, 2023.
[http://dx.doi.org/10.1016/j.metabol.2023.155614] [PMID: 37286128]

[97] Z. Luo, S. Yu, W. Zeng, and J. Zhou, "Comparative analysis of the chemical and biochemical synthesis of keto acids", *Biotechnol. Adv.,* vol. 47, p. 107706, 2021.
[http://dx.doi.org/10.1016/j.biotechadv.2021.107706] [PMID: 33548455]

[98] Y. Kuang, X. Han, M. Xu, Y. Wang, Y. Zhao, and Q. Yang, "Oxaloacetate ameliorates chemical liver injury via oxidative stress reduction and enhancement of bioenergetic fluxes", *Int. J. Mol. Sci.,* vol. 19, no. 6, p. 1626, 2018.
[http://dx.doi.org/10.3390/ijms19061626] [PMID: 29857490]

[99] D.S. Williams, A. Cash, L. Hamadani, and T. Diemer, "Oxaloacetate supplementation increases lifespan in Caenorhabditis elegans through an AMPK/FOXO-dependent pathway", *Aging Cell,* vol. 8, no. 6, pp. 765-768, 2009.
[http://dx.doi.org/10.1111/j.1474-9726.2009.00527.x] [PMID: 19793063]

[100] F. Campos, T. Sobrino, P. Ramos-Cabrer, B. Argibay, J. Agulla, M. Pérez-Mato, R. Rodríguez-González, D. Brea, and J. Castillo, "Neuroprotection by glutamate oxaloacetate transaminase in ischemic stroke: An experimental study", *J. Cereb. Blood Flow Metab.,* vol. 31, no. 6, pp. 1378-1386, 2011.
[http://dx.doi.org/10.1038/jcbfm.2011.3] [PMID: 21266983]

[101] A. Samad, R. Samant, K.V. Rao, V. Bhargava, S.I. Sadique, and R. Yadav, "Oxaloacetate as a Holy Grail Adjunctive Treatment in Gliomas: A Revisit to Metabolic Pathway", *Cureus,* vol. 15, no. 11, p. e48821, 2023.
[http://dx.doi.org/10.7759/cureus.48821] [PMID: 38106701]

[102] Y. Kuang, X. Han, M. Xu, and Q. Yang, "Oxaloacetate induces apoptosis in HepG2 cells via inhibition of glycolysis", *Cancer Med.,* vol. 7, no. 4, pp. 1416-1429, 2018.
[http://dx.doi.org/10.1002/cam4.1410] [PMID: 29533007]

CHAPTER 4

Health Implications of Incomplete Metabolism

Pankaj Kalia[1], **Swati Pundir**[2], **Ashish Sharma**[3], **Rajesh Kumar**[1] and **Amit Aggarwal**[1,*]

[1] *Maharaja Agrasen School of Pharmacy, Maharaja Agrasen University, Solan, Himachal Pradesh, India*

[2] *School of Pharmaceutical Sciences, Shoolini University of Biotechnology and Management Sciences, Solan – 173212, Himachal Pradesh, India*

[3] *Washington University School of Medicine, Washington University in St. Louis, St. Louis, Missouri, United States*

Abstract: The important effects of incomplete metabolism on human health are examined in this work, with particular attention to how it may play a role in the etiology of long-term conditions like diabetes, obesity, and mitochondrial diseases. Nutritional deficits, the buildup of hazardous metabolites, and a disturbance of biochemical equilibrium are all consequences of metabolic pathway disruption. The classic cases of phenylketonuria (PKU) and maple syrup urine disease (MSUD) are the grave neurological and developmental effects of metabolic pathway disruption. G6PD insufficiency is an example of how incomplete drug metabolism, which can be caused by genetic polymorphisms, enzyme deficiencies, or hepatic/renal impairments, can lead to metabolic toxicity and drug buildup. Deficiencies in vital vitamins (B_{12}, D, A, K) and minerals (calcium, iron) result in metabolic inefficiencies and clinical aftereffects, such as anemia and neurotoxicity. Nutrient metabolism is also compromised. Lipid intermediate buildup, oxidative stress, and systemic inflammation are all influenced by lipid metabolic abnormalities, which are especially prevalent in non-alcoholic fatty liver disease (NAFLD) and non-alcoholic steatohepatitis (NASH). By affecting ATP synthesis and redox equilibrium, mitochondrial dysfunction aggravates metabolic diseases further. Oxidative stress aggravates illnesses, such as myopathies and lactic acidosis. Novel therapeutic approaches, such as enzyme replacement therapy (ERT), CRISPR-Cas9 gene therapy, and nanotechnology-based methods, present intriguing paths toward pharmacogenomics-driven enzyme regulation and metabolic function restoration, as well as toward precision medicine.

Keywords: CRISPR-Cas9 gene editing therapy, Hepatic lipid peroxidation, Inborn errors of metabolism, Metabolic dysregulation, Micronutrient malabsorption, Mitochondrial bio-energetic failure, Phenylalanine hydroxylase deficiency, Toxic metabolite accumulation, Xenobiotic metabolism impairment.

* **Corresponding author Amit Aggarwal:** Maharaja Agrasen School of Pharmacy, Maharaja Agrasen University, Solan, Himachal Pradesh, India; E-mail: amitaggarwal78@yahoo.co.in

INTRODUCTION

This chapter titled "Health Implications of Incomplete Metabolism" explores the adverse effects that arise when the body cannot efficiently convert nutrients into energy and essential biochemical substances. Metabolism includes all the chemical processes necessary for sustaining life, encompassing both catabolic reactions that break down molecules for energy and anabolic reactions that use this energy to build cellular components like proteins and nucleic acids [1]. When metabolism is incomplete or disrupted, it can lead to substantial health problems like accumulation of metabolic byproducts, energy and nutrient deficiency, obesity, and metabolic syndromes, along with a few chronic and developmental disorders. The incomplete metabolism of certain substances can cause harmful byproducts to accumulate in the body, leading to various health complications. Phenylketonuria (PKU) is a notable example, where the body fails to process the amino acid phenylalanine, leading to its toxic buildup. This accumulation can cause intellectual disability and other neurological issues if not controlled through dietary management. Inefficient metabolism can lead to insufficient energy production, which is commonly seen in mitochondrial diseases. These conditions, where the mitochondria do not produce enough energy, typically present with symptoms like muscle weakness, fatigue, and various systemic problems [2]. These issues often stem from improper regulation of hormones involved in metabolism, like insulin and leptin, emphasizing the importance of maintaining metabolic health to prevent chronic diseases. Incomplete metabolism can cause nutrient deficiencies even with sufficient dietary intake. Conditions like malabsorption syndromes, including celiac disease, impede the body's ability to digest and absorb nutrients properly, leading to deficiencies in essential vitamins and minerals. These deficiencies can significantly impact overall health and well-being [3]. Prolonged metabolic inefficiencies can also lead to chronic diseases. For example, diabetes mellitus is marked by improper glucose metabolism, resulting in consistently high blood sugar levels and related complications like neuropathy, nephropathy, and cardiovascular diseases. Focusing on metabolic health is essential for preventing and managing these chronic conditions [4, 5]. Incomplete metabolism can impact growth and development, especially in children. Disorders like maple syrup urine disease (MSUD) hinder the breakdown of certain amino acids, causing severe neurological damage and developmental delays if not properly managed. Initial diagnosis and interposition are crucial in reducing the impact of these disorders on development [6]. Recognizing the health implications of incomplete metabolism is vital for the early diagnosis, management, and treatment of metabolic disorders. Ongoing research and medical advancements continue to enhance our ability to identify and address these metabolic issues, improving the quality of life for those affected. Sustained efforts in this field are essential for developing effective strategies to mitigate the health impacts of incomplete metabolism [6, 8, 9, 11].

Metabolic Pathways and Disorders

Metabolic pathways consist of a series of chemical reactions occurring within cells, which are essential for maintaining cellular function and overall health. Disruptions in these pathways can result in metabolic disorders with significant health impacts [7].

For instance, the urea cycle is a pathway that converts ammonia to urea for excretion. Any enzyme defect in the urea cycle can cause an increase in ammonia, characterized by high levels of ammonia in the body, which can lead to severe neurological indications and be fatal if untreated [8, 9].

Phenylketonuria (PKU) is a genetic metabolic syndrome resulting from an insufficiency in the enzyme phenylalanine hydroxylase, which converts phenylalanine into tyrosine. Lacking this enzyme, phenylalanine builds up in the body, potentially causing severe intellectual disability, seizures, and other serious health problems if not managed with a strict low-phenylalanine diet [10].

MSUD is a genetic condition impacting the body's capability to metabolize specific amino acids. The disorder is termed for the sweet-smelling urine of diagnosed persons, similar to maple syrup. Without appropriate treatment, MSUD can result in severe neurological damage, developmental delays, and even death. Treatment generally involves a specialized diet low in these amino acids and close medical monitoring [6].

Mitochondrial disorders comprise a group of metabolic conditions resulting from dysfunctions in the mitochondria, often causing muscle weakness, neurological issues, and various other symptoms. Since mitochondria produce most of the body's energy, defects in these organelles can lead to widespread and debilitating effects [11, 12].

Managing and treating metabolic disorders typically involves dietary changes, medications, and other interventions to address the underlying metabolic imbalance. For instance, people with PKU need a low-phenylalanine diet to prevent its buildup in the body. Those with MSUD require a diet low in the specific amino acids they cannot metabolize. In some cases, enzyme replacement therapy or other medical treatments may be needed to manage symptoms and prevent complications [6].

Future directions in ongoing research in metabolic disorders aim to further understand and treat these conditions. Advances in genetics and molecular biology have identified many genes and pathways involved in metabolic disorders, leading to the development of targeted therapies. Additionally, research into diet, lifestyle, and ecological causes on metabolic health is informing more effective prevention and management strategies.

Metabolic Toxicity and Drug Accumulation

Metabolic toxicity and drug accumulation are key areas in pharmacology and toxicology, focusing on the adverse effects that arise when the body fails to effectively metabolize and eliminate drugs. This inefficiency can lead to harmful levels of the drug or its metabolites in the body, impacting various organs and systems [7]. Metabolic toxicity refers to the adverse effects caused by the buildup of toxic metabolites in the body. This can result from genetic variations, liver dysfunction, or interactions with other drugs that inhibit the enzymes responsible for drug metabolism (Fig. **1**).

Enzyme deficiencies and genetic variations include genetic differences in metabolic enzymes, such as cytochrome P450 enzymes may cause a decrease in or elimination of enzyme activity, which might accumulate harmful metabolites. For instance, many medications may cause hemolytic anemia in people with glucose-6-phosphate dehydrogenase (G6PD) deficiency. The principal location of drug metabolism is the liver.

Liver diseases, such as cirrhosis or hepatitis, can impair the liver's ability to metabolize drugs, leading to the buildup of toxic substances. For example, patients with liver dysfunction may experience increased toxicity from drugs like acetaminophen [13, 14]. Concurrent use of several drugs causes metabolic toxicity if one drug inhibits the metabolism of another, resulting in elevated levels of the latter. For example, ketoconazole, an antifungal drug, is a potent inhibitor of CYP3A4, an enzyme that metabolizes many drugs, leading to increased toxicity when co-administered with drugs metabolized by CYP3A4 [7, 15 - 17].

Drug accumulation occurs when the rate of drug intake surpasses the rate of elimination, leading to progressively higher concentrations in the body. This can result from chronic dosing, impaired clearance mechanisms, or high lipophilicity of the drug. Drugs with a long half-life or those administered frequently can accumulate in the body, especially if the dosing interval is shorter than the time required for elimination. For example, digoxin, a cardiac glycoside with a long half-life, can accumulate and cause toxicity if not carefully managed [18, 19]. Kidneys are crucial for drug elimination. Impaired renal function can lead to reduced clearance and accumulation of drugs primarily excreted by the kidneys, such as aminoglycoside antibiotics, which can lead to nephrotoxicity and ototoxicity. Lipophilic drugs can accumulate in fatty tissues and be released slowly over time, leading to prolonged drug effects and potential toxicity. An example is the antipsychotic drug haloperidol, which can accumulate in fatty tissues and have prolonged sedative effects [20, 21].

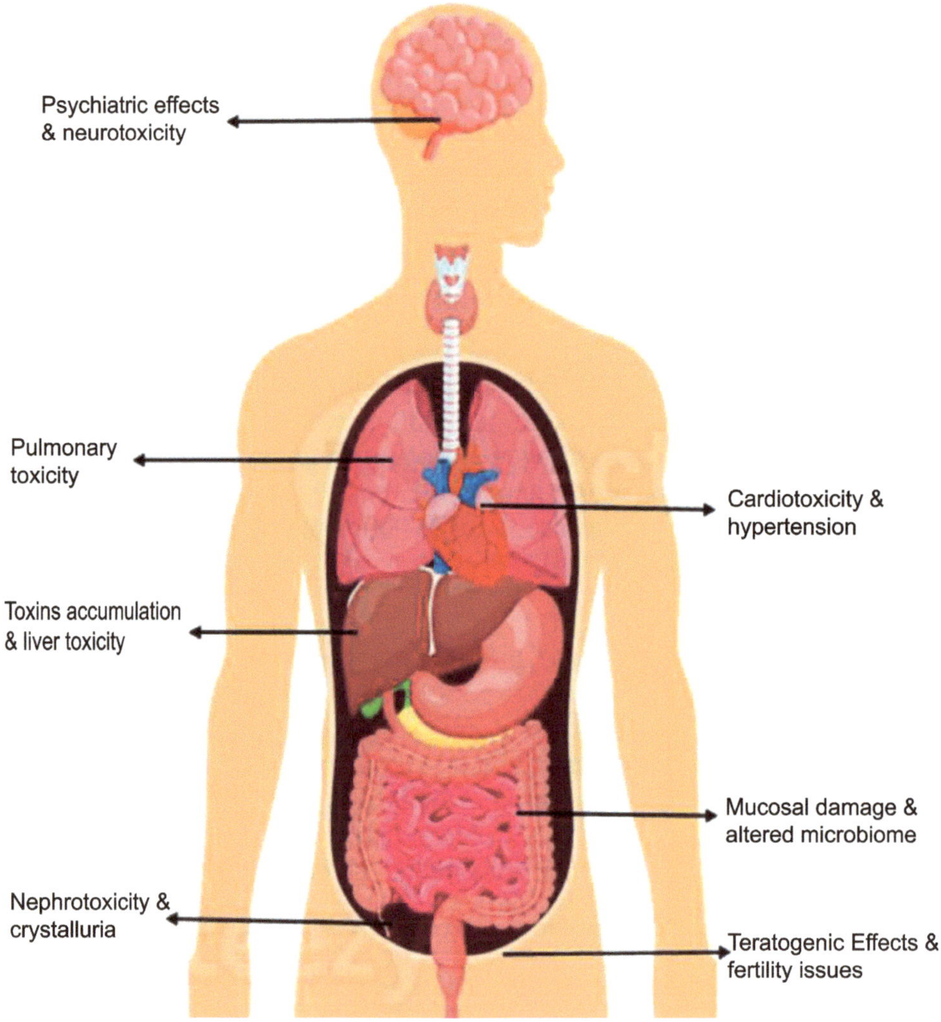

Fig. (1). Effect of incomplete metabolism of drugs and minerals on the human body.

Understanding the mechanisms behind metabolic toxicity and accumulation is essential for maximizing medication administration, avoiding side effects, and ensuring patient safety. Clinicians must consider individual patient factors, potential drug interactions, and the pharmacokinetic properties of drugs when prescribing medications.

Incomplete Metabolism and Related Issues

Metabolic toxicity and drug accumulation become particularly problematic in cases of incomplete metabolism, where the body's ability to process and eliminate drugs is compromised. This can lead to the buildup of drugs and their metabolites, causing toxic effects and adverse reactions [7, 22]. Incomplete metabolism occurs when the body's enzymatic systems fail to fully metabolize drugs or endogenous substances, often due to genetic variations, disease states, or interactions with other medications.

Mechanisms Leading to Metabolic Toxicity and Drug Accumulation

Genetic polymorphisms include genetic variations in metabolic enzymes, such as cytochrome P450 (CYP) enzymes, which can result in poor or non-functioning enzymes. This can hinder drug metabolism, leading to their accumulation and potential toxicity. Individuals with polymorphisms in the CYP2D6 enzyme might experience adverse effects from drugs like codeine or tamoxifen, which require CYP2D6 for activation or clearance. Liver dysfunction conditions, such as cirrhosis or hepatitis, can impair hepatic function, reducing the metabolism and clearance of drugs, which is important to amplify drug levels and noxiousness. For instance, patients with liver dysfunction may be more susceptible to acetaminophen toxicity due to impaired glutathione conjugation [13, 23].

Drug-drug interactions involving the simultaneous administration of medications can cause or inhibit metabolic enzymes, changing how those medications are metabolized. Grapefruit juice and other CYP enzyme inhibitors can raise the plasma levels of medications that these enzymes process, which can be harmful. Conversely, enzyme inducers can decrease drug levels, reducing its efficacy [1]. Renal impairment can reduce drug clearance, leading to accumulation and toxicity. This is particularly relevant for drugs with narrow therapeutic indices, such as digoxin and aminoglycosides [18, 19]. Drugs with long half-lives or those administered frequently (chronic dosing) can accumulate in the body. Chronic dosing of medications like lithium or methotrexate can lead to toxicity if not monitored carefully. Fat-soluble medications or lipophilic drugs may build up in body fat and release slowly, extending their effects and increasing the risk of toxicity, such as with antipsychotics like haloperidol, which can cause lasting sedation [20].

Clinical Implications

Understanding the mechanisms of metabolic toxicity and drug accumulation is essential for augmenting drug treatment and preventing adverse effects. Key strategies include pharmacogenetic testing, which can identify genetic polymorphisms in metabolic enzymes and help tailor drug treatment to individual patients, reducing the threat of toxicity, and monitoring liver and kidney function, where regular assessment of hepatic and renal function can guide dosage adjustments and prevent drug accumulation.

Moreover, awareness of potential drug-drug interactions can help manage the risk of altered drug metabolism. Therapeutic drug monitoring is the clinical practice of measuring specific drugs at designated intervals to maintain a constant concentration in a patient's bloodstream, avoiding toxicity or sub-therapeutic effects [24, 25].

Nutrient Metabolism, Vitamin Deficiencies, and Mineral Imbalance in Incomplete Metabolism

Incomplete metabolism can significantly impact the body's capability to process nutrients, leading to vitamin deficiencies and mineral imbalances. These deficiencies and imbalances can have far-reaching health consequences, affecting various physiological processes and leading to a range of clinical symptoms.

Incomplete Metabolism of Nutrients

Nutrient metabolism refers to the biochemical processes that convert nutrients from food into energy needed for the growth, maintenance, and restoration of the body. This includes the metabolism of carbohydrates, proteins, fats, vitamins, and minerals. Incomplete metabolism can disrupt these processes, resulting in insufficient nutrient availability and utilization. Incomplete metabolism of nutrients refers to the inefficient catabolism of macronutrients, resulting in the accumulation of metabolic intermediates due to enzymatic deficiencies or impaired biochemical pathways.

For instance, in glycolysis, inadequate oxygen availability can lead to the accumulation of lactate *via* anaerobic fermentation rather than complete oxidation through the citric acid cycle. In lipid metabolism, defective β-oxidation may cause excess production of ketone bodies, contributing to ketoacidosis. Genetic disorders like phenylketonuria (PKU) result from mutations in the phenylalanine hydroxylase enzyme, disrupting the normal degradation of phenylalanine, leading to toxic metabolite accumulation and impaired homeostasis [20, 26].

Vitamin Deficiencies in Incomplete Metabolism

Vitamin (Vit.) B12 is essential for DNA synthesis and neurological function. Incomplete metabolism, often due to malabsorption syndromes or pernicious anemia, can lead to B12 deficiency, causing megaloblastic anemia, neuropathy, and cognitive disturbances. Vit. D is crucial for calcium homeostasis and bone health. Incomplete metabolism, often related to kidney disease or malabsorption, can result in vit. D deficit. Vit. A is vital for vision, immune function, and cell growth.

Incomplete metabolism due to liver disease or fat malabsorption can cause vit. A deficiency, leading to night blindness, immune dysfunction, and xerophthalmia. Vit. K is necessary for blood coagulation and bone strength. Incomplete metabolism, particularly due to liver disease or antibiotic use disrupting gut flora, can result in vit. K deficiency, leading to bleeding disorders and reduced bone density. Folate is essential for DNA synthesis and repair. Incomplete metabolism, often seen in alcoholics or individuals with malabsorptive conditions, can lead to folate deficiency, resulting in megaloblastic anemia and increased risk of neural tube defects in fetuses [26, 27].

Mineral Imbalance in Incomplete Metabolism

Calcium is crucial for bone health, muscle growth, and nerve conduction. Incomplete metabolism, often due to vitamin D deficiency or renal disease, can lead to hypocalcemia or hypercalcemia, causing symptoms ranging from muscle cramps and tetany to kidney stones and cardiac arrhythmias. Iron is essential for hemoglobin formation and oxygen transport. Incomplete metabolism, frequently due to gastrointestinal disorders or chronic blood loss, can result in iron deficiency anemia, characterized by fatigue, pallor, and reduced cognitive function. Magnesium is required in numerous enzymatic reactions, as well as in ATP production and muscle function.

Incomplete metabolism due to gastrointestinal disorders or chronic alcoholism can lead to hypomagnesemia, causing muscle weakness, seizures, and cardiac arrhythmias. Incomplete metabolism, often due to malabsorption syndromes or chronic liver disease, can cause zinc deficiency, leading to impaired immune function, skin lesions, and growth retardation in children. Potassium is critical for maintaining cellular function, nerve impulses, and muscle contractions. Incomplete metabolism, particularly in conditions affecting renal function, can lead to hypokalemia or hyperkalemia, causing muscle weakness, arrhythmias, and neuromuscular disturbances [28, 29].

Incomplete Metabolism of Lipids During Fatty Liver Diseases

Incomplete lipid metabolism in fatty liver diseases, such as NAFLD and NASH, is closely related to mitochondrial dysfunction, peroxisomal stress, and impaired lipid export, leading to metabolic overload in hepatocytes. In particular, mitochondrial β-oxidation serves as a crucial factor in fatty acid degradation. However, in NAFLD, studies have demonstrated that mitochondrial activity is significantly impaired. Decreased expression of enzymes like carnitine palmitoyltransferase I (CPT1), which regulates fatty acid entrance into mitochondria, limits the rate of β-oxidation. As a result, hepatocytes accumulate excess free fatty acids (FFAs), leading to incomplete β-oxidation and the formation of lipotoxic intermediates, such as acylcarnitines, diacylglycerol (DAG), and ceramides [30, 31].

Lipotoxic intermediates have been shown to activate stress-related signaling pathways that contribute to insulin resistance and inflammation. For example, ceramides can disrupt insulin signaling by inhibiting Akt phosphorylation, a key step in glucose metabolism. Ceramide accumulation has been linked to the activation of pro-inflammatory pathways, together with the c-Junction N-terminal kinase (JNK) pathway, which exacerbates hepatic inflammation and contributes to hepatocyte apoptosis. Furthermore, the overproduction of reactive oxygen species (ROS) due to mitochondrial inefficiency generates oxidative stress, leading to damage to hepatocyte membranes, proteins, and DNA [32, 33].

Peroxisomal dysfunction further contributes to incomplete lipid metabolism in fatty liver disease. In NAFLD, peroxisomal β-oxidation becomes overwhelmed due to the increased fatty acid load, leading to excess ROS production. Excess ROS from both mitochondria and peroxisomes has been associated with the upregulation of pro-inflammatory cytokines contributing to the progression of simple steatosis to NASH. Moreover, mitochondrial DNA (mtDNA) damage due to oxidative stress has been detected in NAFLD patients, further disrupting mitochondrial function and exacerbating lipid metabolism defects [34].

Normally, triglycerides synthesized within hepatocytes are packaged into VLDL particles for export into circulation. However, in fatty liver disease, microsomal triglyceride transfer protein (MTP), essential for VLDL assembly, is often downregulated. As a result, triglycerides that are not exported accumulate within the liver, contributing to steatosis. Data suggest that NAFLD patients have a 30-40% reduction in VLDL secretion, further exacerbating lipid overload in hepatocytes. The accumulation of intrahepatic triglycerides promotes the formation of harmful lipid intermediates, which perpetuate cellular stress and inflammation, driving the disease progression from NAFLD to NASH and fibrosis [32, 34].

Amino Acid Metabolism and Maple Syrup Urine Disease

A hereditary condition known as MSUD is caused by a lack of branched-chain alpha-keto acid dehydrogenase, an enzyme that is essential for the metabolism of the BCAAs leucine, isoleucine, and valine. This impairment leads to toxic accumulation of BCAAs and their metabolites, causing neurological damage. Characteristic symptoms include sweet-smelling urine, developmental delays, and potential cognitive impairment. Early dietary intervention is essential for managing MSUD and preventing severe complications.

Incomplete Metabolism of Amino Acids in Maple Syrup Urine Disease

A rare genetic metabolic illness known as MSUD is caused by a malfunction in the breakdown of the branched-chain amino acids (BCAAs), which include valine, isoleucine, and leucine. Mutations in the genes encoding the multienzyme complex BCKDH (branched-chain α-ketoacid dehydrogenase) cause this deficiency. BCKDH is responsible for the oxidative decarboxylation of branched-chain α-ketoacids (BCKAs), which are the transamination products of BCAAs. The impaired function of BCKDH results in the incomplete metabolism of these essential amino acids, leading to their pathological accumulation in blood, urine, and tissues in the CNS [35, 36].

In normal physiological conditions, BCAAs undergo transamination by branched-chain amino acid aminotransferase (BCAT) to yield their respective α-ketoacids: α-ketoisocaproate (from leucine), α-keto-β-methylvalerate (from isoleucine), and α-ketoisovalerate (from valine). These α-ketoacids are subsequently oxidatively decarboxylated by the BCKDH complex, which consists of three catalytic components: E1 (α-ketoacid decarboxylase), E2 (dihydrolipoyl transacylase), and E3 (dihydrolipoyl dehydrogenase). In MSUD, mutations in the genes coding for these subunits (BCKDHA, BCKDHB, and DBT) disrupt the complex's activity, leading to the accumulation of BCAAs and their respective ketoacid derivatives [37].

Leucine is particularly neurotoxic when present in excessive concentrations. Elevated leucine levels disrupt normal neurotransmitter synthesis and cause cerebral edema by altering osmotic balance in neurons, contributing to the characteristic neurological manifestations of MSUD, including encephalopathy, seizures, and psychomotor retardation. The accumulation of isoleucine and valine, although less neurotoxic, further exacerbates metabolic disturbances. The accumulation of 2-oxo acids, such as α-ketoisocaproate, contributes to the distinctive “maple syrup” odor in the urine and sweat of affected individuals [38, 39].

The mainstay of treatment is lifelong dietary restriction of BCAAs, preventing the toxic accumulation of these amino acids and their metabolites. Specialized amino acid formulas that exclude BCAAs but provide essential nutrients are crucial to ensure appropriate growth and metabolic balance. In acute metabolic crises, interventions, such as dialysis or hemofiltration, may be necessary to reduce plasma leucine levels. Liver transplantation, which provides a functional source of BCKDH activity, has emerged as a potentially curative option for severe cases. Nonetheless, the incomplete catabolism of BCAAs in MSUD underscores the critical role of the BCKDH complex in amino acid homeostasis and CNS function [40].

Urea Cycle Disorder and Maple Syrup Urine Disease

UCDs and MSUD are both in-born disorders of metabolism, each involving distinct biochemical pathways and resulting in significant metabolic derangements. UCDs arise from enzymatic deficiencies in the urea cycle, the primary pathway for the detoxification of ammonia (NH_3), a neurotoxic byproduct of nitrogen metabolism derived from amino acid catabolism. In individuals with UCDs, mutations in genes encoding enzymes impair the conversion of ammonia to urea. This disruption leads to hyperammonemia, characterized by elevated plasma ammonia levels, which can result in encephalopathy, lethargy, ataxia, seizures, and, in severe cases, cerebral edema and coma. Treatment strategies for UCDs typically involve dietary protein restriction to limit nitrogen intake, administration of nitrogen scavenger medications (*e.g*., sodium phenylbutyrate, sodium benzoate), and, in refractory cases, liver transplantation to restore normal urea cycle function [41].

MSUD, in contrast, is triggered by mutations in the genes encoding an enzyme critical for oxidative decarboxylation, specifically leucine, isoleucine, and valine. Deficiencies in BCKD activity lead to the accumulation of branched-chain α-keto acids and their corresponding amino acids, particularly leucine, which exerts neurotoxic effects by disrupting neurotransmitter synthesis.

The classic form of MSUD presents in neonates with ketoacidosis, poor feeding, nausea, tetchiness, and a sugary scent in the urine due to the buildup of isoleucine derivatives. Without intervention, the disorder can lead to irreversible neurodegeneration, cerebral edema, and death. Management of MSUD includes lifelong dietary restriction of BCAAs, careful monitoring of plasma amino acid levels, and acute metabolic crisis management *via* intravenous glucose, insulin therapy, and specialized amino acid-free formulas. Liver transplantation has emerged as a potentially curative therapy, providing a source of functional BCKD enzyme [42, 43].

While both UCDs and MSUD involve metabolic dysregulation of amino acids, UCDs are centered on impaired nitrogen detoxification and ammonia clearance, whereas MSUD is caused by defective catabolism of branched-chain amino acids, leading to toxic metabolite accumulation. Both disorders require meticulous metabolic control to prevent severe neurological sequelae and ensure long-term patient survival [44].

Incomplete Metabolism of Carbohydrate in Glycogen Disorder Leads to Diabetes Mellitus

Carbohydrate metabolism plays a pivotal role in maintaining energy homeostasis, with glycogen serving as the primary polysaccharide storage form of glucose in hepatic and skeletal muscle tissues. Disruptions in carbohydrate metabolism, particularly incomplete or impaired glycogen metabolism, significantly contribute to various metabolic disorders, including diabetes mellitus.

Glycogen metabolism encompasses two essential biochemical pathways: glycogenesis, the anabolic process of glycogen synthesis from glucose- 6-phosphate, and Glycogenolysis, the catabolic break of glycogen to liberate glucose-1-phosphate. These pathways are intricately regulated by hormones, ensuring glucose availability is matched to the body's metabolic demands. Incomplete metabolism of glycogen can arise due to enzymatic deficiencies, leading to glycogen storage diseases (GSDs), which disrupt normal glycogen storage in tissues. This abnormal storage affects glucose availability, insulin sensitivity, and metabolic regulation, all of which are fundamental to the pathophysiology of diabetes mellitus [45, 46].

Diabetes mellitus, particularly Type 2 diabetes, is regarded as enduring hyperglycemia due to insulin conflict, β-cell dysfunction, or a combination of both. In disorders of glycogen metabolism, insulin signaling pathways are often impaired, resulting in defective glucose uptake, abnormal glycogen synthesis, and compromised glycogenolysis. For instance, deficiencies or mutations result in poor glycogen turnover. In Type 2 diabetes, inefficient glycogen storage exacerbates hyperglycemia, as cells fail to sequester glucose into glycogen, leading to elevated blood glucose levels. This persistent hyperglycemia contributes to a feedback loop of insulin resistance, further driving the progressive nature of diabetes [47].

A specific example is Glycogen Storage Disease Type 0 (GSD 0), a metabolic disorder caused by mutations in the *GYS2* gene encoding glycogen synthase, which marks a diminished capacity to synthesize glycogen. Patients often experience fasting hypoglycemia and postprandial hyperglycemia due to their inability to properly store glucose as glycogen. This dysregulation of glucose homeostasis mirrors some aspects of Type 1 diabetes, where impaired glucose regulation is prominent. Additionally, in other GSDs, such as Glycogen Storage Disease Type I (von Gierke's disease), excessive hepatic glycogen accumulation leads to hepatomegaly, glucose-6-phosphatase dysfunction, and severe insulin resistance, increasing the susceptibility to diabetes.

Incomplete glycogen metabolism also results in a profound cellular energy deficit. In GSDs, impaired glycogenolysis reduces glucose availability for glycolysis, impairing ATP production and cellular bioenergetics. This energy deficiency induces oxidative stress, which is implicated in the development of insulin resistance. Mitochondrial dysfunction, commonly observed in diabetes mellitus, further impairs ATP synthesis, forcing cells to increase reliance on alternative energy substrates, such as fatty acids, which exacerbates metabolic dysregulation and insulin signaling impairment [48, 49].

Mitochondrial Disorder and Oxidative Stress

Mitochondrial disorders and oxidative stress significantly impact drug metabolism and nutrient or vitamin utilization. Mitochondria are critical for cellular energy production and are involved in the metabolism of various drugs and nutrients. Mitochondrial dysfunction, often supplemented by amplified oxidative stress, impairs electron transport and ATP production, leading to altered pharmacokinetics and reduced efficacy of drugs metabolized through mitochondrial pathways. Additionally, mitochondrial damage disrupts the metabolism of essential vitamin, such as vit. B and coenzyme Q10, required for mitochondrial function and energy production. Oxidative stress exacerbates these effects by damaging mitochondrial DNA, proteins, and lipids, further impairing cellular metabolism and overall physiological health [50, 51].

Incomplete Metabolism of Drugs Due to Mitochondrial Disorder

Mitochondrial disorders can significantly impact drug metabolism, resulting in incomplete or altered processing of medications within the body. Mitochondria are essential for cellular energy production, oxidative phosphorylation, and various biosynthetic and catabolic pathways, including drug metabolism. Cytochrome P450 (CYP) enzymes, particularly *CYP2D6*, *CYP3A4*, and *CYP2C19*, are crucial for drug metabolism and are localized in the mitochondrial membrane. Research has reported that patients with mitochondrial disorders exhibit a 30-50% reduction in the activity of these enzymes due to defects in mitochondrial DNA (mtDNA) or oxidative phosphorylation (OXPHOS) machinery. This impaired enzyme action causes incomplete drug metabolism, reducing drug clearance and causing toxic accumulation [52].

The incomplete metabolism of drugs due to mitochondrial dysfunction can lead to several adverse effects. For example, studies have indicated that patients with mitochondrial disorders are up to 60% more likely to experience severe adverse drug reactions (ADRs) when treated with drugs, such as valproic acid, a known mitochondrial toxin. Valproic acid inhibits mitochondrial beta-oxidation. Furthermore, mitochondrial dysfunction has been associated with a three-fold increase in the risk of statin-induced myopathy, as statins reduce coenzyme Q10 levels, a critical component of the electron transport chain, further impairing mitochondrial function [53, 54].

In addition, mitochondrial dysfunction can alter drug pharmacokinetics and pharmacodynamics. Reduced ATP production affects ATP-binding cassette (ABC) transporters, which are essential for drug efflux and distribution. This alteration can reduce drug absorption, increase plasma concentrations, and prolong the half-life of drugs. For instance, in patients with mitochondrial disorders, plasma concentrations of metformin, a drug dependent on mitochondrial uptake, can be elevated by 40%, leading to an increased risk of lactic acidosis [55, 56].

Incomplete Metabolism of Nutrients due to Mitochondrial Disorder

Mitochondrial disorders are a group of heterogeneous genetic conditions characterized by impaired mitochondrial function, particularly affecting oxidative phosphorylation (OXPHOS). Mitochondria are essential organelles responsible for the production of adenosine triphosphate (ATP) through the electron transport chain (ETC) coupled with the Krebs cycle (citric acid cycle). When mitochondrial function is compromised due to genetic mutations or defects in mitochondrial DNA (mtDNA) or nuclear DNA (nDNA), the metabolism of essential nutrients, such as carbohydrates, fatty acids, and amino acids, becomes incomplete, leading to profound physiological and biochemical disturbances [57].

Repercussions on Carbohydrate Metabolism

Mitochondrial dysfunction is very susceptible to the metabolism of carbohydrates. Under typical physiological circumstances, glucose is digested in the cytosol *via* glycolysis, which produces pyruvate. Pyruvate is then carried into the mitochondria by the pyruvate dehydrogenase complex (PDC), where it is transformed into acetyl-CoA.

After entering the Krebs cycle, this acetyl-CoA proceeds through a number of enzymatic processes that produce reducing equivalents ($FADH_2$ and NADH), which power the enzyme transfer chain (ETC) and produce ATP. Defects in the *ETC* or PDC hamper oxidative metabolism in mitochondrial diseases, resulting in pyruvate buildup. Lactate dehydrogenase subsequently converts this pyruvate to lactate, causing lactic acidosis. This poor metabolic transition from aerobic to anaerobic glycolysis lowers ATP output, which can lead to clinical manifestations, such as cardiomyopathy, exercise intolerance, and myopathy [58, 59].

Effect on Fatty Acid Oxidation

An essential process for energy production, fatty acid β-oxidation, is mostly found in the mitochondrial matrix and is particularly important in tissues with high energy requirements, such as the liver, heart, and skeletal muscle. Defects in the enzymes responsible for β-oxidation (such as acyl-CoA dehydrogenase) or the entry of fatty acids into the mitochondria can lead to diseases in the mitochondria (*e.g.*, carnitine palmitoyltransferase deficiency), resulting in incomplete fatty acid oxidation. This impairment leads to the accumulation of toxic acylcarnitine intermediates and free fatty acids, which can disrupt membrane integrity and cellular homeostasis. Furthermore, reduced capacity for fatty acid oxidation leads to increased dependence on glucose metabolism, contributing to hypoketotic hypoglycemia, particularly during fasting or metabolic stress [60, 61].

Significance of Amino Acid Catabolism

Amino acid catabolism is also impacted by mitochondrial diseases, specifically with regard to branched-chain amino acids (BCAAs) like valine, isoleucine, and leucine. Mitochondrial enzymes are necessary for the breakdown of BCAAs, and malfunctions in these processes result in the build-up of hazardous intermediates, such as branched-chain α-keto acids. Additionally, impaired mitochondrial function can compromise the urea cycle, causing hyperammonemia and encephalopathy. Disruption in amino acid metabolism also affects the biosynthesis of critical cofactors and molecules, such as heme, lipoic acid, and glutathione, which are vital for cellular redox balance and mitochondrial function.

Incomplete Metabolism of Vitamins Due to Mitochondrial Disorder

Mitochondrial disorders profoundly affect the incomplete metabolism of vitamins, with significant implications for cellular biochemistry and clinical outcomes. Mitochondria, the cell's principal energy-producing organelles, are pivotal in various biochemical pathways, including the metabolism of critical vitamins. Mitochondrial dysfunction impairs the bioavailability and efficacy of vitamins, leading to a range of metabolic disturbances [62].

The B complex vitamins, including thiamine (B1), riboflavin (B2), niacin (B3), pantothenic acid (B5), and biotin (B7), are essential cofactors in mitochondrial metabolism. The pyruvate dehydrogenase complex, which catalyzes the conversion of pyruvate to acetyl-CoA, a crucial metabolite in the citric acid cycle, includes thiamine as a cofactor. Research indicates that a decrease in thiamine may hinder the generation of ATP in the mitochondria, worsening energy deficiencies associated with diseases, such as mitochondrial myopathy. For many redox processes, including those in the electron transport chain, riboflavin is necessary because it is a precursor to flavin adenine dinucleotide (FAD) [63, 64].

A lack of riboflavin can cause a decrease in the availability of FAD, which can interfere with mitochondrial oxidative phosphorylation and cause symptoms like cardiomyopathy and muscle weakness. As it is essential for the synthesis of NAD^+, niacin is also important for the mitochondrial dehydrogenases that are part of the electron transport chain. Niacin deficiency can hinder the synthesis of NAD^+, which can damage mitochondria and make them more vulnerable to oxidative stress. As a part of coenzyme A, pantothenic acid plays a role in the metabolism of fatty acids and the Krebs cycle. A pantothenic acid deficiency may result in metabolic acidosis and decreased ATP synthesis. Biotin serves as a cofactor for carboxylase enzymes, essential for gluconeogenesis and fatty acid synthesis. Biotin deficiency can impair these pathways, leading to metabolic derangements and skin disorders [65].

Furthermore, the metabolism of fat-soluble vitamins, including A, D, E, and K, might be impacted by mitochondrial diseases. Strong lipophilic antioxidant vitamin E guards against oxidative damage to cell membranes. Decreases in vitamin E levels are frequently the result of oxidative stress being intensified by mitochondrial malfunction. Serum vitamin E levels were found to be considerably lower in patients with mitochondrial diseases, and this was correlated with elevated markers of oxidative stress. Vitamin D is hydroxylated in the liver and kidneys, two organ systems that are susceptible to mitochondrial malfunction and are necessary for maintaining calcium homeostasis. Osteoporosis risk and reduced bone mineralization can result from vitamin D deficiency [66, 67].

Incomplete Metabolism of Drugs Due to Oxidative Stress

The incomplete metabolism of drugs due to oxidative stress poses significant challenges in pharmacology and toxicology. When an imbalance occurs between reactive oxygen species (ROS) and the body's antioxidant defenses, it causes oxidative stress, which, in turn, damages cells and modifies medication metabolism. The liver is the main organ involved in drug metabolism. To change lipophilic medicines into hydrophilic metabolites that may be excreted, the liver uses cytochrome P450 enzymes (CYPs), UDP-glucuronosyltransferases (UGTs), and other metabolic pathways [68].

CYPs, particularly *CYP3A4*, *CYP2D6*, and *CYP2C19*, play critical roles in phase I metabolism. Oxidative stress can impair these enzymes through direct ROS-mediated modifications. For instance, ROS can induce oxidative modifications of *CYP3A4*, leading to enzyme inactivation and altered drug metabolism. Research has indicated that oxidative stress can reduce the activity of *CYP3A4* by up to 50% in certain conditions, which can significantly affect drug clearance [69, 70].

Additionally, oxidative stress can damage liver cells, leading to inflammation and fibrosis, which further impairs metabolic processes. Increased production of pro-inflammatory cytokines and decreased activity of antioxidant enzymes like glutathione peroxidase (GPx) and superoxide dismutase (SOD) are frequently associated with this damage. For instance, there is a noticeable decrease in UGT activity in liver disorders, such as non-alcoholic fatty liver disease (NAFLD), which is linked to increased oxidative stress and affects the glucuronidation of medications. This can lead to the accumulation of drugs and their metabolites, potentially resulting in toxicity or therapeutic failure. Oxidative stress is also implicated in altering drug interactions and pharmacokinetics. For example, in patients with chronic conditions like diabetes or cardiovascular diseases, oxidative stress can lead to significant alterations in drug metabolism. Oxidative stress can increase the expression of CYPs implicated in drug metabolism in diabetic patients, which can change clearance rates and increase the risk of drug interactions [71, 72].

Furthermore, the pharmacokinetics of drugs, such as statins and antihypertensives, can be significantly affected by oxidative stress, necessitating careful dose adjustments and monitoring. Optimizing drug metabolism and reducing side effects require addressing oxidative stress with antioxidant treatments or choosing medications with low oxidative stress impacts. Therapeutic outcomes and safety can be improved by using personalized medicine techniques that take individual differences in oxidative stress and medication metabolism into consideration.

Thus, understanding the intricate connection between medication metabolism and oxidative damage is essential for improving patient care and ensuring the effective use of pharmaceuticals [68, 71].

Incomplete Metabolism of Nutrients Due to Oxidative Stress

Oxidative stress has a significant impact on nutrition metabolism and can result in dysregulated or incomplete metabolic processes. It is defined as an imbalance between the production of reactive oxygen species (ROS) and the cellular antioxidant defense mechanisms. Superoxide anions, hydroxyl radicals, and hydrogen peroxide are examples of ROS that can cause oxidative damage to macromolecules, such as proteins, lipids, and nucleic acids, which can disrupt vital metabolic pathways.

Oxidative stress can hinder the activity of important enzymes in the tricarboxylic acid cycle (TCA cycle), such as pyruvate dehydrogenase and aconitase, which lowers the efficiency of oxidative phosphorylation and the synthesis of adenosine triphosphate (ATP), a crucial step in the metabolism of carbohydrates. As a result, lactate and other partially oxidized metabolic intermediates build up, which exacerbates metabolic acidosis. Such impairment is evident in metabolic diseases like diabetes mellitus, where hyperglycemia-induced oxidative stress exacerbates insulin resistance by modifying glucose transporter (GLUT) proteins and disrupting glucose uptake and glycolysis (Fig. **2**) [73, 74].

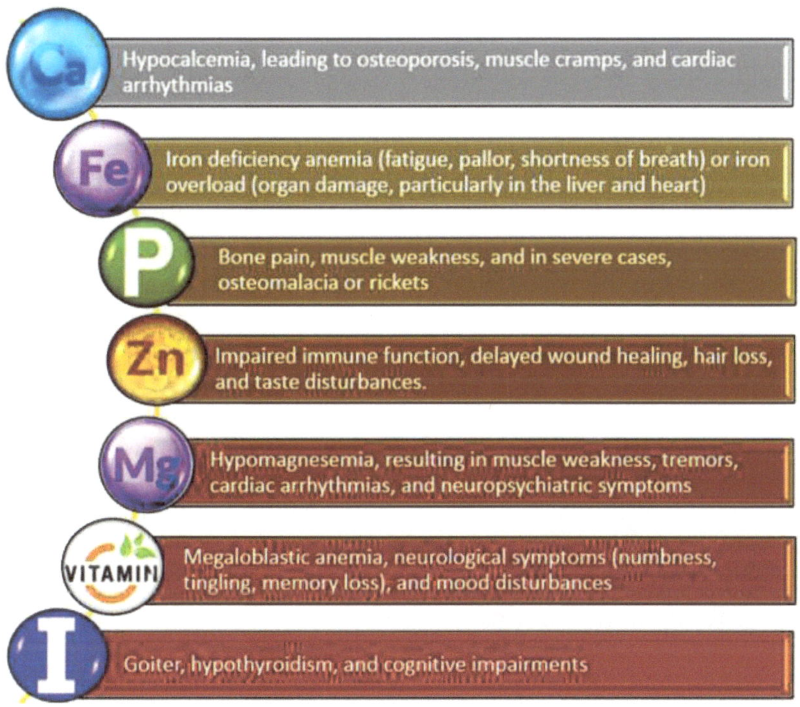

Fig. (2). Disorders and diseases caused by incomplete metabolism of vitamins and nutrients.

Oxidative stress in lipid metabolism encourages polyunsaturated fatty acids (PUFAs) in cellular membranes to peroxide, producing lipid peroxides and reactive aldehydes, such as 4-hydroxynonenal (4-HNE) and malondialdehyde (MDA). Reactive aldehydes have the ability to create adducts with nucleic acids and proteins, which can further impair bioenergetics and cellular function.

Lipid peroxidation also impairs mitochondrial beta-oxidation, where oxidative damage to enzymes, such as Acyl-CoA dehydrogenase, diminishes the breakdown of fatty acids, leading to reduced ATP generation and the accumulation of partially oxidized fatty acid intermediates like acylcarnitines. An accumulation of toxic metabolites like ammonia and impaired protein synthesis due to oxidative damage to ribosomal RNA and translational machinery result from oxidative modifications to amino acids like cysteine and methionine, which can change the conformation and function of enzymes and transporters involved in amino acid catabolism and the urea cycle [75].

Oxidative stress also disrupts the bioavailability and metabolism of essential micronutrients, particularly those that function as cofactors in various enzymatic reactions. For instance, oxidative damage to thiamine, a crucial cofactor for pyruvate dehydrogenase, results in decreased enzymatic activity and impaired carbohydrate metabolism. Similarly, oxidative stress can deplete levels of antioxidant vitamins, such as vitamins A, C, and E, further exacerbating the oxidative environment and perpetuating a cycle of cellular damage. This impaired nutrient metabolism due to oxidative stress contributes to a pathogenic feedback loop, where incomplete metabolic processes enhance ROS production, aggravating cellular damage, inflammation, and the pathogenesis of various metabolic disorders, including obesity, cardiovascular diseases, and neurodegenerative conditions. Gaining an understanding of these pathways is essential to creating treatment plans that lessen the damaging effects of oxidative stress on metabolism, such as tailored antioxidant medications and lifestyle changes [76].

Incomplete Metabolism of Vitamins Due to Oxidative Stress

When the body's antioxidant defenses are overwhelmed by reactive oxygen species (ROS), a condition known as oxidative stress occurs. This greatly hinders the body's ability to metabolize vital vitamins, particularly those that have antioxidant qualities like vitamins A, C, E, and several B vitamins.

The incomplete metabolism of these vitamins under oxidative stress can result in reduced bioavailability, impaired function, and exacerbation of cellular damage. For instance, studies have shown that during oxidative stress, the body's demand for vitamin C (ascorbic acid) increases substantially due to its role in neutralizing

ROS. Research indicates that plasma vitamin C levels can decrease by 70-90% during acute oxidative stress conditions, such as sepsis or intense physical exercise, resulting in inadequate levels for other metabolic functions, such as collagen synthesis and carnitine production. Incomplete metabolism of vitamin C can, therefore, compromise cellular repair mechanisms, leading to tissue damage and impaired immune response [67, 77].

Similarly, vitamin E (tocopherol), a lipid-soluble antioxidant, is crucial for protecting cell membranes from lipid peroxidation. Under oxidative stress, vitamin E is rapidly oxidized to its radical form, which must be reduced back to its active form, a process heavily dependent on the availability of other antioxidants like vitamin C. Studies indicate that oxidative stress situations, like persistent inflammation, can cause a 40–60% decrease in vitamin E levels, hence decreasing its preventive effectiveness. A meta-analysis revealed that low levels of vitamin E are linked to a 30–50% greater risk of coronary artery disease because of elevated lipid peroxidation.

This depletion raises the risk of chronic illnesses. Oxidative stress also has a major impact on the metabolism of B vitamins, including riboflavin (B2) and folate (B9). An established risk factor for cardiovascular disease, *i.e.*, high homocysteine levels result from poor metabolism, when folate is deficient as it is necessary for both DNA synthesis and repair. Clinical study data show that patients with elevated homocysteine levels, up to 25–30% greater than those with lower oxidative stress, have higher levels of oxidative stress markers, such as malondialdehyde (MDA), which raises the risk of cardiovascular problems [78].

Additionally, riboflavin, a cofactor for antioxidant enzymes, such as glutathione reductase, can undergo oxidative degradation. Studies reported that during periods of oxidative stress, riboflavin concentrations in tissues can decrease by up to 30%, weakening the antioxidant defense system.

Moreover, vitamin A (retinoids), vital for vision, immune function, and cell differentiation, is particularly susceptible to oxidative degradation. Oxidative stress can oxidize retinol to form less active retinoid radicals, which may accumulate and contribute to cellular damage.

A study showed that in individuals exposed to high oxidative stress, such as smokers, plasma retinol levels were 20-40% lower than those in non-smokers, correlating with a higher incidence of age-related macular degeneration. Therefore, oxidative stress-induced incomplete metabolism of these vitamins not only diminishes their antioxidant capacity but also leads to a cascade of adverse health effects, highlighting the importance of strategies to mitigate oxidative damage and maintain optimal vitamin function [77].

Enzyme Replacement and Gene Therapy

Enzyme replacement therapy (ERT) and gene therapy represent pivotal strategies in managing inherited metabolic disorders characterized by deficient or aberrant enzyme activity, leading to disrupted drug and nutrient metabolism. ERT involves the exogenous administration of functional enzymes to compensate for endogenous deficiencies, thereby mitigating substrate accumulation and restoring metabolic equilibrium. This therapeutic modality has demonstrated efficacy in lysosomal storage disorders, including Gaucher and Fabry diseases, although immunogenicity and the necessity for chronic administration pose limitations [79].

Gene therapy, by contrast, targets the genetic etiology of these disorders, employing viral vectors like adeno-associated viruses (AAV) to deliver functional gene sequences into affected cells, facilitating endogenous enzyme synthesis. Recent advancements in gene-editing platforms, notably CRISPR-Cas9, enhance precision in correcting metabolic defects. Despite significant potential, challenges, such as vector immunogenicity, off-target effects, and sustained gene expression, remain areas of active investigation. Synergistic approaches integrating ERT and gene therapy may offer superior therapeutic outcomes in metabolic disease management [80].

Enzyme Replacement in Incomplete Metabolism

Enzyme replacement therapy (ERT) has emerged as an important treatment option for metabolic disorders caused by enzyme deficiencies. When enzymes responsible for metabolizing drugs or nutrients are absent or malfunctioning, it can result in incomplete metabolism, leading to the accumulation of harmful substances or impairing the body's ability to process essential compounds effectively. Enzyme replacement offers a targeted way to restore these missing enzymes, potentially improving both drug metabolism and nutrient processing [81].

Drug metabolism relies heavily on enzymes, such as those in the cytochrome P450 family, which are involved in the activation, deactivation, and elimination of various medications. Genetic variations or external factors can lead to deficiencies or malfunctions in these enzymes, causing improper drug metabolism. For example, individuals with mutations in the *CYP2D6* gene may not be able to convert prodrugs like codeine into its active form, morphine, resulting in reduced therapeutic effects. In such cases, enzyme replacement could be a promising solution to enhance drug metabolism, allowing drugs to be processed as intended.

However, achieving this on a systemic level is complex, as enzymes must maintain their activity in the bloodstream and effectively reach the tissues or cells where metabolism occurs [82].

Beyond drugs, enzyme deficiencies can interfere with the metabolism of nutrients, leading to nutritional deficiencies or the accumulation of toxic intermediates. Lactose intolerance, where the enzyme lactase is deficient, is a common example. Without lactase, individuals are unable to break down lactose into glucose and galactose, causing gastrointestinal issues when consuming dairy products. Lactase supplements serve as a form of enzyme replacement therapy, enabling lactose metabolism and preventing discomfort. Similarly, phenylketonuria (PKU), a genetic disorder caused by a deficiency in phenylalanine hydroxylase, leads to an inability to convert phenylalanine into tyrosine. The resulting buildup of phenylalanine can have severe consequences if untreated. Enzyme replacement therapies, such as recombinant phenylalanine ammonia-lyase (PAL), help break down excess phenylalanine, preventing toxicity and improving patient outcomes [83].

Despite the potential of enzyme replacement, significant challenges remain. One major obstacle is the stability of enzymes in the human body. Enzymes can be rapidly degraded by proteolytic enzymes or lose functionality due to pH variations, limiting their therapeutic efficacy. To counter this, enzymes are often modified, such as through PEGylation, which involves attaching polyethylene glycol chains to increase enzyme stability and reduce immune system detection. Additionally, ensuring the enzyme reaches the appropriate tissues or cellular compartments is critical for effective treatment. Lysosomal storage diseases, such as Gaucher's or Fabry's disease, have seen more success with enzyme replacement, as the enzymes can be targeted directly to the lysosomes of cells [84].

The future of enzyme replacement therapy may benefit from advancements in nanotechnology. Using nanocarriers, such as liposomes or nanoparticles, could protect enzymes from degradation and improve targeted delivery to specific cells or organs. Furthermore, gene therapy holds the promise of enabling the body to produce the deficient enzymes internally, potentially offering a long-term solution to metabolic enzyme deficiencies. While enzyme replacement therapy shows great promise for addressing incomplete drug and nutrient metabolism, ongoing research into improving enzyme stability, delivery, and targeting will be key to expanding its therapeutic potential [81, 82, 84].

Gene Therapy in Incomplete Metabolism

Gene therapy represents a transformative approach with immense potential in correcting genetic defects that underlie the incomplete metabolism of drugs, nutrients, and vitamins. Metabolic dysfunctions of this kind are often caused due to mutations in genes encoding enzymes or proteins essential for metabolic pathways. These genetic aberrations can result in the accumulation of toxic intermediates or impair the synthesis of vital compounds, thereby leading to pathological conditions. By targeting the molecular root of these dysfunctions, gene therapy offers a more precise and long-lasting solution than traditional treatments [85].

The cytochrome P450 enzyme family is essential for the biotransformation of xenobiotics, such as medications, in the context of drug metabolism.

Drug metabolism rates have been found to be dramatically altered by genetic polymorphisms in genes like *CYP2D6*, *CYP3A4*, or *CYP2C19*, which can classify people as poor, moderate, extensive, or ultra-rapid metabolizers. Ultra-rapid metabolizers may clear medications too fast, making treatment ineffective, while poor metabolizers have the risk of drug buildup and toxicity. For instance, mutations in the *CYP2D6* gene can disrupt the metabolism of a wide array of drugs, including opioids like codeine, where the metabolic conversion to morphine is essential for its analgesic effect. Through the delivery of functional alleles of these polymorphic genes, gene therapy may be able to minimize the risk of adverse medication reactions, normalize drug clearance rates, and restore normal metabolic activity [86].

Such interventions, while still theoretical, highlight the capability of gene therapy to address personalized pharmacogenomics in ways that current dose-modification strategies cannot. Metabolism of essential nutrients and vitamins is similarly dependent on enzymes that are frequently compromised by genetic mutations. An important component of the folate cycle, for instance, is the enzyme 5,10-methylenetetrahydrofolate reductase (MTHFR), which is encoded by the MTHFR gene. It transforms 5,10-methylenetetrahydrofolate into 5-methyltetrahydrofolate, the bioactive form of folate needed for homocysteine biotransformation. Mutations in the MTHFR gene lead to hyperhomocysteinemia and can increase the risk of cardiovascular diseases, neurological defects, and impaired DNA synthesis. By utilizing gene therapy to introduce a functional MTHFR allele, the normal folate cycle could be reinstated, lowering homocysteine levels and mitigating the associated risks [87].

Phenylketonuria (PKU) offers another striking example of incomplete nutrient metabolism due to genetic mutations. Mutations in the gene encoding phenylalanine hydroxylase (PAH), the enzyme that converts phenylalanine to tyrosine, are the cause of PKU. A defect in this enzyme leads to elevated phenylalanine levels, which are neurotoxic, especially during early development. Current treatments involve strict dietary restrictions, but gene therapy holds the potential to introduce functional PAH genes into hepatocytes, enabling the catabolism of phenylalanine and preventing neurotoxicity. This could lessen the burden of ongoing dietary monitoring and greatly enhance the quality of life for those who have PKU [86].

Vitamin metabolism, especially that of vitamin B12 (cobalamin), provides another context where gene therapy could revolutionize treatment. Vitamin B12 is essential for DNA synthesis and neurological function, and its absorption, transport, and intracellular conversion require a complex interplay of several proteins. Mutations in the genes involved in cobalamin metabolism, such as *MMACHC* (methylmalonic aciduria cblC type with homocystinuria), lead to disorders like methylmalonic acidemia and pernicious anemia, where vitamin B12 cannot be properly utilized. By delivering a functional version of the mutated gene *via* gene therapy, normal B12 metabolism can be restored, reducing the accumulation of toxic intermediates and improving cellular function [87].

Despite the remarkable promise of gene therapy, challenges, such as efficient tissue-specific delivery and immune responses, remain. Given the liver's crucial role in systemic metabolism, hepatocytes in the liver are frequently the primary target for metabolic diseases. Viral vectors, such as adeno-associated viruses (AAVs), are commonly employed for gene delivery but face limitations in terms of immunogenicity and vector size. Recent advances in non-viral delivery methods and genome editing technologies, such as CRISPR-Cas9, offer more precise and potentially safer approaches. CRISPR-Cas9, for example, could correct point mutations *in situ*, offering a more permanent and refined intervention compared to gene replacement strategies.

The application of gene therapy to metabolic disorders also aligns with the broader goals of precision medicine. By tailoring interventions to the patient's unique genetic makeup, therapeutic outcomes can be optimized, reducing the variability seen in response to current treatments. This could be particularly impactful in the management of complex metabolic conditions that involve multifactorial genetic and environmental interactions. As gene therapy continues to evolve, it holds the promise of correcting the molecular basis of incomplete metabolism in a wide range of genetic disorders, thereby improving patient outcomes, reducing the incidence of complications, and offering more sustainable long-term solutions than traditional therapies [88].

Management and Treatment

The incomplete metabolism of drugs, nutrients, and vitamins is a critical issue in pharmacokinetics and pharmacodynamics, often leading to suboptimal therapeutic efficacy, toxicity, or bioavailability reduction. To effectively manage and treat this metabolic inefficiency, a variety of strategies, ranging from pharmacogenomics, enzyme modulation, advanced drug delivery technologies, and prodrug development, are employed. These methods ensure optimized metabolic pathways, efficient absorption, and proper biotransformation, all critical for improving patient outcomes [17, 89].

Pharmacogenomics and Personalized Medicine

Pharmacogenomics leverages individual genetic profiles to predict metabolic responses to drugs and nutrients. Genes encoding for metabolizing enzymes, such as cytochrome P450 enzymes (CYP450), UDP-glucuronosyltransferases (UGT), and N-acetyltransferases (NAT), exhibit polymorphisms that profoundly influence metabolic capacity.

For example, 7–10% of Caucasians are poor metabolizers due to *CYP2D6* polymorphism, which severely hinders their capacity to convert codeine into morphine, the active substance.

This leads to therapeutic inefficacy and, in some cases, drug toxicity due to the accumulation of unmetabolized drugs. Conversely, *CYP2C9* and *CYP2C19* polymorphisms affect the metabolism of anticoagulants like warfarin and anti-platelet agents like clopidogrel, respectively. Adjusting doses based on these polymorphisms has been shown to reduce bleeding complications by 30-50% compared to standard dosing. By using genotyping to identify metabolic phenotypes (for example, individuals with low, middle, extended, or ultra-rapid metabolisms), physicians can optimize drug dosing and reduce adverse effects through personalized treatment regimens [90].

Enzyme Induction or Inhibition

Enzyme modulation is a critical strategy to manage incomplete metabolism, as many drugs rely on specific enzymes for biotransformation.

Enzyme induction: Substrates like statins and oral contraceptives can be metabolized more quickly when certain agents, such as rifampicin, which is a strong *CYP3A4* inducer, are used.

This induction can reduce therapeutic levels, necessitating dose adjustments to maintain efficacy.

Enzyme inhibition: On the other hand, ketoconazole, a strong CYP3A4 inhibitor, can significantly increase plasma concentrations of drugs like midazolam or simvastatin, raising the risk of toxicity. Co-administration with inhibitors often requires reducing the dose of the affected drug to avoid adverse drug reactions (ADRs). The area under the curve (AUC) of midazolam increased tenfold when CYP3A4 inhibitors, such as itraconazole, were administered concurrently in clinical trials, emphasizing the significance of dose modifications while employing inhibitors [91].

Nanotechnology and Drug Delivery Systems

The advent of nanotechnology has revolutionized drug delivery systems, particularly in overcoming challenges related to incomplete metabolism. Nanoparticles (NPs), such as liposomes, polymeric nanoparticles, and micelles, can encapsulate drugs, protecting them from premature metabolism and improving pharmacokinetics. Liposome-based delivery systems, for instance, have been shown to increase the bioavailability of water-insoluble drugs like paclitaxel by 2-3 fold, compared to conventional formulations. Liposomes also provide controlled release and targeted delivery, reducing systemic clearance and enhancing tissue-specific uptake. Studies on PEGylated nanoparticles demonstrated a 50% reduction in clearance rate compared to non-PEGylated systems, increasing circulation half-life and improving therapeutic efficacy [92].

This method can be used for vitamins and nutrients. For example, fat-soluble vitamins (such as vitamins D and E) in nano-emulsion form have much higher bioavailability because of better gastrointestinal absorption and resistance to metabolic breakdown.

Prodrug Design

Prodrugs are chemically modified precursors that undergo enzymatic or chemical transformation into their active form *in vivo*. Prodrug strategies are particularly useful in bypassing first-pass metabolism, enhancing lipophilicity for improved absorption, or reducing toxicity of the active moiety during systemic circulation. Oseltamivir (Tamiflu) is a classic example, where the prodrug is converted by hepatic esterases into its active form to inhibit viral neuraminidase. Prodrug design for drugs with poor bioavailability (*e.g.*, acyclovir) improves bioavailability by up to 4-fold. Similarly, the prodrug fosphenytoin circumvents solubility issues associated with phenytoin, providing a more favorable pharmacokinetic profile with fewer side effects [88].

Nutrient and Vitamin Formulation Adjustments

In cases of incomplete metabolism due to absorption deficiencies, particularly with vitamins, the formulation can be adjusted to enhance bioavailability. For instance, vitamin B12, which requires intrinsic factors for absorption, is often poorly metabolized in individuals with pernicious anemia or gastric atrophy. In such cases, sublingual or intramuscular formulations bypass gastrointestinal absorption challenges, ensuring more efficient systemic uptake. The solubility and absorption of fat-soluble vitamins (A, D, E, and K) can be improved by delivering them in lipid-based carriers, such as self-emulsifying drug delivery systems (SEDDSs). Clinical studies have demonstrated that lipid-based formulations of vitamin E exhibit 2-3 times higher bioavailability compared to non-emulsified versions [75].

Co-Factor Supplementation

Metabolic reactions rely heavily on co-factors, such as vitamins and minerals, to function efficiently. In the case of incomplete metabolism due to co-factor deficiency, supplementation can restore enzyme function and optimize metabolism. For example, the metabolism of medications, such as methotrexate, depends on methylfolate, which is the active form of folic acid. Studies have shown that methylfolate supplementation reduces methotrexate toxicity by up to 60% while preserving therapeutic efficacy by providing adequate methyl group donors for cellular reactions [93].

Metabolic Pathway Modulation

Modulating metabolic pathways to favor alternative, more efficient routes can significantly impact the metabolism of drugs and nutrients. For instance, diverting metabolism away from pathways prone to toxic intermediate formation can reduce adverse effects. For example, CYP2E1 forms the hazardous metabolite N-acetyl-p-benzoquinone imine (NAPQI), which is detoxified by shunting acetaminophen metabolism *via* glutathione conjugation. When administering N-acetyl cysteine after an acetaminophen overdose, glutathione stores are replenished, which aids in detoxification and lessens liver damage [62].

Biodegradable Nanomaterials for Targeted Metabolism

By directing distribution to particular tissues and managing the release of active ingredients, biodegradable nanoparticles have become an innovative way to enhance the digestion of nutrients and medications. After delivering their cargo, these nanocarriers break down, reducing systemic exposure and preventing first-pass metabolism. Due to their biocompatibility and capacity to incorporate both hydrophilic and hydrophobic medications, polylactic-co-glycolic acid (PLGA) nanoparticles have found extensive application. According to a study, PLGA nanoparticles have the potential to be used in targeted delivery because they show a three-fold increase in drug retention at tumor locations when compared to free drug formulations [89].

CONCLUSION

The study on the consequences of incomplete metabolism for health emphasizes how vital effective metabolic processes are to preserving homeostasis and general health. These pathways are disrupted, especially when incomplete metabolism occurs. This results in the build-up of toxic byproducts, nutritional imbalances, and the development of chronic illnesses like diabetes, obesity, and mitochondrial disorders. In this chapter, the serious neurological and developmental repercussions of faulty metabolic pathways are highlighted, with diseases like Phenylketonuria (PKU) and Maple Syrup Urine Disease (MSUD) serving as prime examples.

Medication buildup and metabolic toxicity are among the serious dangers associated with incomplete metabolism. Genetic polymorphisms, liver and kidney problems, and enzyme deficiencies can all cause problems with effective medication metabolism. As seen in diseases like G6PD deficiency and hepatic/renal impairments, this inefficiency can lead to toxic build-up, underscoring the necessity of individualized therapeutic measures. Current developments in therapeutic drug monitoring and pharmacogenetics can help reduce these risks and customize medications based on each patient's unique metabolic profile. Another major concern is the metabolism of nutrients, especially those involving vitamins and minerals. Clinical consequences ranging from anemia to neurological illnesses are caused by deficiencies in key vitamins (*e.g.*, B12, D, A, K) and minerals (*e.g.*, calcium, magnesium, iron). These deficiencies increase metabolic inefficiencies. The accumulation of hazardous lipid intermediates and mitochondrial dysfunction caused by inadequate lipid metabolism, especially in fatty liver diseases like NAFLD and NASH, also contribute to the development of liver damage and systemic inflammation.

In the future, innovative therapeutic techniques, including gene therapy and enzyme replacement therapy (ERT), hold promise for reestablishing metabolic balance. Although gene therapy, using tools like CRISPR-Cas9, provides the possibility of long-term repair of enzyme deficits, ERT has shown effectiveness in treating lysosomal storage disorders. The treatment of metabolic illnesses could be revolutionized by improvements in nanotechnology and gene-editing techniques, notwithstanding present obstacles in delivery and immune reactions.

Furthermore, prodrug design, enzyme regulation, and pharmacogenomics are all types of precision therapies that offer a route toward individualized medicine.

REFERENCES

[1] S. Nagar, U. A. Argikar, and D. J. Tweedie, "Enzyme kinetics in drug metabolism: fundamentals and applications," Methods Mol. Biol., vol. 1113, pp. 1–6, 2014.
PMID: 24523105 DOI: 10.1007/978-1-62703-758-7_1

[2] N. Barzilai, D.M. Huffman, R.H. Muzumdar, and A. Bartke, "The critical role of metabolic pathways in aging", *Diabetes,* vol. 61, no. 6, pp. 1315-1322, 2012.
[http://dx.doi.org/10.2337/db11-1300] [PMID: 22618766]

[3] A. Fasano and C. Catassi, "Celiac disease," N. Engl. J. Med., vol. 367, no. 25, pp. 2419–2426, 2012.
[http://dx.doi.org/10.1056/NEJMcp1113994]

[4] N.A. ElSayed, G. Aleppo, V.R. Aroda, R.R. Bannuru, F.M. Brown, D. Bruemmer, B.S. Collins, J.L. Gaglia, M.E. Hilliard, D. Isaacs, E.L. Johnson, S. Kahan, K. Khunti, J. Leon, S.K. Lyons, M.L. Perry, P. Prahalad, R.E. Pratley, J.J. Seley, R.C. Stanton, and R.A. Gabbay, "2. Classification and Diagnosis of Diabetes: Standards of Care in Diabetes—2023", *Diabetes Care,* vol. 46, no. January, suppl. Suppl. 1, pp. S19-S40, 2023.
[http://dx.doi.org/10.2337/dc23-S002] [PMID: 36507649]

[5] M. Adeva-Andany, M. López-Ojén, R. Funcasta-Calderón, E. Ameneiros-Rodríguez, C. Donapetry-García, M. Vila-Altesor, and J. Rodríguez-Seijas, "Comprehensive review on lactate metabolism in human health", *Mitochondrion,* vol. 17, pp. 76-100, 2014.
[http://dx.doi.org/10.1016/j.mito.2014.05.007] [PMID: 24929216]

[6] L.C. Burrage, S.C.S. Nagamani, P.M. Campeau, and B.H. Lee, "Branched-chain amino acid metabolism: from rare Mendelian diseases to more common disorders", *Hum. Mol. Genet.,* vol. 23, no. R1, pp. R1-R8, 2014.
[http://dx.doi.org/10.1093/hmg/ddu123] [PMID: 24651065]

[7] C.R. Scriver, "Does hereditary metabolic disease modulate senescence and ageing?", *J. Inherit. Metab. Dis.,* vol. 25, no. 3, pp. 235-251, 2002.
[http://dx.doi.org/10.1023/A:1015650517020] [PMID: 12137233]

[8] O.E. Owen, S.C. Kalhan, and R.W. Hanson, "The key role of anaplerosis and cataplerosis for citric acid cycle function", *J. Biol. Chem.,* vol. 277, no. 34, pp. 30409-30412, 2002.
[http://dx.doi.org/10.1074/jbc.R200006200] [PMID: 12087111]

[9] H.A. Krebs, "The history of the tricarboxylic acid cycle", *Perspect. Biol. Med.,* vol. 14, no. 1, pp. 154-172, 1970.
[http://dx.doi.org/10.1353/pbm.1970.0001] [PMID: 4923349]

[10] A.I. Zuñiga Vinueza, "Recent Advances in Phenylketonuria: A Review", *Cureus,* vol. 15, no. 6, p. e40459, 2023.
[http://dx.doi.org/10.7759/cureus.40459] [PMID: 37456395]

[11] S.B. Vafai, and V.K. Mootha, "Mitochondrial disorders as windows into an ancient organelle", *Nature,* vol. 491, no. 7424, pp. 374-383, 2012.
[http://dx.doi.org/10.1038/nature11707] [PMID: 23151580]

[12] S. DiMauro and E. A. Schon, "Mitochondrial respiratory-chain diseases," *N. Engl. J. Med.*, vol. 348, no. 26, pp. 2656–2668, 2003.

[13] W.M. Lee, "Acetaminophen (APAP) hepatotoxicity—Isn't it time for APAP to go away?", *J. Hepatol.,* vol. 67, no. 6, pp. 1324-1331, 2017.
[http://dx.doi.org/10.1016/j.jhep.2017.07.005] [PMID: 28734939]

[14] R. Kumar, Y. Singh, M. Piplani, P. Bhateja, R. Garg, and B. Kumar, "The Chemistry and

Pharmacological Study of Copaifera spp. Oils", *Chem. Insid. Spices Herbs Res. Dev.,* no. Jul, pp. 79-116, 2024.
[http://dx.doi.org/10.2174/9789815196832124040006]

[15] A. Kumar, P. Singh, R. Kumar, and S. Dutt, "Novel drug delivery systems for inflammatory bowel disease," In: *Nanobiotechnology and Artificial Intelligence in Gastrointestinal Diseases*, 2024, V. K. Chaturvedi, A. K. Singh, J. Singh, and D. P. Yadav, Eds., pp. 4-1, 4-24.
[http://dx.doi.org/10.1088/978-0-7503-6134-7ch4]

[16] R. Kumar, S. Dutt, A. Goyal, A. Kumar, and B. Kumar, "Nanotechnology in gastrointestinal endoscopy," In: Nanobiotechnology and Artificial Intelligence in Gastrointestinal Diseases, 2024, V. K. Chaturvedi, A. K. Singh, J. Singh, and D. P. Yadav, Eds., pp. 5-1, 5-32.
[http://dx.doi.org/10.1088/978-0-7503-6134-7ch5]

[17] A. Hamid, R. Kumar, V. K. Chaturvedi, and S. Dutt, "Artificial intelligence (AI) based colonoscopy," In: *Nanobiotechnology and Artificial Intelligence in Gastrointestinal Diseases*, ch. 12, 2024, V. K. Chaturvedi, A. K. Singh, J. Singh, and D. P. Yadav, Eds., IOP Publishing Ltd, pp. 5-1, 5-32.
[http://dx.doi.org/10.1088/978-0-7503-6134-7ch12]

[18] B. Beermann, and M. Groschinsky-Grind, "Clinical pharmacokinetics of diuretics", *Clin. Pharmacokinet.,* vol. 5, no. 3, pp. 221-245, 1980.
[http://dx.doi.org/10.2165/00003088-198005030-00003] [PMID: 6993083]

[19] K. Lecointre, L. Pisanté, F. Fauvelle, and S. Mazouz, "Digoxin toxicity evaluation in clinical practice with pharmacokinetic correlations", *Clin. Drug Investig.,* vol. 21, no. 3, pp. 225-232, 2001.
[http://dx.doi.org/10.2165/00044011-200121030-00009] [PMID: 27517553]

[20] T. Fukao, G.D. Lopaschuk, and G.A. Mitchell, "Pathways and control of ketone body metabolism: on the fringe of lipid biochemistry", *Prostaglandins Leukot. Essent. Fatty Acids,* vol. 70, no. 3, pp. 243-251, 2004.
[http://dx.doi.org/10.1016/j.plefa.2003.11.001] [PMID: 14769483]

[21] A. Singh, L.B. Prasad, K. Shiv, R. Kumar, and S. Garai, "Synthesis, characterization, and in vitro antibacterial and cytotoxic study of Co(II), Ni(II), Cu(II), and Zn(II) complexes of N- (4-methoxybenzyl) N-(phenylethyl) dithiocarbamate ligand", *J. Mol. Struct.,* vol. 1288, no. May, p. 135835, 2023.
[http://dx.doi.org/10.1016/j.molstruc.2023.135835]

[22] A. Singh, R. Kumar, K. Shiv, S.K. Pandey, M.K. Bharty, R.J. Butcher, and L.B. Prasad, "Synthesis, crystal structure and screening for anticonvulsant and antianxiety activities of three new Ni(II), Cu(II), and Zn(II) dithiocarbamate complexes", *J. Mol. Struct.,* vol. 1298, no. P2, p. 137052, 2024.
[http://dx.doi.org/10.1016/j.molstruc.2023.137052]

[23] "A. Singh, R. Kumar, R. Patel, Trishna, R. N. Gautam, M. K. Bharty, and L. B. Prasad, Dithiocarbamate-based novel anti-histaminic agents: synthesis, characterization, crystal structure and thermal study", *Dalton Trans.*, no. 33, 2024.
[http://dx.doi.org/10.1039/D4DT01777C]

[24] M.V. Relling, and T.E. Klein, "CPIC: Clinical pharmacogenetics implementation consortium of the pharmacogenomics research network", *Clin. Pharmacol. Ther.,* vol. 89, no. 3, pp. 464-467, 2011.
[http://dx.doi.org/10.1038/clpt.2010.279] [PMID: 21270786]

[25] A. Kumar, B. Kumar, R. Kumar, A. Kumar, M. Singh, V. Tiwari, A. Trigunayat, P. Paul, and P. Singh, "Acute and subacute toxicity study of ethanolic extract of Calotropis procera (Aiton) Dryand flower in Swiss albino mice", *Phytomed. Plus,* vol. 2, no. 2, p. 100224, 2022.
[http://dx.doi.org/10.1016/j.phyplu.2022.100224]

[26] D. Pathology, and B. Boston, "Tissue Changes in Vitamin Deficiencies", *Nutr. Rev.,* vol. 1, no. 7, pp. 203-205, 1943.
[http://dx.doi.org/10.1111/j.1753-4887.1943.tb08050.x]

[27] R. Dhiman, "Novel LC-MS/MS bioanalytical method validation for gabapentin in human plasma in

accordance with current regulatory guidances", *Int J Pharm Chem Sci*, vol. 4, no. 4, pp. 462-468, 2015.
Available from: http://www.ijpcsonline.com/files/08-10-15/07-1035.pdf

[28] F. N. Ali and C. B. Langman, "Disorders of mineral metabolism," In: *Clinical Pediatric Nephrology*, 3rd ed., K. Kher, H. W. Schnaper, and L. A. Greenbaum, Eds. Boca Raton, FL, USA: CRC Press, 2016, pp. 205–233.
[http://dx.doi.org/10.1201/9781315382319]

[29] "H. Pandey, S. Srivastava, R. Kumar, and Y. B. Tripathi, Preclinical acute and repeated dose toxicity of Pueraria tuberosa (PTWE) on Charles Foster rats", *J Pharm Sci Res.*, pp. 4572–4581, 2018. Available from: https://ijpsr.com/bft-article/preclinical-acute-and-repeated-dose-toxicity-of--ueraria-tuberosa-ptwe-on-charles-foster-rats/

[30] J. Chen, C. Ding, Y. Chen, W. Hu, C. Yu, C. Peng, X. Feng, Q. Cheng, W. Wu, Y. Lu, H. Xie, L. Zhou, J. Wu, and S. Zheng, "ACSL4 reprograms fatty acid metabolism in hepatocellular carcinoma via c-Myc/SREBP1 pathway", *Cancer Lett.,* vol. 502, pp. 154-165, 2021.
[http://dx.doi.org/10.1016/j.canlet.2020.12.019] [PMID: 33340617]

[31] R. Dhiman, D. Kumar, B. Kumar, and B.L. Pandey, "Quantitative determination of atorvastatin, ortho-hydroxy atorvastatin, para-hydroxy atorvastatin in human plasma using rosuvastatin as internal standard by LC-MS / MS", *Int. J. Pharm. Chem. Sci.,* vol. 4, no. 4, pp. 487-500, 2016.

[32] O. Govaere, S. Cockell, D. Tiniakos, R. Queen, R. Younes, M. Vacca, L. Alexander, F. Ravaioli, J. Palmer, S. Petta, J. Boursier, C. Rosso, K. Johnson, K. Wonders, C.P. Day, M. Ekstedt, M. Orešič, R. Darlay, H.J. Cordell, F. Marra, A. Vidal-Puig, P. Bedossa, J.M. Schattenberg, K. Clément, M. Allison, E. Bugianesi, V. Ratziu, A.K. Daly, and Q.M. Anstee, "Transcriptomic profiling across the nonalcoholic fatty liver disease spectrum reveals gene signatures for steatohepatitis and fibrosis", *Sci. Transl. Med.,* vol. 12, no. 572, p. eaba4448, 2020.
[http://dx.doi.org/10.1126/scitranslmed.aba4448] [PMID: 33268509]

[33] M. Singh, B. Kumar, A. K Singh, C. Shekhar Azad, M. K Yadav, A. Kumar, R. Kumar, A. Kumar, and P. Chaudhary, "Pharmacokinetic screening to Estimate the drug likeliness characteristics of selected Herbal Anticancer Drugs", *Res. J. Pharm. Technol.,* vol. 16, no. 7, pp. 3422-3426, 2023.
[http://dx.doi.org/10.52711/0974-360X.2023.00566]

[34] M. Dudek, D. Pfister, S. Donakonda, P. Filpe, A. Schneider, M. Laschinger, D. Hartmann, N. Hüser, P. Meiser, F. Bayerl, D. Inverso, J. Wigger, M. Sebode, R. Öllinger, R. Rad, S. Hegenbarth, M. Anton, A. Guillot, A. Bowman, D. Heide, F. Müller, P. Ramadori, V. Leone, C. Garcia-Caceres, P. A. Knolle, "Auto-aggressive $CXCR6^+$ CD8 T cells cause liver immune pathology in NASH," *Nature*, vol. 592, pp. 444–449, 2021.
[http://dx.doi.org/10.1038/s41586-021-03233-8]

[35] R. Elango, R.O. Ball, and P.B. Pencharz, "Amino acid requirements in humans: with a special emphasis on the metabolic availability of amino acids", *Amino Acids,* vol. 37, no. 1, pp. 19-27, 2009.
[http://dx.doi.org/10.1007/s00726-009-0234-y] [PMID: 19156481]

[36] B. Kumar, R. Kumar, H. Patel, and A. Kumar, "Development and validation of an RP-HPLC method for simultaneous estimation of amlodipine besylate and indapamide in bulk and tablet dosage form", *Indo Am. J. Pharm. Sci.,* vol. 4, no. 11, pp. 4278-4287, 2017.

[37] H. Huang, P. Zhou, J. Wei, L. Long, H. Shi, Y. Dhungana, N.M. Chapman, G. Fu, J. Saravia, J.L. Raynor, S. Liu, G. Palacios, Y.D. Wang, C. Qian, J. Yu, and H. Chi, "In vivo CRISPR screening reveals nutrient signaling processes underpinning CD8+ T cell fate decisions", *Cell,* vol. 184, no. 5, pp. 1245-1261.e21, 2021.
[http://dx.doi.org/10.1016/j.cell.2021.02.021] [PMID: 33636132]

[38] G. Kahl, ""Msud," Dict. Genomics.", *Transl. Proteom.,* vol. 99, no. 4, pp. 1-1, 2015.
[http://dx.doi.org/10.1002/9783527678679.dg07970]

[39] "B. Kumar, R. Kumar, A. Kumar, and R. Patel, "Simultaneous RP-HPLC method development and

validation of fenofibrate and rosuvastatin calcium in bulk and tablet dosage form,"", *Indo Am. J. Pharm. Sci.,* vol. 4, no. 11, pp. 4288-4297, 2017.

[40] A.J.M. Wagenmakers, "Tracers to investigate protein and amino acid metabolism in human subjects", *Proc. Nutr. Soc.,* vol. 58, no. 4, pp. 987-1000, 1999. [http://dx.doi.org/10.1017/S0029665199001305] [PMID: 10817167]

[41] M.A. Dahpy, T.H. Saleem, O.M. El-Asheer, A.A. ELrasoul, and A.M. Abo Elgeit, "Clinical, biochemical, molecular, and therapeutic analysis of maple syrup urine disease in upper Egypt", *J. Pediatr. Genet.,* vol. 10, no. 2, pp. 116-125, 2021. [http://dx.doi.org/10.1055/s-0040-1715111] [PMID: 33996182]

[42] S. Rauf, T. Almas, I. Ullah, N. Usman, and M. Irfan, "Maple syrup urine disease masquerading as urea cycle disorder: A tale of two clinical mimics", *Cureus,* vol. 12, no. 8, p. e9706, 2020. [http://dx.doi.org/10.7759/cureus.9706] [PMID: 32944429]

[43] K. Ajay, N. Manmath Kumar, K. Brijesh, K. Ashutosh, K. Rajesh, K. Vikas, and S. Ambrish Kumar, "Toxicity (acute and subacute) assessment and in-vivo antiurolithiatic activity of ethanolic extract of Caesalpinia bonducella seed in albino Wistar rat", *J. Appl. Pharm. Sci.,* vol. 12, no. 2, pp. 187-197, 2022. [http://dx.doi.org/10.7324/JAPS.2022.120220]

[44] S. Kalkan Ucar, M. Coker, S. Habif, E. Ulas Saz, B. Karapinar, H. Ucar, O. Kitis, and M. Duran, "The first use of N-carbamylglutamate in a patient with decompensated maple syrup urine disease", *Metab. Brain Dis.,* vol. 24, no. 3, pp. 409-414, 2009. [http://dx.doi.org/10.1007/s11011-009-9155-4] [PMID: 19688253]

[45] G. Boden, X. Chen, and T. P. Stein, "Gluconeogenesis in moderately and severely hyperglycemic patients with type 2 diabetes mellitus", *Am J Physiol Endocrinol Metab,* vol. 280, no. 1, pp. 23-30, 2001. [http://dx.doi.org/10.1152/ajpendo.2001.280.1.e23] [http://dx.doi.org/10.1152/ajpendo.2001.280.1.E23]

[46] R. Kumar, B. Kumar, and A. Kumar, "Stability indicating method development and validation of abiraterone in human plasma by using LC-MS/MS",

[47] A. Rossi, M. Ruoppolo, P. Formisano, G. Villani, L. Albano, G. Gallo, D. Crisci, A. Moccia, G. Parenti, P. Strisciuglio, and D. Melis, "Insulin-resistance in glycogen storage disease type Ia: linking carbohydrates and mitochondria?", *J. Inherit. Metab. Dis.,* vol. 41, no. 6, pp. 985-995, 2018. [http://dx.doi.org/10.1007/s10545-018-0149-4] [PMID: 29435782]

[48] I. Glucose, T. As, and C. Of, "Impaired glucose transport as a cause of decreased insulin-stimulated muscle glycogen synthesis in type 2 diabetes", *N Engl J Med,* 1999.

[49] R. Kumar, B. Kumar, and A. Kumar, "Pharmacokinetic profile of antitubercular drugs and their clinical applications in therapeutic drug monitoring", *Asian J Pharm,* vol. 3, 2017. no. 5, pp. 145-154, 2017. Available from: http://ajpp.in/uploaded/p71.pdf

[50] A. Federico, E. Cardaioli, P. Da Pozzo, P. Formichi, G.N. Gallus, and E. Radi, "Mitochondria, oxidative stress and neurodegeneration", *J. Neurol. Sci.,* vol. 322, no. 1-2, pp. 254-262, 2012. [http://dx.doi.org/10.1016/j.jns.2012.05.030] [PMID: 22669122]

[51] A. Kumar, B. Kumar, R. Kumar, A. Kumar, M. Singh, P. Singh, V. Tiwari, and A. Trigunayat, "Assessment of pain-ameliorative potential of ethanolic extract of Calotropis procera and its active components characterization by HPTLC and GC-MS analysis," *Syst. Rev. Pharm., vol. 13, no. 9, pp. 581–586, 2022.*

[52] K. Begriche, J. Massart, M.A. Robin, A. Borgne-Sanchez, and B. Fromenty, "Drug-induced toxicity on mitochondria and lipid metabolism: Mechanistic diversity and deleterious consequences for the liver", *J. Hepatol.,* vol. 54, no. 4, pp. 773-794, 2011. [http://dx.doi.org/10.1016/j.jhep.2010.11.006] [PMID: 21145849]

[53] G.L. Nicolson, "Mitochondrial dysfunction and chronic disease: Treatment with natural supplements", *Integr. Med. (Encinitas),* vol. 13, no. 4, pp. 35-43, 2014. [PMID: 26770107]

[54] R.D. Vijaya Durga, "Determination of S-amlodipine in human plasma using tizanidine as internal standard by Lc/Ms/Ms method", *J. Bioanal. Biomed.,* vol. 7, no. 3, pp. 91-96, 2015. [http://dx.doi.org/10.4172/1948-593X.1000130]

[55] N. Kaplowitz, "Hepatotoxicity due to mitochondrial injury," In: *Drug-Induced Liver Disease*, 2002, N. Kaplowitz and L. D. DeLeve, Eds., pp. 55–84. [http://dx.doi.org/10.3109/9781420021141-8]

[56] A. Kumar, M.K. Nandi, B. Kumar, A. Kumar, R. Kumar, A.K. Singh, and M. Singh, "Antiurolithiasis, antioxidant, anti-inflammatory, analgesic, and diuretic activity of ethanolic extract of seeds of Caesalpinia bonducella", *Int. J. Pharm. Investig.,* vol. 11, no. 3, pp. 306-311, 2021. [http://dx.doi.org/10.5530/ijpi.2021.3.54]

[57] M. Liesa, and O.S. Shirihai, "Mitochondrial dynamics in the regulation of nutrient utilization and energy expenditure", *Cell Metab.,* vol. 17, no. 4, pp. 491-506, 2013. [http://dx.doi.org/10.1016/j.cmet.2013.03.002] [PMID: 23562075]

[58] J. McInnes, "Mitochondrial-associated metabolic disorders: foundations, pathologies and recent progress", *Nutr. Metab. (Lond.),* vol. 10, no. 1, p. 63, 2013. [http://dx.doi.org/10.1186/1743-7075-10-63] [PMID: 24499129]

[59] R. Dhiman, V. Durga, K. Jayasankar, and S. Raj, "Solid phase extraction: Potential carbocisteine bioanalysis using carbocisteine 13c3 stable isotope: LC-MS/MS technique with pharmacokinetics on healthy human subjects", *Pharm. Lett.,* vol. 7, no. 8, pp. 126-134, 2015.

[60] P.E. Beitune, G. Duarte, M.C. Foss, R.M. Montenegro Jr, P. Spara, S.M. Quintana, E.A. Figueiró-Filho, A.G. da Costa, and F.M. Filho, "Effect of antiretroviral agents on carbohydrate metabolism in HIV-1 infected pregnant women", *Diabetes Metab. Res. Rev.,* vol. 22, no. 1, pp. 59-63, 2006. [http://dx.doi.org/10.1002/dmrr.576] [PMID: 16021650]

[61] R. Dhiman, V. Durga, and A. J. Rio, "Bioanalytical method for lornoxicam determination in human plasma by using piroxicam as internal standard by LC-MS/MS", *J Chem Pharm Res,* vol. 7, no. 8, pp. 54-60, 2015.

[62] J. M. Smellie, "High Basal Metabolic Rate: Exophthalmic Goitre.," *Proceedings of the Royal Society of Medicine*, vol. 26, no. 12. p. 1545, Oct. 1933.

[63] H. McNulty, and J. M. Scott, "Intake and status of folate and related B-vitamins: considerations and challenges in achieving optimal status", *Br J Nutr,* vol. 99, no. suppl. 3, pp. 48-54, 2008. [http://dx.doi.org/10.1017/S0007114508006855]

[64] R. Dhiman, B. Kumar, D. Kumar, and A. Kumar, "Liquid-liquid extraction method for ziprasidone (ZRS) bioanalysis by using ZRS-D 8 (stable isotope) as internal standard", *J Chem Pharm Res,* vol. 8, no. 3, pp. 722-727, 2016.

[65] P. Fricker, "Nutrition in sport," *Br. J. Sports Med.*, vol. 36, p. 154, 2002. [http://dx.doi.org/10.1136/bjsm.36.2.154-b]

[66] R.E. Frye, R. DeLaTorre, H. Taylor, J. Slattery, S. Melnyk, N. Chowdhury, and S.J. James, "Redox metabolism abnormalities in autistic children associated with mitochondrial disease", *Transl. Psychiatry,* vol. 3, no. 6, p. e273, 2013. [http://dx.doi.org/10.1038/tp.2013.51] [PMID: 23778583]

[67] R. Kumar, A. Kumar, A. Kumar, and M. Singh, "HPTLC and GC-MS analyses of biologically active extracts and fractions from *Premna latifolia* Roxb leaves reveal anti-inflammatory and analgesic effects," *Syst. Rev. Pharm.*, vol. 13, no. 8, 2021.

[68] P. Ghezzi, V. Jaquet, F. Marcucci, and H.H.H.W. Schmidt, "The oxidative stress theory of disease:

levels of evidence and epistemological aspects", *Br. J. Pharmacol.,* vol. 174, no. 12, pp. 1784-1796, 2017.
[http://dx.doi.org/10.1111/bph.13544] [PMID: 27425643]

[69] B.K. Yamamoto, and J. Raudensky, "The role of oxidative stress, metabolic compromise, and inflammation in neuronal injury produced by amphetamine-related drugs of abuse", *J. Neuroimmune Pharmacol.,* vol. 3, no. 4, pp. 203-217, 2008.
[http://dx.doi.org/10.1007/s11481-008-9121-7] [PMID: 18709468]

[70] R. Garg, K. Dobhal, B. Kumar, and A. K. Gupta, "COVID-19: a challenge towards the sustainability of health in platform era", In: *NeuroQuantology*, vol. 20. , 2022, no. 11, pp. 2503-2519.

[71] R.K.J. Olsen, N. Cornelius, and N. Gregersen, "Redox signalling and mitochondrial stress responses; lessons from inborn errors of metabolism", *J. Inherit. Metab. Dis.,* vol. 38, no. 4, pp. 703-719, 2015.
[http://dx.doi.org/10.1007/s10545-015-9861-5] [PMID: 26025548]

[72] R. Kumar, B. Kumar, A. Kumar, A. Kumar, and M. Singh, "GC-MS analysis of Phytocomponents in the Methanol Extract of Premna latifolia Roxb", *Pharmacognosy Res.,* vol. 14, no. 1, pp. 19-23, 2021.
[http://dx.doi.org/10.5530/pres.14.1.4]

[73] K.E. Wellen, and C.B. Thompson, "Cellular metabolic stress: considering how cells respond to nutrient excess", *Mol. Cell,* vol. 40, no. 2, pp. 323-332, 2010.
[http://dx.doi.org/10.1016/j.molcel.2010.10.004] [PMID: 20965425]

[74] H. Pandey, S. Srivastava, A. Kumar, R. Kumar, and Y. Bhusan Tripathi, "Pharmacokinetics study of puerarin absorption in blood after consumption of Pueraria tuberosa water extract (PTWE) by rats", *Explor. Anim. Med. Res.,* vol. 9, no. 2, pp. 188-196, 2019.

[75] B.L. Tan, M.E. Norhaizan, and W.P.P. Liew, "Nutrients and oxidative stress: Friend or foe?", *Oxid. Med. Cell. Longev.,* no. 1, p. 9719584, 2018.
[http://dx.doi.org/10.1155/2018/9719584] [PMID: 29643982]

[76] A.A. Petti, C.A. Crutchfield, J.D. Rabinowitz, and D. Botstein, "Survival of starving yeast is correlated with oxidative stress response and nonrespiratory mitochondrial function", *Proc. Natl. Acad. Sci. USA,* vol. 108, no. 45, pp. E1089-E1098, 2011.
[http://dx.doi.org/10.1073/pnas.1101494108] [PMID: 21734149]

[77] A.K. Panda, and G. Cherian, "Role of vitamin E in counteracting oxidative stress in poultry", *J. Poult. Sci.,* vol. 51, no. 2, pp. 109-117, 2014.
[http://dx.doi.org/10.2141/jpsa.0130134]

[78] F.J. Rupérez, D. García-Martínez, B. Baena, N. Maeso, A. Cifuentes, C. Barbas, and E. Herrera, "Evolution of oxidative stress parameters and response to oral vitamins E and C in streptozotocin-induced diabetic rats", *J. Pharm. Pharmacol.,* vol. 60, no. 7, pp. 871-878, 2008.
[http://dx.doi.org/10.1211/jpp.60.7.0008] [PMID: 18549673]

[79] D.A. Carbonaro, X. Jin, X. Wang, X.J. Yu, N. Rozengurt, M.L. Kaufman, X. Wang, D. Gjertson, Y. Zhou, M.R. Blackburn, and D.B. Kohn, "Gene therapy/bone marrow transplantation in ADA-deficient mice: roles of enzyme-replacement therapy and cytoreduction", *Blood,* vol. 120, no. 18, pp. 3677-3687, 2012.
[http://dx.doi.org/10.1182/blood-2012-02-408591] [PMID: 22833548]

[80] D. Concolino, F. Deodato, and R. Parini, "Enzyme replacement therapy: efficacy and limitations", *Ital J Pediatr,* vol. 44, no. Suppl 2, 2018.
[http://dx.doi.org/10.1186/s13052-018-0562-1]

[81] M. Kalivianakis, D.M. Minich, C.M.A. Bijleveld, W.M.C. van Aalderen, F. Stellaard, M. Laseur, R.J. Vonk, and H.J. Verkade, "Fat malabsorption in cystic fibrosis patients receiving enzyme replacement therapy is due to impaired intestinal uptake of long-chain fatty acids", *Am. J. Clin. Nutr.,* vol. 69, no. 1, pp. 127-134, 1999.
[http://dx.doi.org/10.1093/ajcn/69.1.127] [PMID: 9925134]

[82] H. Naz, A. Islam, A. Waheed, W.S. Sly, F. Ahmad, and M.I. Hassan, "Human β-glucuronidase: structure, function, and application in enzyme replacement therapy", *Rejuvenation Res.,* vol. 16, no. 5, pp. 352-363, 2013.
[http://dx.doi.org/10.1089/rej.2013.1407] [PMID: 23777470]

[83] M. Ries, "Enzyme replacement therapy and beyond—in memoriam Roscoe O. Brady, M.D. (1923–2016)", *J. Inherit. Metab. Dis.,* vol. 40, no. 3, pp. 343-356, 2017.
[http://dx.doi.org/10.1007/s10545-017-0032-8] [PMID: 28314976]

[84] E. Shemesh, "Enzyme replacement and substrate reduction therapy for Gaucher disease", *Cochrane Database Syst. Rev.,* vol. 2013, no. 1, 2013.
[http://dx.doi.org/10.1002/14651858.CD010324] [PMID: 25812601]

[85] M.P. Cicalese, F. Ferrua, L. Castagnaro, R. Pajno, F. Barzaghi, S. Giannelli, F. Dionisio, I. Brigida, M. Bonopane, M. Casiraghi, A. Tabucchi, F. Carlucci, E. Grunebaum, M. Adeli, R.G. Bredius, J.M. Puck, P. Stepensky, I. Tezcan, K. Rolfe, E. De Boever, R.R. Reinhardt, J. Appleby, F. Ciceri, M.G. Roncarolo, and A. Aiuti, "Update on the safety and efficacy of retroviral gene therapy for immunodeficiency due to adenosine deaminase deficiency", *Blood,* vol. 128, no. 1, pp. 45-54, 2016.
[http://dx.doi.org/10.1182/blood-2016-01-688226] [PMID: 27129325]

[86] A. Aiuti, and M.G. Roncarolo, "Ten years of gene therapy for primary immune deficiencies", *Hematology (Am. Soc. Hematol. Educ. Program),* vol. 2009, no. 1, pp. 682-689, 2009.
[http://dx.doi.org/10.1182/asheducation-2009.1.682] [PMID: 20008254]

[87] G.M. Rubanyi, "The future of human gene therapy", *Mol. Aspects Med.,* vol. 22, no. 3, pp. 113-142, 2001.
[http://dx.doi.org/10.1016/S0098-2997(01)00004-8] [PMID: 11470139]

[88] J. Rautio, H. Kumpulainen, T. Heimbach, R. Oliyai, D. Oh, T. Järvinen, and J. Savolainen, "Prodrugs: design and clinical applications", *Nat. Rev. Drug Discov.,* vol. 7, no. 3, pp. 255-270, 2008.
[http://dx.doi.org/10.1038/nrd2468] [PMID: 18219308]

[89] C.F. He, C-F. He, S-H. Wang, Y-J. Yu, H-Y. Shen, Y. Zhao, H-L. Gao, H. Wang, L-L. Li, H-Y. Liu, S-H. Wang, Y-J. Yu, H-Y. Shen, Y. Zhao, H-L. Gao, H. Wang, L-L. Li, and H-Y. Liu, "Advances in biodegradable nanomaterials for photothermal therapy of cancer", *Cancer Biol. Med.,* vol. 13, no. 3, pp. 299-312, 2016.
[http://dx.doi.org/10.20892/j.issn.2095-3941.2016.0052] [PMID: 27807498]

[90] E. Cecchin, and G. Stocco, "Pharmacogenomics and personalized medicine", *Genes (Basel),* vol. 11, no. 6, p. 679, 2020.
[http://dx.doi.org/10.3390/genes11060679] [PMID: 32580376]

[91] M. Barry, and J. Feely, "Enzyme induction and inhibition", *Pharmacol. Ther.,* vol. 48, no. 1, pp. 71-94, 1990.
[http://dx.doi.org/10.1016/0163-7258(90)90019-X] [PMID: 2274578]

[92] S.S. Suri, H. Fenniri, and B. Singh, "Nanotechnology-based drug delivery systems", *J. Occup. Med. Toxicol.,* vol. 2, no. 1, p. 16, 2007.
[http://dx.doi.org/10.1186/1745-6673-2-16] [PMID: 18053152]

[93] R. Remington, C. Bechtel, D. Larsen, A. Samar, L. Doshanjh, P. Fishman, Y. Luo, K. Smyers, R. Page, C. Morrell, and T.B. Shea, "A Phase II Randomized Clinical Trial of a Nutritional Formulation for Cognition and Mood in Alzheimer's Disease", *J. Alzheimers Dis.,* vol. 45, no. 2, pp. 395-405, 2015.
[http://dx.doi.org/10.3233/JAD-142499] [PMID: 25589719]

CHAPTER 5

Case Studies and Patient Stories

Devinder Kumar[1,2], **Brijesh Kumar**[4], **Rajesh Kumar**[2,*] and **Pratistha Singh**[3]

[1] *Amar Shaheed Baba Ajit Singh Jujhar Singh Memorial College of Pharmacy, Bela, Ropar, Punjab, India*

[2] *Maharaja Agrasen School of Pharmacy, Maharaja Agrasen University, Solan, Himachal Pradesh, India*

[3] *Department of Ophthalmology, University of California, Los Angeles, California, USA*

[4] *Department of Pharmacology, Institute of Medical Sciences, Banaras Hindu University, Varanasi, Uttar Pradesh, India*

Abstract: Pharmacological metabolism issues put patient safety at risk by raising the possibility of side effects and inadequate treatment results. This summary attempts to explain the challenges associated with drug metabolism, particularly long-term polypharmacy with metabolites - when an unintended change has direct and clinically meaningful effects. This is done through a series of cases and patient stories. Metabolism typically consists of a series of enzymatic reactions that occur primarily in the liver and result in drugs being converted into water-soluble forms for excretion. Nevertheless, it is a factor that can be disrupted by genetic polymorphisms, liver diseases affecting its expression or function, drug interactions, and age-related changes in physiological mechanisms. Metabolites can lead to bioaccumulation of a drug, poor therapeutic effects, or adverse events if metabolism is incomplete. A patient with a CYP2D6 enzyme deficiency may convert codeine to morphine more slowly, resulting in inadequate pain relief. Another example is a distant psychiatric patient with underlying liver disease who experiences significant inadequacy of benzodiazepine metabolism, leading to prolonged sedation and respiratory depression. There is also an example of polypharmacy, in which a patient taking multiple medications, including a strong CYP3A4 inhibitor, receives a toxic amount of a normally safe dose because the metabolic clearance rate is reduced. They are important examples of incomplete drug metabolism and underscore the imperative incorporation of precise, personalized medicine techniques, including genetic testing, close monitoring of liver function tests (which would account for gender differences in response), and thorough consideration of possible interactions between pharmaceutical agents that contribute to individual variability. This, in turn, allows healthcare professionals to provide more personalized treatment, minimize the risk of side effects, and improve patient outcomes.

* **Corresponding author Rajesh Kumar:** Maharaja Agrasen School of Pharmacy, Maharaja Agrasen University, Solan, Himachal Pradesh, India; E-mail: rajdhiman60@gmail.com

Keywords: Disease, Drug interactions, Drug metabolism, Personalized medicine techniques, Pharmacogenomics.

INTRODUCTION

Metabolism, the process by which our bodies convert food into energy, involves a complex series of chemical reactions occurring in every cell. While it's essential for sustaining life, this process can sometimes encounter disruptions, leading to various metabolic issues [1]. Unfortunately, some other factors can lead to (partially) inefficient nutrient metabolism, such as genetic diseases, changes in the environment, and food-drug interactions. Investigating these interactions is essential to making meaningful progress in the prevention and treatment of metabolic disorders [2]. Impairment of the body or organ's metabolic process through drug interactions leads to undesirable and serious health consequences, including the accumulation of toxic substances and even therapeutic failure [3]. These may be due to genetic factors, comorbidities, high glucose food intake, and increased consumption of multiple medications. The case studies highlight the need for individualized care and judicious use of medications in patients at risk of side effects [4].

This disruption can result in the incomplete metabolism of drugs, as well as endogenous or xenobiotic substances, which in turn alters their pharmacokinetics and/or pharmacodynamics. Such issues become especially concerning when diseases impair key metabolic organs like the liver and kidneys. The case studies in this article highlight the challenges of disease-related incomplete metabolism and demonstrate how these complications can significantly affect patient care [5]. Enzyme deficiencies or variations are also of great relevance as they can influence drug therapy, nutrition, and health in general. Enzymes are important for the metabolism of substances *in vivo*, and their abnormality can potentially trigger drug accumulation, acute toxicity, or treatment failure [6]. Here, we present case studies that illustrate the difficulties of incomplete metabolism in relation to enzyme systems. CIIIM is a phenomenon in which exogenous chemicals (*e.g.*, environmental toxins, drugs, or dietary components) disrupt the normal metabolic process and instead produce different results. This disorder can lead to the accumulation of unmetabolized elements, leading to toxicity or treatment failure. These case studies demonstrate the importance of understanding and controlling these interactions in clinical practice [7]. In addition, environmental factors greatly influence metabolism by preventing the complete breakdown of nutrients, drugs, or toxins, thus leading to the accumulation of toxic substances in harmful concentrations. This also results in ineffective drug action and disruption of various biochemical pathways. Environmental factors play an important role in metabolic impairment by highlighting case studies that help drive greater environmental awareness and targeted treatment (Fig. **1**).

Moreover, environmental factors significantly affect metabolism by hindering the full breakdown of nutrients, drugs, or toxins, resulting in the buildup of toxic substances at harmful levels. This can lead to ineffective drug action and the disruption of critical biochemical pathways. Case studies emphasizing the role of environmental factors in metabolic impairment help raise awareness of these issues and promote the development of more targeted treatments and strategies for mitigating environmental impacts on health [8, 9].

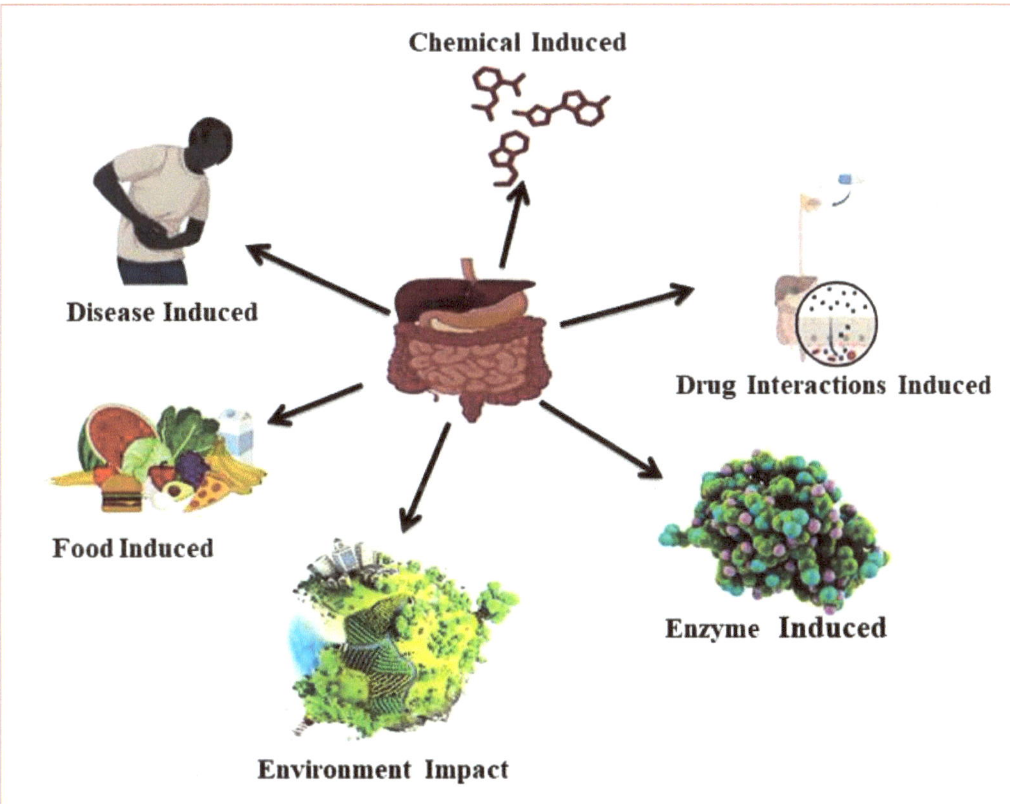

Fig. (1). Incomplete metabolism.

ROLE OF FOOD INTERACTION IN INCOMPLETE METABOLISM

Food-drug interactions are critical to drug metabolism and can result in incomplete or altered drug processing. Components of the diet, including macronutrients, vitamins, phytochemicals, and alcohol, can alter the activity of metabolic enzymes, thereby affecting the bioavailability, effectiveness, and safety of medications. Understanding these interactions is crucial for optimizing drug therapy and minimizing adverse effects [10, 11].

Drug metabolism occurs primarily in the liver, where a family of enzymes, particularly those from the cytochrome P450 (CYP) group, play a crucial role in converting drugs into more water-soluble forms that are more easily excreted by the kidneys [6]. Dietary proteins can also modulate drug metabolism. A high-protein diet may increase the activity of certain CYP450 enzymes, resulting in accelerated drug metabolism. This effect is particularly important for drugs that undergo extensive first-pass metabolism, in which a significant portion of the drug is metabolized by the liver before entering the systemic circulation. More rapid metabolism in these cases may reduce the therapeutic effect of the drug and require dosage adjustment (Fig. **2**) [7]. These changes in enzyme activity caused by dietary protein intake can result in incomplete metabolism, causing the drug to be metabolized either too quickly or too slowly, which may affect both its safety and effectiveness [10, 12]. On the other hand, some phytochemicals, for example in cruciferous vegetables such as broccoli and Brussels sprouts, can induce detoxifying enzymes such as CYP1A2. This could increase the metabolism of certain drugs and reduce their effectiveness. These interactions are particularly critical for drugs with narrow therapeutic windows, where even minor changes in drug levels can result in suboptimal outcomes or toxicity [13, 14].

Case Study 1: Lactose 2 and Dairy Products

Lactose intolerance is one of the most common metabolic diseases worldwide, affecting an estimated 65% of people [15]. It is a result of a lack of lactase, the protein that separates milk sugar, or what we absorb in the body with lactose. Unabsorbed lactose enters the large intestine, where bacteria ferment it, causing unpleasant gas, bloating, and sometimes diarrhea [16, 17].

Patient Story: Maria's Struggle with Lactose Intolerance

The 35-year-old nurse has always struggled with digestive problems. After consuming dairy products, she would usually experience gas, bloating, and stomach cramps, followed by diarrhea. At first, she thought the symptoms were due to stress and sporadic eating during her hectic work schedule. But when the symptoms got worse, she saw a gastroenterologist. After a test, she was diagnosed with lactose intolerance. Her doctor told her to drink the dairy products cold (which she did at my request) or just always consume them in moderation. Maria struggled to completely eliminate dairy from her diet as she had consumed milk most of her life, so lactase enzyme treatments had a positive effect on the digestion of lactose and resulted in fewer symptoms. Over time, she was able to control her symptoms by limiting dairy intake and taking enzyme supplements as needed. The present case highlights the need to consider food-drug interactions in the context of metabolic disorders such as lactose intolerance, where incomplete metabolism can cause significant stress and lifestyle disruption [18, 19].

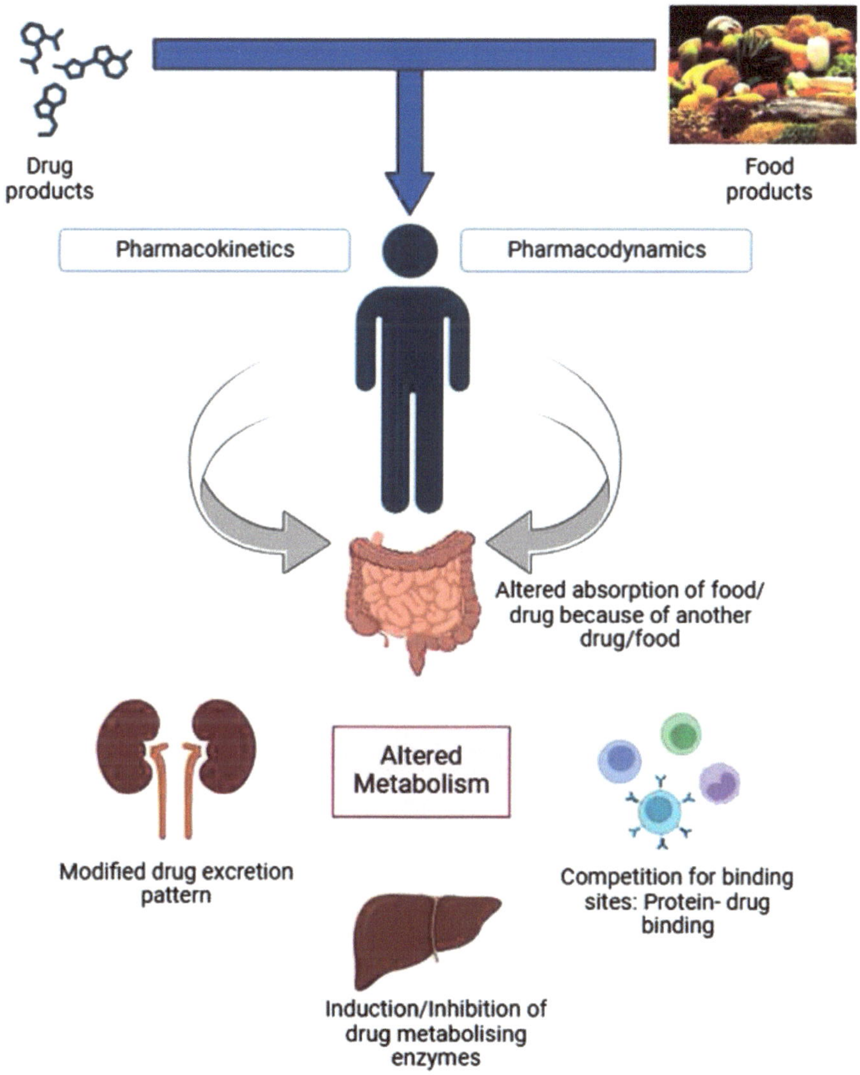

Fig. (2). Drug and food interaction in incomplete mechanism.

Case Study 2: Grapefruit Juice and Medication Metabolism

Grapefruit juice can interact with a variety of medications, sometimes causing serious or even fatal consequences [20]. This interaction is well documented and is primarily due to compounds called furanocoumarins, which are found in grapefruit. Drugs may have a longer half-life and higher concentrations in the bloodstream due to enzyme inhibitors, increasing the risk of toxicity [10, 21].

Patient Story: John's Experience with Statins and Grapefruit Juice

John, a 60-year-old pensioner who had a history of high cholesterol, was prescribed statins. He was a fan of grapefruit juice and drank some for breakfast every day. However, after taking the drug for several months, muscle pain and weakness occurred with exertion without relief at rest (consistent with statin-induced myopathy). He became alarmed and went to see his doctor, who examined his diet more closely — and it was easy to see where the problem likely lay: Grapefruit juice appeared to interact with this patient's statin drug in a way that led to excessively high levels of the medicine. He stopped taking grapefruit juice along with the statins, and his symptoms improved over time. This case highlights the need to understand food-drug interactions as possible causes of incomplete metabolism and serious side effects [13, 22].

Case Study 3: Vitamin B12 Deficiency and Dietary Choices

Vitamin B12 is essential for red blood cell production, DNA synthesis and nerves. Since this nutrient is primarily found in animal sources, the absorption of this particular nutrient requires the intrinsic factor protein, which is produced during production in your stomach. Still, some medical conditions and dietary habits as well as certain medications commonly prescribed for acid reflux symptoms, such as hormone replacement therapy or the diabetes drug metformin-can hinder the absorption of vitamin B12 and lead to a deficiency that causes other health problems [23, 24].

Patient Story: Linda's Battle with Vitamin B12 Deficiency

After experiencing some worrying symptoms, 45-year-old vegetarian Linda experienced fatigue, numbness (especially in her hands and feet), and memory difficulties. Worried that her health was deteriorating, she consulted a doctor who recommended a series of blood tests. The results were revealing: Linda was severely deficient in vitamin B12. Because she didn't eat animal products, which are the main source of this vitamin. Adding to the problem, Linda was also taking (PPIs) for acid reflux, which further impaired her ability to absorb vitamin B12 through lower stomach acids *via* knock-out *H. Pylori*. Linda's doctor recommended that she take vitamin B12 supplements and include fortified foods like grains and plant-based milk in her diet. Eventually, Linda's symptoms disappeared, and her energy and mental acuity returned. Her story highlights the need for an understanding of how diet can impact nutrient deficiencies and, in many metabolic states, lead to inadequate processing of certain nutrients in cells, meaning that while these nutrients are still energetically present, they are not metabolized properly [25, 26].

Case Study 4: Vitamin D Deficiency and Calcium Absorption

Calcium absorption, bone health. The body can produce vitamin D itself with sufficient exposure to sunlight, but it can also be absorbed through certain foods and dietary supplements [27, 28]. Surprisingly, however, vitamin D deficiency is not uncommon, and the incidence continues to rise worldwide - especially in areas where sunlight exposure is limited. Our bones crave vitamin D, low concentrations of which can hinder calcium absorption, leading to brittle bones and a higher risk of fractures [29, 30].

Patient Story: Anna's Struggle with Vitamin D Deficiency

Anna, a 50-year-old who lived in the northern part of the country where she did not get much sun exposure, presented with joint pain and bone fractures that were all too common. She took calcium supplements to strengthen her bones, but despite all the work and thousands of dollars she put into it, she wanted nothing in return. After additional testing, which included a bone density test that measured the strength of Anna's bones, it was eventually determined that she had osteoporosis. However, some other blood tests revealed that she was extremely low in vitamin D deficiency. As a result, although Linda was consuming plenty of calcium through her diet, the lack of sufficient vitamin D impaired its absorption, ultimately affecting her bone health. Her doctor gave her a high-dose vitamin D prescription with instructions to get out in the sun more. The alternative practitioner recommended that she eat more foods with vitamin D, such as fatty fish. Incorporate fortified dairy products and egg yolks into their diet. Over time, Anna's bone density improved, and she broke fewer bones. The same earlier case explains that vitamin D is an essential piece of the puzzle for calcium metabolism (and therefore bone health) and thus contextualizes situations where dietary sources are insufficient to allow full normal metabolic activity, *i.e.*, rickets [31, 32].

Case Study 5: Hyperuricemia and Purine-Rich Foods

Gout is an inflammatory arthritis that causes sudden, severe joint pain and is often caused by hyperuricemia or high levels of uric acid in the blood [33, 34]. Uric acid is produced when the body breaks down purine, a component of certain foods such as red meat and packaged meals. An excess of uric acid or poor disposal can lead to deposits in the joint area and, thus, to acute gout attacks [35, 36].

Patient Story: George's Battle with Gout

George, a 55-year-old businessman, had been suffering from regular gout attacks for several years. His daily life was constantly plagued by severe pain that interrupted his work, casual greetings, and social activities. However, the gout continued to worsen despite taking medication to normalize serum uric acid concentration. George's diet consisted primarily of purine-rich foods such as shellfish and red meat. George learned from his nutritionist that these purine-rich foods were actually making the condition worse. He was encouraged to limit the amount of it and instead eat more low-purine foods such as fruits, vegetables, and whole grains. He also started drinking more water to help get rid of some of the uric acid overload. After a while, he noticed that his gout occurred less frequently and became significantly milder [37, 38].

DRUG INTERACTIONS WITH REFERENCE TO INCOMPLETE METABOLISM

Drug interactions represent an important area of research in pharmacology due to their profound impact on the safety and effectiveness of drugs. Pharmacokinetic and pharmacodynamics drug-drug interactions are caused by one drug acting against another drug and interfering with multiple processes that affect the ADME (absorption, distribution, metabolism, and excretion) of individual drugs. This can lead to concentration changes and resulting pharmacological effects [39, 40]. In fact, among these interactions, those that alter the way drugs are metabolized in the body are of great importance since they can lead to incomplete metabolism and thus reduce the effectiveness of drugs. Most metabolisms are mediated by cytochrome P450 enzymes (CYP) in the liver, which convert drugs into metabolites that are excreted. Drug interactions may affect mechanisms that control the activity of these enzymes. When drug interactions disrupt the metabolism of xenobiotics, alliance-low reactions can interfere and lead to the accumulation of pharmacologically active concentrations in the parent or toxic metabolite form, which would ultimately lead to therapeutic failure or side effects [41, 42]. Incomplete metabolism most commonly, drug interactions result in inhibition or induction of metabolic enzymes due to incomplete metabolism. Particularly members of the cytochrome p450 (CYP450) family. Among drug interactions, enzymatic interactions occur when an enzyme inhibitor inactivates an enzyme responsible for metabolizing another drug (the substrate), leading to enzyme inhibition [43]. Such reduction mechanisms may decrease the clearance of the substrate drug, increase its plasma levels, prolong its half-life, and prolong or potentiate pharmacological effects that may result in toxicity. An example of a real interaction is that between warfarin (a CYP2C9 substrate) and fluconazole (a CYP2C9 inhibitor). Fluconazole, when administered concurrently with warfarin, may significantly inhibit the metabolism of warfarin, which in turn increases warfarin plasma concentrations and the risk of serious bleeding [44]. However, care must be taken to avoid incomplete metabolism by appropriately managing drug interactions in clinical pharmacology. The effects can be very variable and depend on genetic polymorphisms, liver function, and the specific pharmacokinetic characteristics of both drugs. To mitigate these risks and optimize therapeutic potential, therapeutic drug monitoring and interpretation by clinical pharmacologists with advanced knowledge of metabolic pathways and individual patient factors are required [45].

Case Study 1: Warfarin and Vitamin K-Rich Foods

Warfarin is a commonly prescribed anticoagulant that helps prevent blood clots. However, the effectiveness of warfarin depends very much on food intake and especially on foods that have a high vitamin K content, such as: B. green leafy vegetables [46]. Vitamin K, which is necessary for blood clotting and can, therefore, counteract the anticoagulant effect of warfarin, reduces its migration time before it is fully metabolized to achieve sufficient blood thinning [47, 48].

Patient Story: Margaret's Warfarin Therapy Complications

Margaret, 68, was taking warfarin to prevent a stroke due to atrial fibrillation. The trio initially told her that she needed to eat vitamin K-rich foods regularly. Margaret does not eat red meat but regularly eats large salads and green smoothies; these two high-intake groups resulted in fluctuating warfarin levels. The imbalance was evident in numerous hospital visits - either because of excessive bleeding or because of a clotting problem that also required treatment. Read Margaret's Vitamin K story. After nutritional advice, Margret managed to balance her vitamin K intake with medication that stabilized her INR value. This highlights the need to monitor dietary interactions in warfarin-treated patients [49, 50].

Case Study 2: Antiepileptic Drugs and Oral Contraceptives

The metabolism of oral contraceptives is known to be induced by certain antiepileptic drugs (AEDs), including phenytoin and carbamazepine. This not only reduces the effectiveness of contraception but also increases the risk of an unwanted pregnancy [51].

Patient Story: Lisa's Unplanned Pregnancy

Lisa, a 25-year-old woman diagnosed with epilepsy, was taking carbamazepine to prevent seizures and an oral contraceptive for contraception. Lisa adhered to a strict medication schedule but still found herself suddenly pregnant. Her neurologist told her that carbamazepine had increased the metabolism of her contraceptive and reduced its effectiveness. As a result, she was switched to another AED that did not interact with her contraceptive and achieved good seizure control. We would like to point out that the present case should serve as a reminder to any patient who requires simultaneous anticonvulsant therapy and hormonal contraception [51, 52].

Case Study 3: Digoxin and Amiodarone

Amiodarone, an antiarrhythmic drug, and digoxin, a cardiac glycoside, are both used to treat atrial fibrillation or heart failure. Digoxin toxicity may also arise from the combined action of amiodarone and digoxin, which inhibits the renal clearance of the drug at follow-up [53, 54].

Patient Story: Michael's Digoxin Toxicity

Digoxin and amiodarone were administered to a 70-year-old man who had a history of heart failure and atrial fibrillation. Soon, symptoms such as nausea, dizziness, blurred vision, *etc.* developed. This finding confirmed the toxicity of digoxin by high concentrations of the drug in his blood analysis. Michael had his digoxin dose adjusted by the cardiologist, and kidney function was monitored more carefully. He was placed on a cardiac monitor, and eventually, symptoms improved, and his underlying cardiac condition remained stable. This case highlights the importance of careful dose titration and monitoring when using combinations with known drug interactions [55, 56].

Case Study 4: Fluoroquinolones and Theophylline

Fluoroquinolones, a class of antibiotics, can interfere with the metabolism of the actual bronchodilator theophylline, which is used in the treatment of COPD and other lung diseases. One of the consequences can be an increased level of theophylline in the blood, which can lead to seizures or fatal arrhythmias and other symptoms of intoxication [57, 58].

Patient Story: Tom's Theophylline Overdose

Tom, a 65-year-old man with COPD, received treatment with ciprofloxacin for an acute respiratory infection. However, at the time of treatment, he was taking her chronic lung medication, theophylline. Tom was then quickly rushed to the hospital with severe shaking, brain fog, and heart palpitations. His blood contained toxic levels of theophylline. Withdrawal of ciprofloxacin and a reduction in his theophylline levels resulted in improvement in Tom's symptoms. This case highlights the risk of drug interactions associated with prescribing antibiotics to patients on chronic therapies [57].

Case Study 5: Rifampin and Oral Anticoagulants

Rifampin, an antibiotic commonly prescribed to treat tuberculosis, has a very strong enzyme inducer of CYP3A4. Warfarin and similar warfarin, which is also an oral anticoagulant, may be reduced in effectiveness by induction of metabolism [49].

Patient Story: Alan's Stroke Risk

Alan was a 63-year-old man with long-standing deep vein thrombosis on warfarin therapy. After being diagnosed with tuberculosis, he was given rifampicin. In the weeks that followed, his INR fell to dangerously low levels. This put him at significant risk for blood clot formation—and, paradoxically, for life-threatening complications such as uncontrolled bleeding, including intracranial hemorrhage, due to unstable anticoagulation. It turned out that the new dose of warfarin was interacting with something, and a doctor's referral to monitor INR levels had eventually stabilized his dosage. Alan's case demonstrates the complexity of managing drug interactions with multimorbidity or multiple comorbidities as he escaped a potentially catastrophic ischemic cerebrovascular event [59].

DISEASE-INDUCED INCOMPLETE METABOLISM

Drug metabolism can be significantly affected by disease, making partial metabolism of drugs a major concern in clinical pharmacology. Drug metabolism is mostly carried out by the liver. Therefore, any type of liver injury, including a disease such as cirrhosis, often leads to impairment of liver enzymes and disruption of drug biotransformation [7]. They may also change your metabolism in ways that affect how medications are metabolized by your body, but this also applies to other conditions such as kidney disease or heart failure. These can impair the metabolic clearance of drugs, thereby overpowering them from therapeutic ranges to toxic concentrations relative to their parent drug or active metabolites, resulting in a lack of efficacy or toxicity [60]. Treatment planning for these patients requires an understanding of how changes in disease status affect drug metabolism due to their significant comorbidities. Disease-specific factors that control the extent of incomplete metabolism are defined by the metabolic pathways involved, and these effects are related to the type and severity of the disease (Fig. **3**). It is critical to reduce the likelihood of incomplete metabolism in these patients, and the previously mentioned strategies of dose titration, monitoring drug levels, and individualized treatment regimens may be useful [61].

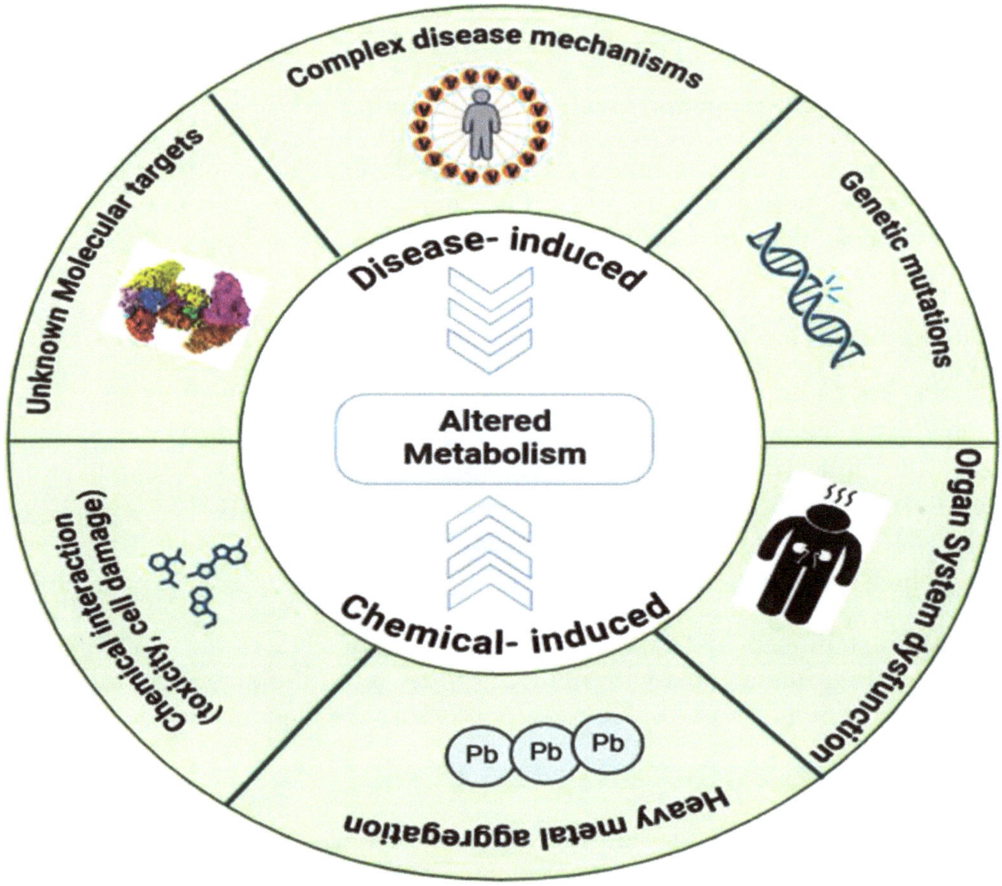

Fig. (3). Disease and chemical-induced incomplete metabolism.

Case Study 1: Liver Cirrhosis and Drug Metabolism

Liver scarring impairs the liver's ability to perform its regular functions in cirrhosis, a progressive disease. A significant portion of drug metabolism is also carried out by the liver, primarily *via* the cytochrome P450 cycle. Additionally, because the enzymes responsible for phase II reactions have low activity, patients with liver cirrhosis or those taking medications that require phase II reactions may also be at risk [62].

Patient Story: Sarah's Sedative Overdose

Diazepam was prescribed for Sarah, a 52-year-old woman with a history of alcoholic liver cirrhosis, to treat her anxiety. Despite receiving the usual dose, Sarah suffered from severe sedation, disorientation, and respiratory depression. Chris Hahn, the staff physician who first examined Chavez that day at Salinas Valley State Prison, knew immediately. For some reason, her liver had become less effective at processing the toxicity of the diazepam drug. Sarah began to feel better as her dose was reduced and the severity of her condition was monitored. It highlights the need for dose adjustments and close monitoring in patients with liver cirrhosis to prevent partial metabolism from causing adverse drug reactions [63].

Case Study 2: Chronic Kidney Disease and Drug Clearance

Chronic kidney disease (CKD) impairs the kidneys' ability to remove waste products from the blood, allowing drugs and their metabolites to be retained at elevated levels. This condition alters the pharmacokinetics of several drugs, particularly those that are predominantly eliminated *via* the kidneys [64].

Patient Story: Michael's Digoxin Toxicity

Digoxin is clinically indicated for the treatment of Michael's atrial fibrillation because he has stage 4 chronic kidney disease, and his age-ischemia score could not be determined. Over time, he developed symptoms of digoxin poisoning in the form of nausea and visual disturbances, as well as cardiac arrhythmias. A blood test revealed high levels of digoxin, which his nephrologist attributed to decreased renal clearance because the patient had kidney disease. Michael felt better after his digoxin dose was reduced. This highlights the importance of monitoring serum levels and appropriate dosage adjustments in CKD patients to avoid drug-induced toxicity due to decreased renal clearance [65].

Case Study 3: Congestive Heart Failure and Hepatic Congestion

Hepatic congestion may occur in CHF patients when venous return decreases. Drug toxicity may result from incomplete first-pass clearance of this congestion, which could impair the liver's ability to metabolize drugs [66].

Patient Story: Alice's Warfarin Sensitivity

Alice, a 75-year-old patient with CHF receiving warfarin for stroke prevention. Even at a low dose, her INR (a measure of blood clotting) remained elevated, meaning she was at risk of bleeding. Her cardiologist suspected that liver congestion from CHF caused warfarin to be metabolized more slowly, resulting in over-therapeutic levels. Alice, who received warfarin, was then successfully treated with dose adjustments and careful monitoring of her INR. This case highlights the need for individualized treatment in patients with CHF and possible changes in pharmacokinetics [67].

Case Study 4: Hypothyroidism and Drug Metabolism

Hypothyroidism leads to a reduction in thyroid hormones, which leads to reduced metabolic processes throughout the body. High doses or long-term use of medications can affect drug metabolism. Lower blood flow rates can slow the absorption of drugs from transdermal patches, prolonging drug release but also increasing the risk of side effects [68].

Patient Story: Emma's Prolonged Sedation

After surgery, a 60-year-old woman named Emma (with untreated hypothyroidism) was given morphine for pain. She had remained heavily sedated for an unusually long time despite receiving a standard dose. In this case, her endocrinologist found that drug metabolism had decreased due to hypothyroidism. By treating her hypothyroidism and changing her pain medication, Emma was able to overcome the side effects of sedation. This case highlights the need to consider thyroid function in patient treatment, particularly with sedating medications [69].

Case Study 5: Inflammatory Bowel Disease and Drug Absorption

Inflammatory bowel disease (IBD): Crohn's disease and ulcerative colitis are examples of inflammatory bowel disease (IBD), in which our ability to absorb nutrients and, therefore, medications are limited. A ruptured and inflamed intestinal mucosa may also be responsible for the presence of drugs with variable bioavailability or being partially absorbed [70].

Patient Story: John's Inconsistent Response to Oral Medications

John, a 35-year-old man with Crohn's disease, had difficulty adhering to his oral corticosteroids and maintaining consistent therapeutic levels. Even if he took his medication diligently, Zaldivar said inflammatory symptoms would appear without warning. Crohn's disease interfered with medication absorption, his gastroenterologist said. John's symptoms were more effectively controlled after switching from topical corticosteroids to an injectable formulation. Furthermore, this case highlights the difficulties of oral drug administration in IBD patients and the possible role of non-oral formulation when absorption is in question [71].

ENZYME-BASED INCOMPLETE METABOLISM

Enzyme-related factors are an important area of pharmacokinetics due to incomplete metabolism and the resulting significant impact on drug therapy outcomes. The metabolism of many drugs involves enzymes that convert them into more easily excreted metabolites, which helps maintain therapeutic drug levels and minimize toxicity [72]. In addition, some factors can influence the activity of an enzyme (*e.g.*, differences in genotype from patient to patient, concomitant drug administration, and environmental influences) that can lead to incomplete metabolism. This may lead to the accumulation of parenteral drugs and inactive metabolites in the body, which may lead to worsening side effects and/or treatment failure (Fig. **4**). Of particular importance in this context are the cytochrome P450 (CYP 450) enzymes, which represent a cornerstone for phase I metabolic reactions such as oxidation and reduction [6]. Each CYP450 enzyme recognizes substrates and ensures that specific drugs are metabolized by specific enzymes. Incomplete metabolism may be due to differences in enzyme activity, typically due to genetic polymorphisms. For example, patients with genetic differences that result in such low CYP2D6 activity that they become "very slow metabolizers" may experience longer drug effects and a higher risk of toxicity if they take commonly prescribed medications that rely on this enzyme metabolism can be influenced. In contrast, ultrarapid metabolizers may process these drugs too quickly to be effective [73].

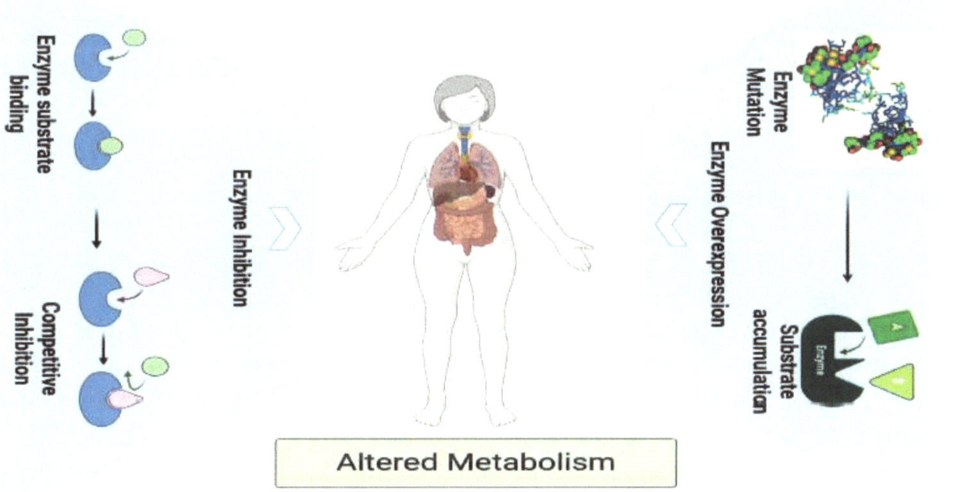

Fig. (4). Enzyme-based incomplete metabolism.

Case Study 1: Cytochrome P450 2d6 Deficiency and Codeine Toxicity

Cytochrome P450 2D6 (CYP2D6) is an enzyme that metabolizes about a quarter of all medications, including codeine, a commonly prescribed opioid for pain relief [74]. Its effect is based on the enzyme converting codeine into its active form, morphine, thereby relieving pain. On the other hand, poor metabolizers without CYP2D6 activity may not receive adequate analgesia, and ultra-rapid metabolizers can convert codeine to morphine very quickly, so even children receiving recommended doses of codeine-containing medications may experience toxicity [75].

Patient Story: Emily's Codeine Overdose

After having her tonsils removed, 6-year-old Emily is given codeine to ease the pain. Despite receiving the correct dose, Emily fainted and had difficulty breathing, resulting in her having to go to the emergency room at the last minute. In this case, tests showed that Emily was an extremely rapid metabolizer of CYP2D6, resulting in a greatly accelerated conversion of codeine (to morphine with excessive opioid toxicity as the end result). The diagnosis and treatment with codeine were discontinued, and the acute treatment was carried out by the doctors. Emily has thankfully made a full recovery but cautions that genetic testing and an assessment of enzyme activity are required when prescribing codeine [76].

Case Study 2: Glucose-6-Phosphate Dehydrogenase Deficiency and Hemolytic Anemia

An enzyme called Glucose-6-phosphate dehydrogenase that prevents the oxidation of red blood cells. G6PD deficiency predisposes to Hemolytic anemia following drugs/diet & in some infections where oxidative stress occurs as a response [77].

Patient Story: Ahmed's Hemolytic Crisis

Ahmed, a 25-year-old man with G6PD deficiency, was prescribed primaquine for malaria prophylaxis before traveling to an endemic area. About a month after taking the drug, he experienced typical jaundice (yellowing of the skin/eyes), dark urine, and fatigue - typical symptoms of hemolytic anemia. His doctors were quickly able to link his symptoms to his G6PD deficiency. The patient was diagnosed with hemolysis, and primaquine treatment was discontinued. This case highlights an important aspect of prior screening for G6PD deficiency prior to administration of certain medications, particularly in communities where this enzyme is deficient [78].

Case Study 3: Thiopurine S-Methyltransferase (TPMT) Deficiency and Azathioprine Toxicity

Thiopurine S-methyltransferase (TPMT) and its relationship to the metabolism of thiopurine drugs, *e.g.*, azathioprine in autoimmune diseases and some cancer treatments, are discussed here. Likewise, patients with low or deficient TPMT enzyme activity are at risk of severe bone marrow toxicity when standard doses of thiopurines are administered [79].

Patient Story: Rachel's Life-Threatening Myelosuppression

Azathioprine was recommended for the treatment of Crohn's disease in 32-yea--old Rachel. A few weeks after starting treatment, she developed a severe drop in white blood cells - known as myelosuppression - which increased her risk of infection and could have been life-threatening. Genetic testing ultimately revealed that Rachel was among the 10% of patients with TPMT deficiency, which led to a buildup of toxic thiopurine metabolites. Her medications were withheld, and given supportive care until her blood counts recovered. This case provides a strong argument for initial clearance prior to thiopurine ingestion to avoid potentially fatal events [80].

Case Study 4: Dihydropyrimidine Dehydrogenase (DPD) Deficiency And 5-FU Toxicity

Dihydropyrimidine dehydrogenase (DPD) is the enzyme that breaks down 5-fluorouracil (5-FU), a chemotherapy drug used in cancer. Defective metabolic degradation of 5-FU by circulating DPD can lead to drug accumulation and increased severe toxicity [81].

Patient Story: John's Severe 5-FU Toxicity

John, a 58-year-old man diagnosed with colorectal cancer (Dukes C, T4N2M0), presented for treatment with adjuvant therapy consisting of 5-FU chemotherapy. Within hours of his first infusion, he developed significant diarrhea, mucositis, and severe fatigue. The patient experienced severe toxicity, and his oncologist suspected 5-FU toxicity, leading to testing that revealed DPD deficiency. Chemotherapy was stopped immediately, and this patient, John, received much supportive care. This case highlights the need to perform a DPD deficiency test before administering 5-FU to avoid life-threatening toxicity [82].

Case Study 5: Aldehyde Dehydrogenase 2 (ALDH2) Deficiency and Alcohol Intolerance

Aldehyde dehydrogenase 2 plays a central role in alcohol metabolism [83]. One benefit of genetic variation may be ALDH2 deficiency, which affects many East Asians and causes acetaldehyde to build up after drinking alcohol, leading to symptoms such as facial flushing or possibly even nausea due to poor processing of consumed alcohol [84].

Patient Story: Mei's Social Struggles with Alcohol Intolerance

Mei, a 24-year-old Chinese woman, felt fearful of her belonging at social events that included drinking alcohol. Something as simple as a glass of alcohol would cause her face to turn red, give her a headache and make her feel terribly dizzy after just a single drink or so - which has a visible impact on her ability to engage in such events. A genetic test of Mei revealed mutations in ALDH2, which seemed to explain her symptoms. Although it was not life-threatening, it had a devastating impact on her social lifestyle and self-esteem. Mei's case highlights the impact of enzyme defects on lifestyle and social interactions, as well as psychosocial impacts, including stigmatization in communities with little education about these types of diseases [85].

CHEMICAL-INDUCED INCOMPLETE METABOLISM

The incomplete metabolism of chemicals is a major problem for both pharmacology and toxicology. Many exogenous chemicals, such as other drugs, environmental pollutants, food, *etc*., can lead to alteration in drug metabolism due to improper functioning of enzymes, especially from the cytochrome P450 (CYP450) family [86]. Competition for different CYP enzymes is also common when taking drugs that are substrates of the same CYP enzyme when their activity is reduced in some way, as nanomolecular drugs or congeners aid metabolism. This leads to insufficient elimination of the drug by metabolism and leads to an increase in the fraction excreted unchanged or increased metabolite production, *i.e*., h. an increase in the parent drug and active/toxic compound. These have been identified as potentially disruptive forces with deleterious consequences or could lead to therapeutic ineffectiveness [87]. This highlights the importance for healthcare providers to consider potential chemical interactions when prescribing medications, particularly in patients with multiple comorbidities or those exposed to environmental toxins. Understanding this is important to prevent the success of desired therapeutic outcomes and appropriate events [88].

Case Study 1: Grapefruit Juice and Statin Toxicity

Grapefruit juice is known to block the enzyme cytochrome P450 3A4 (CYP3A4), which metabolizes a variety of medications, including many statins used to lower cholesterol. This blockade of CYP3A4 by grapefruit juice leads to an increase in blood statins and increases the risk of muscle toxicity, among other things [10].

Patient Story: Mark's Rhabdomyolysis

Mark's doctor prescribed the statin atorvastatin to control his high cholesterol. He also loved drinking his grapefruit juice every day. Over the next few weeks, Mark began experiencing muscle pain (myalgia), which worsened over time. Over time, he was hospitalized and suffered from extreme rhabdomyolysis - a syndrome in which muscle tissue breaks down and releases harmful substances into the bloodstream. After his doctors discovered that the grapefruit juice had blocked the metabolism of atorvastatin, causing toxic levels to develop in his body. Mark's case also serves as a good reminder to warn patients about drug-food interactions and the consequences of eating certain foods while taking certain medications [89].

Case Study 2: Alcohol And Acetaminophen-Induced Liver Toxicity

Acetaminophen (paracetamol) is a widely used analgesic that relieves pain and fever. However, paracetamol in higher doses or in combination with alcohol can cause life-threatening liver damage. Alcohol stimulates the cytochrome P-450 enzyme CYP2E1 to produce a hepatotoxic metabolite from acetaminophen [90].

Patient Story: Sarah's Acute Liver Failure

A 35-year-old woman, Sarah, who often self-treated her chronic headaches with paracetamol. She was also a frequent drinker. One particularly difficult week, she had such severe headaches and was so anxious that, for the first time in years, she took paracetamol along with alcohol to ease the discomfort. Days later, Sarah showed signs of jaundice as well as vomiting and abdominal pain. Their case highlights the dangers of concomitant use of alcohol with paracetamol, particularly in public officials who take it over a long period of time, and points out that patients need to be warned about events caused by the use, even if it results in one Liver toxicity is simulated [91].

Case Study 3: Smoking and Theophylline Toxicity

A bronchodilator used to treat asthma and chronic obstructive pulmonary disease is theophylline. Smoking increases the activity of the enzyme CYP1A2, which helps break down theophylline. Theophylline is predominantly metabolized by

CYP1A2. However, when a patient stops smoking, the activity of this enzyme decreases, and plasma concentrations of theophylline increase, leading to an increased risk of toxicity [92].

Patient Story: John's Theophylline Overdose

John was a 60-year-old man with COPD who had taken theophylline for many years and was a heavy smoker. John stopped smoking and noticed an increased number of symptoms, such as nausea, heart palpitations, and headaches. A blood test revealed he had toxic levels of theophylline. His theophylline levels had become too high because his CYP1A2 enzyme was less active after he stopped smoking. To rule out other possibilities, John's medications were also adjusted to ensure he did not become toxic. This highlights the need to monitor drug levels in patients who experience significant changes, such as B. Avoiding tobacco consumption, which can lead to incomplete metabolism caused by chemicals and a subsequent side effect [68].

Case Study 4: Herbal Supplements and Warfarin Interactions

These effects are primarily driven by environmental factors, largely through alterations in the activity of drug-metabolizing enzymes—particularly those of the cytochrome P450 (CYP450) family [93].

Patient Story: Linda's Uncontrolled Bleeding

Linda, a 70-year-old woman with atrial fibrillation, was taking warfarin to prevent blood clots. She regularly took ginseng to maintain her energy. Over time, she developed severe bruising and nosebleeds. It was discovered that her blood was far too thin according to the INR (a measure of blood clotting). This was related to inhibition of the metabolism of warfarin from ginseng, resulting in excessive anticoagulant effects. They changed Linda's warfarin dosage and told her to stop taking ginseng. The case is a reminder for health professionals to ask patients, even educated patients with comorbidities and rational reasons that are not linked in their minds, whether they are also taking herbal supplements [94].

Case Study 5: Pesticide Exposure and Reduced Drug Metabolism

Organophosphates, possibly with the exception of organothiophosphate insecticides, are capable of inhibiting and altering the activities of cytochrome P450. Continued exposure to these substances may lead to the accumulation of substrates for metabolism by the enzymes involved, which could increase toxicity when used as drugs [86].

Patient Story: Carlos's Unexplained Seizures

Phenytoin was recommended to Carlos, a 45-year-old farmworker, to treat epilepsy. He suffered from frequent seizures and symptoms of phenytoin poisoning, such as ataxia and confusion, despite taking his prescription as directed. Further investigation revealed that Carlos' phenytoin metabolism was inhibited by organophosphate pesticides to which he was exposed at work. Measures were taken to reduce his exposure to pesticides and change his medications. Carlos' situation serves as an example of how environmental chemicals affect drug metabolism and highlights the importance of considering occupational exposures when treating patients experiencing unexplained drug toxicity [95].

ENVIRONMENTAL IMPACT ON INCOMPLETE METABOLISM

Environmental factors such as exposure to pollutants, dietary components, lifestyle habits, and occupational xenobiotics, in addition to genetic predisposition, may contribute to the exhaustion of drug metabolism [96]. Many of these effects are attributed to environmental exposures altering drug-metabolizing enzyme activity, primarily CYP450 (cytochrome P450) enzymes. These changes result in the alteration of drug clearance or metabolism and the accumulation of toxic metabolites, which can lead to either therapeutic failure or toxicity (Fig. **5**) [43]. Healthcare providers need to be aware of environmental factors that can influence drug metabolism to optimize treatment plans, adjust dosages, or prevent adverse events, especially in patients with high environmental exposures and/or complex medication regimens [97].

Case Study 1: Air Pollution and Cytochrome P450 Enzyme Induction

Heavy exposure to vehicle exhaust and industrial emissions impairs the body's ability to metabolize drugs. Exposure to polycyclic aromatic hydrocarbons (PAHs), which are substrates of certain cytochrome P450 enzymes, especially CYP1A1 and CYP1B1, have also been shown to be determinant risk factors. They can unregulated enzymes that break down certain drugs, thereby reducing the available amount of the inactive drug and potentially reducing its pharmacological effects by increasing their clearance from the body [98]. Healthcare providers need to be aware of these biochemical interactions, especially in patients who are frequently exposed to high levels of air pollution [99].

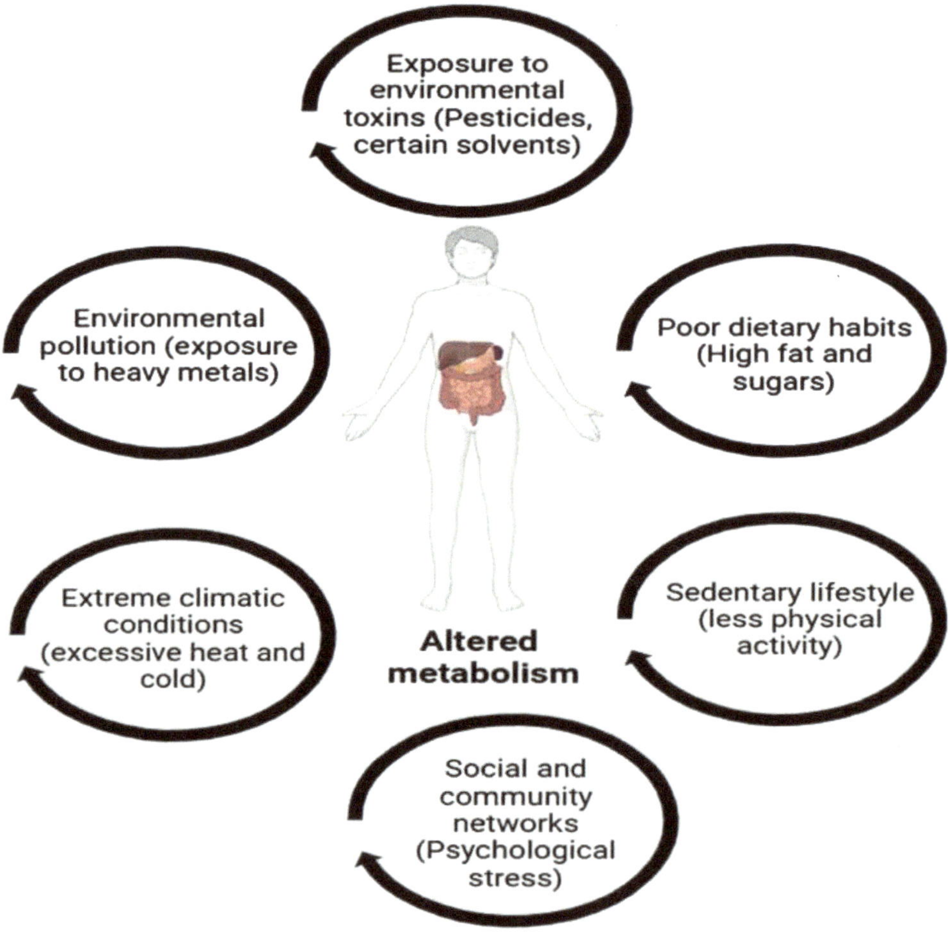

Fig. (5). Environmental impact on incomplete metabolism.

Patient Story: Sarah's Struggle with Asthma Control

Sarah is a 42-year-old senior city resident who works in healthcare and suffers from asthma. Although Sarah took her inhaled corticosteroids and bronchodilators according to specific instructions, she continued to experience severe asthma attacks. Still, control of her asthma began to worsen, but only when she was exposed to air pollution so high that even healthy people living in many cities across India were unable to escape. There are several studies conducted on air pollutants induced by Cytochrome P450 (CYP450) enzymes, including CYP1A1 and CYP1A2, with exposure to particulate matter (PM), Polycyclic aromatic hydrocarbons (PAHs). These enzymes metabolize a variety of medications, including some prescribed to treat asthma. Sarah's activation of these enzymes caused her to metabolize the asthma medications so quickly that they had significantly less effect on her lungs, making Sarah more susceptible to severe asthma attacks. Sarah's results are a good reminder for the clinician to consider many factors, including environmental factors such as air pollution when assessing poorly controlled patients with chronic diseases. White had her medication dosed and was advised to stay indoors when air pollution was high and to see if it would be worth changing the area to get better air [96].

Case Study 2: Pesticide Exposure and Inhibition Of Metabolic Enzymes

Pesticide exposure, particularly from organophosphates, can inhibit key metabolic enzymes like cytochrome P450 (CYP450) [100]. This inhibition can slow down the metabolism of drugs, prolonging their activity in the body. Consequently, there's a higher risk of toxicity or adverse effects from medications that rely on these enzymes for processing. Chronic exposure to pesticides is also associated with liver dysfunction, further worsening these metabolic disturbances [101].

Patient Story: Carlos's Unexplained Seizures

30-year-old farmworker Carlos was the one who came to work with unexplained seizures despite having no history or diagnosis of any illness. Later, they not only occurred more frequently but also with regular use of standard antiepileptic drugs. Further investigation revealed that Carlos was heavily exposed to organophosphate pesticides during construction work. Organophosphates were known to block acetylcholinesterase, the enzyme required for the breakdown of AcA in the nervous system. They also have an inhibitory effect on the cytochrome P450 enzyme, although to a lesser extent than their anti-epilepsy activity. This inhibition of metabolism was the reason for the accumulation of all these drugs in Carlos' system and, thus, the seizures. Carlos' case is an example of considering occupational exposures in patients with otherwise unexplained symptoms, particularly neurological in nature. His measures included minimizing pesticide exposure, changing his medications, and regularly assessing his neurological status [102].

Case Study 3: Heavy Metal Exposure and Disruption of Nutrient Metabolism

Exposure to heavy metals such as lead and cadmium disrupts nutrient metabolism by hindering the absorption of essential minerals like zinc, iron, and calcium. This disruption can cause deficiencies that impair vital cellular functions and enzyme activity, leading to health problems like anemia, osteoporosis, and developmental issues. Furthermore, heavy metals can compromise the liver and kidneys' ability to process nutrients, worsening metabolic imbalances [103].

Patient Story: Linda's Chronic Fatigue and Anemia

Linda, a fifty-year-old woman, complained of headaches, chronic fatigue, and difficulty concentrating. She lived near an industrial area. A blood test revealed that although her diet was rich in iron, she had depleted it all and now has an anemic set of red blood cells called iron deficiency anemia. Cadmium is a heavy metal found in industrial exhaust gases. It turned out that Linda had been exposed to dangerous levels of the material. Cadmium also interferes with the kidneys' release of erythropoietin, whose sole purpose is to promote red blood cell production. In Linda's case, cadmium exposure resulted in incomplete iron metabolism, manifesting as "iron deficiency anemia," even though she was consuming a diet with sufficient oral intake. Intervention strategies included chelation therapy to reduce cadmium levels in the body, nutritional counseling (improving iron absorption), and reducing environmental sources of cadmium [104].

Case Study 4: Endocrine Disruptors and Thyroid Hormone Metabolism

Endocrine disruptors, including bisphenol A (BPA) and phthalates, interfere with thyroid hormone metabolism by altering their synthesis, transport, and clearance. This interference can lead to conditions like hypothyroidism and developmental issues, particularly impacting children's growth and neurological development. Over time, chronic exposure to these chemicals can disrupt thyroid homeostasis, raising the likelihood of thyroid diseases and metabolic syndrome [105].

Patient Story: James's Weight Gain and Depression

James, a 38-year-old man, presented with unexplained weight gain, fatigue, and depression. Laboratory tests revealed that he had hypothyroidism, with low thyroid hormone levels despite normal thyroid-stimulating hormone (TSH) levels. Given his medical history and otherwise healthy lifestyle, the classic etiology of hypothyroidism was less likely. However, his work exposed him to very high levels of BPA and phthalates, which are endocrine disruptors commonly found in plastics. These chemicals affect the synthesis, binding, and metabolism of thyroid hormones, resulting in symptoms of hypothyroidism. A case from James highlights the influence of environmental chemicals on thyroid hormone metabolism. He was treated with measures to reduce exposure to endocrine disruptors and administration of thyroid hormones. If necessary, he will be monitored to adjust the therapeutic dose [106].

Case Study 5: Environmental Toxins and Mitochondrial Dysfunction

Environmental toxins like dioxins and polychlorinated biphenyls (PCBs) impair mitochondrial function by reducing energy production and increasing oxidative stress. This mitochondrial dysfunction results in decreased ATP production and increased cell death, contributing to various diseases, including neurodegenerative disorders, metabolic syndrome, and cancer. Prolonged exposure to these toxins can lead to ongoing mitochondrial damage, progressively worsening organ function and cellular resilience [107].

Patient Story: Emily's Mitochondrial Myopathy

Emily, a twelve-year-old girl, was diagnosed with mitochondrial myopathy. The results explained incidental muscle weakness and fatigue associated with exercise intolerance in this disorder. Even the most comprehensive examination could not identify a cause for her symptoms. Over time, it was discovered that Emily had been poisoned in her own home by neurotoxins such as pesticides and heavy metals. These toxins can damage mitochondria, the fuel-synthesizing organelles in a cell, leading to incomplete nutrient metabolism and energy failure in affected tissues. The Emily case highlights the impact of environmental toxins on dysfunctional mitochondria and highlights the importance of identifying and reducing exposure to potential environmental factors in patients who have not been diagnosed with a metabolic or neuromuscular disorder [108].

CONCLUSION

These case reports and medical histories illustrate the many factors that hinder complete metabolism and how these can subsequently impact health. One of the most important aspects in the control of metabolic disorders in patients is the interaction of food with genetic, hygienic, biological, or drug factors. By understanding these interactions, everyone involved, the doctors and the patients, can formulate the appropriate nutritional measures in relation to the patient's metabolism. This targeted approach not only improves health-related quality of life but also serves as a preventative measure against more advanced health threats. Regarding the mechanisms of drug interactions, genetic factors, comorbidities, or polypharmacy can significantly alter the metabolic course and response to therapies. Achieving these goals requires coordinated strategies, such as: *e.g.*, increasing patient awareness of their medications and nutritional supplements, comprehensive patient education, and optimal management practices. This approach helps reduce stress and improve the effectiveness of treatment. In addition, the presence of a disease-related metabolic disorder adds further complexity to the treatment of patients and requires tailored approaches to idiopathic metabolic disorders, including consideration of drug interactions, treatment of underlying diseases, as well as "on-site" patient cases, how important pharmacogenomics and genetic testing serve the diagnosis and help shape the therapeutic development of the individual, the application of which can increase the safety and effectiveness of the treatment. In addition, interactions with chemicals, such as pollutants, medications, or even food, can be particularly damaging to metabolism and lead to various disruptive consequences. Extensive knowledge of these interactions helps physicians manage and eliminate adverse events and improve care strategies. Pollution and constant poisoning are just some of the environmental factors that affect metabolism and can, therefore, cause a whole range of health problems. Managing these environmental exposures is important as it contributes to proper diagnosis and appropriate treatment. This presents a compelling argument for early modalities, comprehensive education, and prevention. These case studies highlight the growing clinical importance of personalized healthcare and the need to address metabolic disorders more comprehensively, taking into account individual patient factors and specific environmental influences that contribute to such diseases.

REFERENCES

[1] M. Agana, J. Frueh, M. Kamboj, D.R. Patel, and S. Kanungo, "Common metabolic disorder (inborn errors of metabolism) concerns in primary care practice," *Annals of Translational Medicine*, vol. 6, no. 24, 2018.
[http://dx.doi.org/10.21037/atm.2018.12.34]

[2] C.M. Phillips, "Nutrigenetics and metabolic disease: Current status and implications for personalised nutrition," *Nutrients*, vol. 5, no. 1, pp. 32–57, 2013.
[http://dx.doi.org/10.3390/nu5010032]

[3] S. E. Leucuta and L. Vlase, "Pharmacokinetics and metabolic drug interactions," *Curr. Clin. Pharmacol.,* vol. 1, no. 1, pp. 5–20, 2006.
[http://dx.doi.org/10.2174/157488406775268183]

[4] P. P. Gorski, D. C. Turner, J. Iraki, J. P. Morton, A. P. Sharples, and J. L. Areta, "Human skeletal muscle methylome after low-carbohydrate energy-balanced exercise.," *Am. J. Physiol. Endocrinol. Metab*., vol. 324, no. 5, pp. E437–E448, May 2023,
[http://dx.doi.org/10.1152/ajpendo.00029.2023]

[5] "Enzyme replacement combinational therapy: Effective treatments for mucopolysaccharidoses," *Expert Opin. Biol. Ther.,* vol. 21, no. 9, pp. 1181–1197, 2021.
[http://dx.doi.org/10.1080/14712598.2021.1895746]

[6] U.M. Zanger, and M. Schwab, "Cytochrome P450 enzymes in drug metabolism: Regulation of gene expression, enzyme activities, and impact of genetic variation", *Pharmacol. Ther.,* vol. 138, no. 1, pp. 103-141, 2013.
[http://dx.doi.org/10.1016/j.pharmthera.2012.12.007] [PMID: 23333322]

[7] R.K. Verbeeck, "Pharmacokinetics and dosage adjustment in patients with hepatic dysfunction", *Eur. J. Clin. Pharmacol.,* vol. 64, no. 12, pp. 1147-1161, 2008.
[http://dx.doi.org/10.1007/s00228-008-0553-z] [PMID: 18762933]

[8] B. A. Neel and R. M. Sargis, "The paradox of progress: Environmental disruption of metabolism and the diabetes epidemic," Diabetes, vol. 60, no. 7, pp. 1838–1848, 2011.
[http://dx.doi.org/10.2337/db11-0153]

[9] A. Singh, L.B. Prasad, K. Shiv, R. Kumar, and S. Garai, "Synthesis, characterization, and in vitro antibacterial and cytotoxic study of Co(II), Ni(II), Cu(II), and Zn(II) complexes of N-(--methoxybenzyl) N-(phenylethyl) dithiocarbamate ligand", *J. Mol. Struct.,* vol. 1288, no. May, p. 135835, 2023.
[http://dx.doi.org/10.1016/j.molstruc.2023.135835]

[10] M. Chen, S. Zhou, E. Fabriaga, P. Zhang, and Q. Zhou, "Food-drug interactions precipitated by fruit juices other than grapefruit juice: An update review", *J. Food Drug Anal.,* vol. 26, no. 2, pp. S61-S71, 2018.
[http://dx.doi.org/10.1016/j.jfda.2018.01.009] [PMID: 29703387]

[11] A. Kumar, B. Kumar, R. Kumar, A. Kumar, M. Singh, V. Tiwari, A. Trigunayat, P. Paul, and P. Singh, "Acute and subacute toxicity study of ethanolic extract of *Calotropis procera* (Aiton) Dryand flower in Swiss albino mice," *Phytomedicine Plus*, vol. 2, no. 2, p. 100224, May 2022.
[http://dx.doi.org/10.1016/j.phyplu.2022.100224]

[12] A. Singh, R. Kumar, K. Shiv, S.K. Pandey, M.K. Bharty, R.J. Butcher, and L.B. Prasad, "Synthesis, crystal structure and screening for anticonvulsant and antianxiety activities of three new Ni(II), Cu(II), and Zn(II) dithiocarbamate complexes", *J. Mol. Struct.,* vol. 1298, no. P2, p. 137052, 2024.
[http://dx.doi.org/10.1016/j.molstruc.2023.137052]

[13] J. Maskalyk, "Grapefruit juice: potential drug interactions", *CMAJ,* vol. 167, no. 3, pp. 279-280, 2002.
[PMID: 12186177]

[14] R. Dhiman, D. Kumar, B. Kumar, and B.L. Pandey, "Quantitative determination of atorvastatin, ortho-hydroxy atorvastatin, para-hydroxy atorvastatin in human plasma using rosuvastatin as internal standard by LC-MS / MS", *Int. J. Pharm. Chem. Sci.,* vol. 4, no. 4, pp. 487-500, 2016.

[15] M.B. Heyman, and P. Care, "Lactose intolerance in infants, children, and adolescents", *Pediatrics,* vol. 118, no. 3, pp. 1279-1286, 2006.
[http://dx.doi.org/10.1542/peds.2006-1721] [PMID: 16951027]

[16] C. L. Storhaug, S. K. Fosse, and L. T. Fadnes, "RETRACTED: Country, regional, and global estimates for lactose malabsorption in adults: a systematic review and meta-analysis," *Lancet Gastroenterol. Hepatol.,* vol. 2, no. 10, pp. 738–746, Jul. 2017.
[http://dx.doi.org/10.1016/S2468-1253(17)30154-1]

[17] R.D. Vijaya Durga, "Determination of S-amlodipine in Human Plasma Using Tizanidine as Internal Standard by LC/MS/MS Method", *J. Bioanal. Biomed.,* vol. 7, no. 3, pp. 91-96, 2015.
[http://dx.doi.org/10.4172/1948-593X.1000130]

[18] B. Misselwitz, M. Butter, K. Verbeke, and M.R. Fox, "Update on lactose malabsorption and intolerance: pathogenesis, diagnosis and clinical management," *Gut,* vol. 68, pp. 2080–2091, 2019.

[19] A. Kumar, M.K. Nandi, B. Kumar, A. Kumar, R. Kumar, A.K. Singh, and M. Singh, "Antiurolithiasis, antioxidant, anti-inflammatory, analgesic, and diuretic activity of ethanolic extract of seeds of Caesalpinia bonducella", *Int. J. Pharm. Investig.,* vol. 11, no. 3, pp. 306-311, 2021.
[http://dx.doi.org/10.5530/ijpi.2021.3.54]

[20] D. G. Bailey, G. Dresser, and J. M. Arnold, "Grapefruit-medication interactions: forbidden fruit or avoidable consequences?" *CMAJ*, vol. 185, no. 4, pp. 309–316, 2013,

[21] H. Pandey, S. Srivastava, A. Kumar, R. Kumar, and Y. Bhusan Tripathi, "Pharmacokinetics study Of puerarin absorption in blood after consumption Of Pueraria tuberosa water extract (PTWE) by rats", *Explor. Anim. Med. Res.,* vol. 9, no. 2, pp. 188-196, 2019.

[22] R. Garg, K. Dobhal, B. Kumar, and A. K. Gupta, COVID-19: a challenge towards the sustainability of health in platform era", In: NeuroQuantology vol. 20, no. 11, pp. 2503–2519, 2022.

[23] R. Green, L.H. Allen, A.L. Bjørke-Monsen, A. Brito, J.L. Guéant, J.W. Miller, A.M. Molloy, E. Nexo, S. Stabler, B.H. Toh, P.M. Ueland, and C. Yajnik, "Vitamin B12 deficiency", *Nat. Rev. Dis. Primers,* vol. 3, no. 1, p. 17040, 2017.

[http://dx.doi.org/10.1038/nrdp.2017.40] [PMID: 28660890]

[24] K. Ajay, N. Manmath Kumar, K. Brijesh, K. Ashutosh, K. Rajesh, K. Vikas, and S. Ambrish Kumar, "Toxicity (acute and subacute) assessment and in-vivo antiurolithiatic activity of ethanolic extract of Caesalpinia bonducella seed in albino Wistar rat", *J. Appl. Pharm. Sci.,* vol. 12, no. 2, pp. 187-197, 2022.
[http://dx.doi.org/10.7324/JAPS.2022.120220]

[25] Y. Chen, M. Michalak, and L. B. Agellon, "Importance of Nutrients and Nutrient Metabolism on Human Health.," *Yale J. Biol. Med.*, vol. 91, no. 2, pp. 95–103, Jun. 2018.

[26] R. Dhiman, V. Durga, and A. J. Rio, "Bioanalytical method for lornoxicam determination in human plasma by using piroxicam as internal standard by LC-MS/MS", *J Chem Pharm Res,* vol. 7, no. 8, pp. 54-60, 2015.

[27] J. J. Cannell and B. W. Hollis, "Use of Vitamin D in clinical practice," *Altern. Med. Rev.,* vol. 13, no. 1, pp. 6–20, Mar. 2008.

[28] A. Singh, R. Kumar, R. Patel, Trishna, R. N. Gautam, M. K. Bharty, and L. B. Prasad, "Dithiocarbamate-based novel anti-histaminic agents: synthesis, characterization, crystal structure and thermal study," *Dalton Trans*., no. 33, 2024.
[http://dx.doi.org/10.1039/D4DT01777C]

[29] I.H. de Boer, "Vitamin D deficiency," In: *Chronic Kidney Disease*, Dialysis, and Transplantation, 3rd ed., J. Himmelfarb and M. H. Sayegh, Eds. Oxford, UK: Academic Press, 2010, pp. 115–127.
[http://dx.doi.org/10.1016/B978-1-4377-0987-2.00009-1]

[30] R. Kumar, B. Kumar, A. Kumar, A. Kumar, and M. Singh, "GC-MS analysis of Phytocomponents in the Methanol Extract of Premna latifolia Roxb", *Pharmacognosy Res.,* vol. 14, no. 1, pp. 19-23, 2021.
[http://dx.doi.org/10.5530/pres.14.1.4]

[31] A. Prentice, "Vitamin D deficiency: a global perspective", *Nutr Rev,* vol. 66, no. Suppl.2, pp. 153-164, 2008.
[http://dx.doi.org/10.1111/j.1753-4887.2008.00100.x]

[32] R. Dhiman, B. Kumar, D. Kumar, and A. Kumar, "Liquid-liquid extraction method for ziprasidone (ZRS) bioanalysis by using ZRS-D 8 (stable isotope) as internal standard", *J Chem Pharm Res,* vol. 8, no. 3, pp. 722-727, 2016.

[33] N. Dalbeth, H.K. Choi, L.A.B. Joosten, P.P. Khanna, H. Matsuo, F. Perez-Ruiz, and L.K. Stamp, "Gout", *Nat. Rev. Dis. Primers,* vol. 5, no. 1, p. 69, 2019.
[http://dx.doi.org/10.1038/s41572-019-0115-y] [PMID: 31558729]

[34] A. Hamid, R. Kumar, V. K. Chaturvedi, S. Dutt, G. Sulabh, V. Kumar, and D. P. Yadav, "Artificial intelligence (AI) based colonoscopy," In: *Nanobiotechnology and Artificial Intelligence in Gastrointestinal Diseases*, 2024, V. K. Chaturvedi, A. K. Singh, J. Singh, and D. P. Yadav, Eds., pp. 12-1,12-23.
[http://dx.doi.org/10.1088/978-0-7503-6134-7ch12]

[35] E. Roddy and M. Doherty, "Epidemiology of gout," *Arthritis Res. Ther.,* vol. 12, no. 6, p. 223, 2010.

[36] R. Kumar, Y. Singh, M. Piplani, P. Bhateja, R. Garg, and B. Kumar, "The Chemistry and Pharmacological Study of Copaifera spp. Oils", *Chem. Insid. Spices Herbs Res. Dev.,* pp. 79-116, 2024.
[http://dx.doi.org/10.2174/9789815196832124040006]

[37] M. A. B. Siddiq, "Evidence-based management of gout – A systematic search and review," *Preprints*, 2020.
[http://dx.doi.org/10.20944/preprints202012.0156.v1]

[38] A. Kumar, P. Singh, R. Kumar, and S. Dutt, "Novel drug delivery systems for inflammatory bowel disease," in *Nanobiotechnology and Artificial Intelligence in Gastrointestinal Diseases*, Chapter 4, IOP Publishing Ltd, Aug. 2024, pp. 4-1, 4-24.

[http://dx.doi.org/10.1088/978-0-7503-6134-7ch4]

[39] J. Punt, and J. Owen, "Choose your partner and aggregate", *Trends Immunol.,* vol. 23, no. 11, p. 520, 2002.
[http://dx.doi.org/10.1016/S1471-4906(02)02334-7]

[40] M. Singh, B. Kumar, A. K Singh, C. Shekhar Azad, M. K Yadav, A. Kumar, R. Kumar, A. Kumar, and P. Chaudhary, "Pharmacokinetic screening to Estimate the drug likeliness characteristics of selected Herbal Anticancer Drugs", *Res. J. Pharm. Technol.,* vol. 16, no. 7, pp. 3422-3426, 2023.
[http://dx.doi.org/10.52711/0974-360X.2023.00566]

[41] R. H. Elbekai, H. M. Korashy, and A. O. S. El-Kadi, "The effect of liver cirrhosis on the regulation and expression of drug metabolizing enzymes," *Res. J. Pharm. Technol.,* vol. 5, no. 2, pp. 157–167, 2004.
[http://dx.doi.org/10.2174/1389200043489054]

[42] A. Kumar, B. Kumar, R. Dhiman, A. Kumar, M. Singh, P. Singh, V. Tiwari, and A. Trigunayat, "Assessment of pain-ameliorative potential of ethanolic extract of Calotropis procera and its active components characterization by HPTLC and GC-MS analysis," *Res. J. Pharm. Technol.*, July 2022.

[43] A. Daly, A. Rettie, D. Fowler, and J. Miners, "Pharmacogenomics of CYP2C9: Functional and clinical considerations", *J. Pers. Med.,* vol. 8, no. 1, pp. 1-31, 2017.
[http://dx.doi.org/10.3390/jpm8010001] [PMID: 29283396]

[44] C.R. Lee, J.A. Goldstein, and J.A. Pieper, "Cytochrome P450 2C9 polymorphisms: a comprehensive review of the in-vitro and human data", *Pharmacogenetics,* vol. 12, no. 3, pp. 251-263, 2002.
[http://dx.doi.org/10.1097/00008571-200204000-00010] [PMID: 11927841]

[45] M. Gex-Fabry, A.E. Balant-Gorgia, and L.P. Balant, "Therapeutic drug monitoring databases for postmarketing surveillance of drug-drug interactions", *Drug Saf.,* vol. 24, no. 13, pp. 947-959, 2001.
[http://dx.doi.org/10.2165/00002018-200124130-00002] [PMID: 11735651]

[46] M. Wadelius, and M. Pirmohamed, "Pharmacogenetics of warfarin: current status and future challenges," *Pharmacogenomics J.,* vol. 7, pp. 99–111, 2007.
[http://dx.doi.org/10.1038/sj.tpj.6500417]

[47] A. M. Holbrook, J. A. Pereira, R. Labiris, H. McDonald, J. D. Douketis, M. Crowther, and P. S. Wells, "Systematic overview of warfarin and its drug and food interactions," *Arch. Intern. Med.*, vol. 165, no. 10, pp. 1095–1106, 2005.

[48] R. Kumar, B. Kumar, A. Kumar, A. Kumar, and M. Singh, "HPTLC and GC-MS analyses of biologically active extracts and fractions from Premna latifolia Roxb leaves reveal anti-inflammatory and analgesic effects," *Syst. Rev. Pharm.,* vol. 13, no. 58, pp. 544–551, 2022.

[49] E. Nutescu, I. Chuatrisorn, and E. Hellenbart, "Drug and dietary interactions of warfarin and novel oral anticoagulants: an update," *J. Thromb. Thrombolysis*, vol. 31, pp. 326–343, 2011.
[http://dx.doi.org/10.1007/s11239-011-0561-1]

[50] R. Kumar, B. Parashar, and P. Sharma, "Review on Premna Species," *World J. Pharmacol. Res. Technol.*, vol. 2, no. 2, pp. 19–28, 2014.

[51] A. Reimers, E. Brodtkorb, and A. Sabers, "Interactions between hormonal contraception and antiepileptic drugs: Clinical and mechanistic considerations", *Seizure,* vol. 28, pp. 66-70, 2015.
[http://dx.doi.org/10.1016/j.seizure.2015.03.006] [PMID: 25843765]

[52] R. Kumar, B. Kumar, and A. Kumar, "Pharmacokinetic profile of antitubercular drugs and their clinical applications in therapeutic drug monitoring", *Asian J Pharm,* vol. 3, 2015no. 5, pp. 145-154. Available from: http://ajpp.in/uploaded/p71.pdf

[53] M. Paxton, "Current challenges with supply-chain integrity and the threat to the quality of marketed drugs", *Clin. Pharmacol. Ther.,* vol. 89, no. 2, pp. 316-319, 2011.
[http://dx.doi.org/10.1038/clpt.2010.288] [PMID: 21178987]

[54] R. Kumar, S. Dutt, A. Goyal, A. Kumar, and B. Kumar, “Nanotechnology in gastrointestinal endoscopy,” In: *Nanobiotechnology and Artificial Intelligence in Gastrointestinal Diseases*, Chapter 5, IOP Publishing Ltd, Aug. 2024, pp. 5-1–5-32.
[http://dx.doi.org/10.1088/978-0-7503-6134-7ch5]

[55] P.E. Fenster, N.W. White Jr, and C.D. Hanson, "Pharmacokinetic evaluation of the digoxin-amiodarone interaction", *J. Am. Coll. Cardiol.,* vol. 5, no. 1, pp. 108-112, 1985.
[http://dx.doi.org/10.1016/S0735-1097(85)80091-7] [PMID: 3964797]

[56] R. Dhiman, V. Durga, K. Jayasankar, and S. Raj, "Solid phase extraction: Potential carbocisteine bioanalysis using carbocisteine 13c3 stable isotope: LC-MS/MS technique with pharmacokinetics on healthy human subjects", *Pharm. Lett.,* vol. 7, no. 8, pp. 126-134, 2015.

[57] N. Gogan, C.D. Bayliff, N.A.M. Paterson, and J.D. Spence, “Clinical consequences of quinolone-theophylline interactions,” *Le J. Can. Pharm. Hosp.,* vol. 48, no. 6, p. 366, Dec. 1995.

[58] R. Dhiman, “Novel LC-MS/MS bioanalytical method validation for gabapentin in human plasma in accordance with current regulatory guidances,” *Int. J. Pharm. Chem. Sci.,* vol. 4, no. 4, pp. 462–468, 2015.
Available from: http://www.ijpcsonline.com/files/08-10-15/07-1035.pdf

[59] M.C. Herink, Y.F. Zhuo, C.D. Williams, and T.G. DeLoughery, "Clinical Management of Pharmacokinetic Drug Interactions with Direct Oral Anticoagulants (DOACs)", *Drugs,* vol. 79, no. 15, pp. 1625-1634, 2019.
[http://dx.doi.org/10.1007/s40265-019-01183-0] [PMID: 31440911]

[60] D. Telles-Correia, A. Barbosa, H. Cortez-Pinto, C. Campos, N. B. F. Rocha, and S. Machado, “Psychotropic drugs and liver disease: A critical review of pharmacokinetics and liver toxicity,” *World J. Gastrointest. Pharmacol. Ther.,* vol. 8, no. 1, pp. 26–38, 2017.
[http://dx.doi.org/10.4292/wjgpt.v8.i1.26]

[61] T. He, M.G. Priebe, H.J.M. Harmsen, F. Stellaard, X. Sun, G.W. Welling, and R.J. Vonk, "Colonic fermentation may play a role in lactose intolerance in humans", *J. Nutr.,* vol. 136, no. 1, pp. 58-63, 2006.
[http://dx.doi.org/10.1093/jn/136.1.58] [PMID: 16365059]

[62] M.A. Pena, J.F. Horga, and P. Zapater, “Variations of pharmacokinetics of drugs in patients with cirrhosis,” *Expert Rev. Clin. Pharmacol.,* vol. 9, no. 3, pp. 441–458, 2016.
[http://dx.doi.org/10.1586/17512433.2016.1135733]

[63] R.A. Weersink, D.M. Burger, K.L. Hayward, K. Taxis, J.P.H. Drenth, and S.D. Borgsteede, "Safe use of medication in patients with cirrhosis: pharmacokinetic and pharmacodynamic considerations", *Expert Opin. Drug Metab. Toxicol.,* vol. 16, no. 1, pp. 45-57, 2020.
[http://dx.doi.org/10.1080/17425255.2020.1702022] [PMID: 31810397]

[64] T.N. Lea-Henry, J.E. Carland, S.L. Stocker, J. Sevastos, and D.M. Roberts, “Clinical pharmacokinetics in kidney disease: Fundamental principles,” *Clin. J. Am. Soc. Nephrol.,* vol. 13, no. 7, pp. 1085–1095, 2018.
[http://dx.doi.org/10.2215/CJN.00340118] [PMID: 29934432]

[65] P. E. Marik, and L. Fromm, “A case series of hospitalized patients with elevated digoxin levels,” *Am. J. Med.,* vol. 105, no. 2, pp. 110–115, 1999.
[http://dx.doi.org/10.1016/S0002-9343(98)00195-8]

[66] S. Møller, and J.H. Henriksen, “Cardiovascular complications of cirrhosis,” *Gut*, vol. 57, pp. 268–278, 2008.
[http://dx.doi.org/10.1136/gut.2006.112177]

[67] T. H. Self, A. B. Reaves, C. S. Oliphant, and C. Sands, “Does heart failure exacerbation increase response to warfarin? A critical review of the literature,” *Curr. Med. Res. Opin.,* vol. 22, no. 11, pp. 2089–2094, 2006.

[http://dx.doi.org/10.1185/030079906X132479]

[68] Y. Kashiwagura, S. Uchida, S. Tanaka, H. Watanabe, M. Masuzawa, T. Sasaki, and N. Namiki, "Clinical efficacy and pharmacokinetics of levothyroxine suppository in patients with hypothyroidism," *Biol. Pharm. Bull.,* vol. 37, no. 4, pp. 666-670, 2014.

[69] Z. Miyake, K. Ishii, and A. Tamaoka, "Hypothyroidism induced by phenytoin and gabapentin: A case report,", vol. 97, no. 43, p. e12938, 2018.

[70] L.J.J. Derijks, D.R. Wong, D.W. Hommes, and A.A. van Bodegraven, "Clinical Pharmacokinetic and Pharmacodynamic Considerations in the Treatment of Inflammatory Bowel Disease", *Clin. Pharmacokinet.,* vol. 57, no. 9, pp. 1075-1106, 2018. [http://dx.doi.org/10.1007/s40262-018-0639-4] [PMID: 29512050]

[71] A. Effinger, C.M.O. Driscoll, M. Mcallister, and N. Fotaki, "Impact of gastrointestinal disease states on oral drug absorption – implications for formulation design – a PEARRL review," *J. Pharm. Pharmacol.,* vol. 71, no. 4, pp. 674–698, 2019. [http://dx.doi.org/10.1111/jphp.12928]

[72] G. Guarnieri *et al.*, "Simple methods for nutritional assessment in hemodialyzed patients.," *Am. J. Clin. Nutr.*, vol. 33, no. 7, pp. 1598–1607, Jul. 1980. [http://dx.doi.org/10.1093/ajcn/33.7.1598]

[73] J. K. Lamba, Y. S. Lin, E. G. Schuetz, and K. E. Thummel, "Genetic contribution to variable human CYP3A-mediated metabolism", *Adv Drug Deliv Rev,* vol. 64, pp. 256-269, 2012. [http://dx.doi.org/10.1016/j.addr.2012.09.017]

[74] Y. Gasche, Y. Daali, M. Fathi, A. Chiappe, S. Cottini, P. Dayer, and J. Desmeules, "Codeine intoxication associated with ultrarapid CYP2D6 metabolism," *N. Engl. J. Med.,* vol. 351, no. 27, pp. 2827–2831, 2004.

[75] K.R. Crews, A. Gaedigk, H.M. Dunnenberger, T.E. Klein, D.D. Shen, J.T. Callaghan, E.D. Kharasch, and T.C. Skaar, "Clinical Pharmacogenetics Implementation Consortium (CPIC) guidelines for codeine therapy in the context of cytochrome P450 2D6 (CYP2D6) genotype", *Clin. Pharmacol. Ther.,* vol. 91, no. 2, pp. 321-326, 2012. [http://dx.doi.org/10.1038/clpt.2011.287] [PMID: 22205192]

[76] I. Cascorbi, "Pharmacogenetics of cytochrome P4502D6: genetic background and clinical implication," Eur. J. Clin. Invest., vol. 33, Suppl. 2, pp. 17–22, 2003.

[77] M.D. Cappellini, and G. Fiorelli, "Glucose-6-phosphate dehydrogenase deficiency", *Lancet,* vol. 371, no. 9606, pp. 64-74, 2008. [http://dx.doi.org/10.1016/S0140-6736(08)60073-2] [PMID: 18177777]

[78] G. Bancone, and C.S. Chu, "G6PD variants and haemolytic sensitivity to primaquine and other drugs," *Front. Pharmacol.,* vol. 12, 2021. [http://dx.doi.org/10.3389/fphar.2021.638885]

[79] L. Lennard, "TPMT in the treatment of Crohn's disease with azathioprine," *Gut*, vol. 51, no. 2, pp. 143–146, 2002.

[80] E. A. Fargher, K. Tricker, W. Newman, R. Elliott, S. A. Roberts, J. L. Shaffer, I. Bruce, and K. Payne, "Current use of pharmacogenetic testing: a national survey of thiopurine methyltransferase testing prior to azathioprine prescription," *J. Clin. Pharm. Ther.,* vol. 32, no. 2, pp. 187–195, 2007.

[81] U. Amstutz, S. Farese, S. Aebi, and C. R. Largiadèr, "Dihydropyrimidine dehydrogenase gene variation and severe 5-fluorouracil toxicity: A haplotype assessment," *Pharmacogenomics*, vol. 10, no. 6, pp. 931–944, 2009. [http://dx.doi.org/10.2217/pgs.09.28]

[82] J. Latchman, A. Guastella, and C. Tofthagen, "5-Fluorouracil toxicity and dihydropyrimidine dehydrogenase enzyme: implications for practice," *Clin. J. Oncol. Nurs.,* vol. 18, no. 5, pp. 581–585, 2014.

[http://dx.doi.org/10.1188/14.CJON.581-585.5-Fluorouracil]

[83] D.W. Crabb, M. Matsumoto, D. Chang, and M. You, "Overview of the role of alcohol dehydrogenase and aldehyde dehydrogenase and their variants in the genesis of alcohol-related pathology," *Proc. Nutr. Soc.*, vol. 63, no. 1, pp. 49–63, 2004.
[http://dx.doi.org/10.1079/PNS2003327]

[84] C.-H. Chen, B. R. Kraemer, and D. Mochly-Rosen, "ALDH2 variance in disease and populations.," *Dis. Model. Mech.*, vol. 15, no. 6, Jun. 2022
[http://dx.doi.org/10.1242/dmm.049601]

[85] H. Edenberg, "The genetics of alcohol metabolism: Role of alcohol dehydrogenase and aldehyde dehydrogenase variants," *Alcohol Res. Health*, vol. 30, no. 1, pp. 5–13, 2007.

[86] R.H. Abdou, M. Elbadawy, W.F. Khalil, T. Usui, K. Sasaki, and M. Shimoda, "Effects of several organophosphates on hepatic cytochrome P450 activities in rats," *Toxicology*, vol. 82, no. 5, pp. 598–606, 2020.
[http://dx.doi.org/10.1292/jvms.19-0452]

[87] T. Mushiroda, Y. Ohnishi, S. Saito, A. Takahashi, Y. Kikuchi, S. Saito, H. Shimomura, Y. Wanibuchi, T. Suzuki, N. Kamatani, and Y. Nakamura, "Association of VKORC1 and CYP2C9 polymorphisms with warfarin dose requirements in Japanese patients", *J. Hum. Genet.,* vol. 51, no. 3, pp. 249-253, 2006.
[http://dx.doi.org/10.1007/s10038-005-0354-5] [PMID: 16432637]

[88] S.M.J. Mortazavi, G. Mortazavi, and M. Paknahad, "Comment on sundseth et al. Global sources and pathways of mercury in the context of human health, ", *Int. J. Environ. Res. Public Health,* vol. 14, no. 5, p. 481, 2017.
[http://dx.doi.org/10.3390/ijerph14050481] [PMID: 28475133]

[89] F.Y. Khan, and W. Ibrahim, "Rosuvastatin induced rhabdomyolysis in a low risk patient: a case report and review of the literature," *Curr. Clin. Pharmacol.,* vol. 4, no. 1, pp. 1–3, 2009.

[90] G.G. Gyamlani, and C.R. Parikh, "Acetaminophen toxicity: suicidal *vs.* accidental," *Crit. Care*, vol. 6, no. 2, pp. 155–159, 2002.

[91] T. Ishizaki, Y. Horai, T. Sasaki, et al., "Bioavailability and pharmacokinetics of theophylline following plain uncoated and sustained-release dosage forms in relation to smoking habit," *Eur. J. Clin. Pharmacol.,* vol. 24, pp. 361–369, 1983.

[92] Y. J. Ma, D. Q. Jiang, J. X. Meng, M. X. Li, H. H. Zhao, Y. Wang, and L. Q. Wang, "Theophylline: a review of population pharmacokinetic analyses," J. Clin. Pharm. Ther., vol. 41, no. 6, 2016, pp. 594-601.
[http://dx.doi.org/10.1111/jcpt.12435]

[93] J. A. Patel and K. J. Gohil, "Herb-drug interactions: A review and study based on assessment of clinical case reports in literature," *Indian J. Pharmacol.,* vol. 39, no. 3, 2007.

[94] M. Wadelius, and M. Pirmohamed, "Pharmacogenetics of warfarin: current status and future challenges", *Pharmacogenomics J.,* vol. 7, no. 2, pp. 99-111, 2007.
[http://dx.doi.org/10.1038/sj.tpj.6500417] [PMID: 16983400]

[95] A.F. Hernández, T. Parrón, A.M. Tsatsakis, M. Requena, R. Alarcón, and O. López-Guarnido, "Toxic effects of pesticide mixtures at a molecular level: Their relevance to human health", *Toxicology,* vol. 307, pp. 136-145, 2013.
[http://dx.doi.org/10.1016/j.tox.2012.06.009] [PMID: 22728724]

[96] M. Ibrahim, E. M. MacFarlane, G. Matteo, M. P. Hoyeck, K. R. C. Rick, S. Farokhi, C. M. Copley, S. O'Dwyer, and J. E. Bruin, "Functional cytochrome P450 1A enzymes are induced in mouse and human islets following pollutant exposure," *Diabetologia*, vol. 63, no. 1, pp. 162–178, 2020.

[97] B. Weiss, "Vulnerability of children and the developing brain to neurotoxic hazards", *Environ Health Perspect,* vol. 108, no. SUPPL. 3, pp. 375-381, 2000.

[http://dx.doi.org/10.2307/3454523]

[98] M. Ingelman-Sundberg, "Human drug metabolising cytochrome P450 enzymes: properties and polymorphisms", *Naunyn Schmiedebergs Arch. Pharmacol.,* vol. 369, no. 1, pp. 89-104, 2004. [http://dx.doi.org/10.1007/s00210-003-0819-z] [PMID: 14574440]

[99] C. A. Pope III, R. T. Burnett, G. D. Thurston, M. J. Thun, E. E. Calle, D. Krewski, and J. J. Godleski, "Cardiovascular mortality and long-term exposure to particulate air pollution: Epidemiological evidence of general pathophysiological pathways of disease," *Circulation,* vol. 109, no. 1, 2003. [http://dx.doi.org/10.1161/01.CIR.0000108927.80044.7F]

[100] A. K. Daly, "Pharmacogenetics: a general review on progress to date," *Br. Med. Bull.,* vol. 124, no. 1, pp. 65–79, 2017.
[http://dx.doi.org/10.1093/bmb/ldx035]

[101] E. Hodgson and P. E. Levi, "Pesticides: an important but underused model for the environmental health sciences," *Environ. Health Perspect.,* vol. 104, Suppl. 1, pp. 97–106, 1996.

[102] K. Jaga, and C. Dharmani, *Sources of exposure to and public health implications of organophosphate pesticides* vol. 14. , 2003, no. 3, pp. 171-185.
[http://dx.doi.org/10.1590/S1020-49892003000800004]

[103] P.B. Tchounwou, C.G. Yedjou, A.K. Patlolla, and D.J. Sutton, "Heavy metal toxicity and the environment," *Exp. Suppl.,* vol. 101, pp. 133–164, 2012.
[http://dx.doi.org/10.1007/978-3-7643-8340-4]

[104] R.A. Bernhoft, "Cadmium toxicity and treatment", *Sci. World J.,,* vol. 2013, no. 1, p. 394652, 2013. [http://dx.doi.org/10.1155/2013/394652] [PMID: 23844395]

[105] A. Pinson, J. P. Bourguignon, and A. S. Parent, "Exposure to endocrine disrupting chemicals and neurodevelopmental alterations," Andrology, vol. 4, no. 4, pp. 706-722, 2016.
[http://dx.doi.org/10.1111/andr.12211]

[106] M. Boas, U. Feldt-Rasmussen, and K.M. Main, "Thyroid effects of endocrine disrupting chemicals," *Mol. Cell. Endocrinol.,* vol. 355, no. 2, pp. 240–248, 2012.
[http://dx.doi.org/10.1016/j.mce.2011.09.005]

[107] A. Reddam, S. McLarnan, and A. Kupsco, "Environmental Chemical Exposures and Mitochondrial Dysfunction: a Review of Recent Literature.," *Curr. Environ. Heal. reports*, vol. 9, no. 4, pp. 631–649, Dec. 2022.
[http://dx.doi.org/10.1007/s40572-022-00371-7]

[108] D.C. Wallace, "A mitochondrial bioenergetic etiology of disease," *J. Clin. Invest.,* vol. 123, no. 4, pp. 1405–1412, 2013.

CHAPTER 6

Metabolic Biomarkers as Diagnostic Approaches During Metabolism

Monika Rana[1], **Prabhsharan Kaur**[2,*] and **Varinder Singh**[2,*]

[1] *Department of Pharmacognosy, Laureate Institute of Pharmacy,, Kangra, Himachal Pradesh, India*

[2] *Department of Pharmaceutical Sciences and Technology, Maharaja Ranjit Singh Punjab Technical University, Bhatinda, Punjab, India*

Abstract: Over the past ten years, there has been a noticeable growth in the use of biomarkers as early warning systems in the assessment of disease risk. Indicators of normal biological processes, pathological processes, or pharmacological responses to treatment are known as biomarkers. Biochemical indicators are important tools for guiding illness care since they provide important information about a patient's overall functional status and inflammatory state. The last few decades have seen the development and establishment of nanotechnology, which has led to the rise of new methods for the extraction of nanoscale biomarkers. This overview highlights the definitions, background, features, classifications, and the identification, verification, and standardization of biomarkers. Also discussed are the numerous new technologies in discovery and the possible use of biomarkers in the previous 10 years for the diagnosis, prognosis, and therapy of different diseases. The primary objective of this review is to encourage readers to investigate novel directions in biomarker development and research.

Keywords: Biomarkers, Diagnostic biomarkers, Imaging biomarkers, Technology, Validation.

INTRODUCTION

A biological molecule that denotes a healthy or unhealthy process is called a biomarker. It exists in blood, lymph, and other body fluids (such as urine, tears, saliva, sweat, *etc.*) or tissues. It changes into a form that is molecular, cellular, or biochemical, making a measurement predictable and accurate. It is useful for monitoring and diagnosing physiological, pathological, and responses to drug treatment processes [1]. It is also referred to as a signature molecule and

* **Corresponding authors Prabhsharan Kaur and Varinder Singh:** Department of Pharmaceutical Sciences and Technology, Maharaja Ranjit Singh Punjab Technical University, Bhatinda, Punjab, India;
E-mails: prabhsharanhayer@gmail.com; varinderjassal17@gmail.com

molecular marker. Biomarkers are quantifiable and do not reflect an individual's feelings or functions. Physiological parameters (*e.g.*, blood pressure, heart rate), biochemical markers (*e.g.*, alanine transferase, alkaline phosphatase, and bilirubin), x-ray results, angiography and magnetic resonance imaging, and intricate histologic and genetic testing of blood, genes, DNA, RNA, platelets, enzymes, hormones, and biomolecules, such as carbohydrates, proteins, lipids, and other tissues, are a few examples of biomarkers [2]. Biomarkers usually provide information about the progression of a disease, an early warning of the condition before symptoms appear, and/or treatment recommendations [3].

The failure rate is one of the main issues facing drug development today. During that phase, nine out of ten might not succeed. A new generation of biomarkers is required to increase the success rate and efficiency of drug development. These biomarkers should be more informative and able to alert developers early on to potential toxicity or lack of efficacy in their drug, giving them an early indication of what is likely to work and fail. Thus, those that are still in the development stage are considered biomarkers. In the past few years, biomarkers have gained importance in drug discovery as they can help determine the mechanism of action of a treatment, look into efficacy and toxicity signals early in the development process, and identify patients who are likely to respond to therapy [4].

HISTORY OF BIOMARKER

It's possible to argue that the earliest biomarker was discovered by Hindu scholars in 1500 B.C. in India, when they observed that the high sugar content of diabetes patients' urine attracted ants. However, with the development of modern medicine much later, in the 20th century, came the term “biomarker”. The term “biochemical markers” and related terms originated in scientific journals in the 1940s and 1950s [5]. The first papers identifying biomarkers with a clinical application—cancer—were published in the 1970s. The term “biomarker” refers to a biological marker, as defined by the National Institutes of Health Biomarkers Definitions Working Group in 2000. As the internationally recognized definition of a biomarker in clinical pharmacology, this one has gained widespread acceptance. The development of modern medicine in the 20th century, however, introduced with it the term “biomarker” and numerous synonyms [6]. Some highlights are included in Fig. (**1**).

Selection and Validation of Biomarkers

The selection and validation of a biomarker involve several criteria that need to be taken into account. It takes multiple stages to choose the best biomarker:

- The selection and identification of the interest's endpoint.

- The identification of the biomarker(s) unique to the outcome of interest, taking into account the biomarker's sensitivity, specificity, and significance with respect to health outcome or pathological change over time.
- The setup of a database to record the connection between the endpoint, potential biomarkers, and chemical exposure.
- A preference for non-invasive methods and consideration of specimens that might be available for analysis, with a focus on maintaining the specimen's integrity between collection and analysis.
- Assessment or forecast of the risk to the health of people, either for the broader public or a specific subgroup [6, 7].

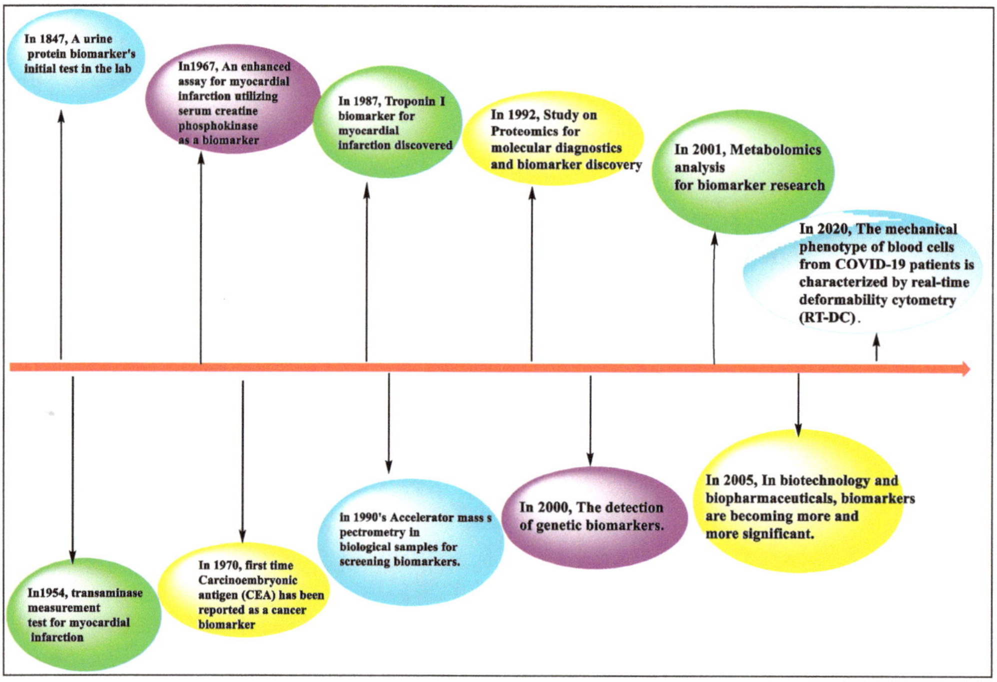

Fig. (1). Historical turning points in the discovery of biomarkers in contemporary biomedicine.

Types of Biomarkers

Biomarkers can be classified based on different parameters: their features, clinical uses, and techniques from the fields of genetics and molecular biology (Fig. **2**).

As per the genetic and molecular biology approach, there are three different kinds of biomarkers: Type 0, Type 1, and Type 2. Natural history biomarkers may be referred to as Type 0 biomarkers, drug activity biomarkers, or Type 1 biomarkers; they show the result of medication intervention, and Type 2 biomarkers are surrogate markers, or surrogate endpoints, that can be used to predict the outcome of a therapeutic intervention as well as a disease's clinical outcome in place of a clinical outcome. Based on their features, biomarkers can be categorized into three main types: imaging, cellular, and molecular biomarkers. Biomarkers can be categorized as diagnostic, prognostic, or therapeutic based on how they are used in various phases of disease [8].

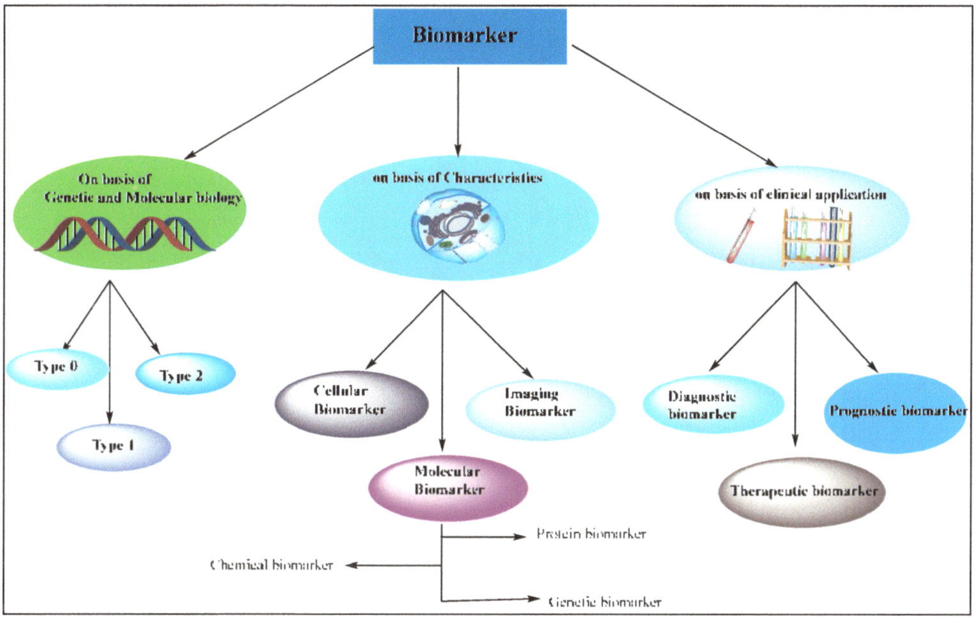

Fig. (2). Classification of biomarkers.

What is an Ideal Biomarker?

- Should be reliable, sensitive, and specific.
- Should be safe and simple to measure.
- Must be modifiable with treatment.
- Must have high prediction accuracy, as well as ease of measurement and reproducibility.

Diagnostic Biomarkers

A biomarker that can be used to find or validate the presence of a disease of interest or to identify individuals with a certain illness subtype. For example, hemoglobin A1c (HbA1c) or blood sugar levels can be used as biomarkers of diagnosis to identify patients with Type 2 Diabetes Mellitus (DM). The biomarker known as glomerular filtration rate, or GFR, is useful in the diagnosis of chronic renal illness [9].

ROLE IN DISEASE DIAGNOSIS

Accurate diagnosis of illnesses and ailments is necessary for medical practice. When determining whether a patient has a condition for which therapy may be recommended or whether they are eligible for a clinical trial investigating a specific disease, diagnostic biomarkers are crucial [10]. The previously identified biomarkers are employed in the diagnosis of various diseases, including cardiac troponin, which is used to diagnose cardiac muscle injury, distributions of 3-hydroxy-fatty acids for planctomycetes, a set of glycans as cancer biomarkers, glutamate for visceral obesity and altered metabolism, catestatin for psychological stress response linked to an increased mortality rate among heart patients, cystatin-C for glomerular filtration, and Liver-type Fatty Acid-Binding Protein (L-FABP) as a diagnostic biomarker. By assisting in early diagnosis, disease prevention, drug target identification, treatment response, and other areas, these biomarkers put the future in our hands. It has been determined that gene-based biomarkers are reliable and valid in the current scientific community [11].

CASE STUDIES OF KEY DIAGNOSTIC BIOMARKERS

"Rapid diagnosis of tuberculosis by detection of mycobacterial Lipoarabinomannan (LAM) in urine" was the subject of a study conducted by B. Hamasur *et al.* Two methods are used to measure LAM concentrations in urine: sandwich ELISA and dipstick approach. Both methods of investigation revealed that LAM was excreted in the urine of patients with active tuberculosis [12].

Zeisler, H., performed an observational study on the Predictive Value of the sFlt-1:PlGF Ratio in Women with Suspected Preeclampsia. In order to determine and validate a serum sFlt-1 to PlGF ratio that would be predictive of the short-term presence or absence of preeclampsia in women with singleton pregnancies in whom preeclampsia was suspected, they conducted a prospective, multicenter observational study (24 weeks 0 days to 36 weeks 6 days of gestation). Finding out if low sFlt-1:PlGF ratios (at or below a defined threshold) indicate preeclampsia absence within 1 week of the first visit and high ratios (above the cutoff) predict preeclampsia presence within 4 weeks were the main goals of the study [13, 14].

Rajkumar T. *et al.* studied the "Identification and validation of plasma biomarkers for diagnosis of breast cancer in South Asian women." Using antibody array and methylation specific PCR, fifteen out of 56 prospective protein markers and three candidate epigenetic markers were first validated in primary breast tumors, and then in plasma samples of patients (N = 202 invasive, 16 ductal carcinoma *in-situ* (DCIS)] and controls [N = 203 healthy, 37 benign). They conducted research with the goal of creating a multi-marker, minimally invasive Breast Cancer diagnostic approach that would be more sensitive and unique to South Asian women. The effectiveness of both single and combination indicators in diagnosis was evaluated. In accordance with research, the combination of *SOSTDC1, DACT2*, and *WIF1* proved effective in distinguishing between benign breast diseases and healthy individuals, as well as between non-invasive and invasive breast cancer. This combination may also be used as a supplemental diagnostic tool for breast cancer [15, 16].

Atik A. *et al* studied the "Alpha-Synuclein as a Biomarker for Parkinson's Disease." A common, progressive, and chronic neurological illness that affects people in the general population between the ages of 60 and 80 is Parkinson's Disease (PD). After Alzheimer's Disease (AD), it is the second most prevalent major neurodegenerative illness. Misdiagnosis of the disease emphasizes the need for disease-specific and early-stage biomarkers, as clinical diagnosis of Parkinson's Disease (PD) is difficult. Lewy bodies are microscopic markers for Parkinson's disease that are found within brain cells [17]. Lewy bodies contain the protein known as alpha-synuclein, or a-synuclein. A substantial amount of data indicates that the very small presynaptic protein -synuclein (-syn) is crucial to the pathophysiology of Parkinson's Disease (PD). The Red Blood Cell (RBC) levels of a multiple system atrophy patient, a Parkinson's Disease patient, and a control subject were assessed using an enzyme-linked immunosorbent test in this investigation. The current results support the idea that elevated-syn oligomers in plasma and cerebrospinal fluid could be a useful diagnostic biomarker for Parkinson's Disease (PD), as they are consistent with prior research indicating an elevation of these oligomers in PD patients' plasma and cerebrospinal fluid as compared to non-PD patients [18, 19].

Masson S. *et al*. evaluated the Circulating presepsin (soluble CD14 subtype) as a marker of host response in patients with severe sepsis or septic shock: data from the multicenter, randomized ALBIOS trial. They assessed the relationship between its concentration in circulation, the host's reaction, the suitability of antibiotic treatment, and the death rate in patients suffering from severe sepsis. In the multicenter Albumin Italian Outcome Sepsis (ALBIOS) experiment, plasma presepsin was assessed 1, 2, and 7 days after the patients with severe sepsis or septic shock were enrolled. From day 28 of randomization, patients were randomly assigned to receive 20% albumin and crystalloids or crystalloids alone. Research indicates that in sepsis patients, sepsin serves as an early predictor of both host response and death. Variations in concentrations over time appear to indicate whether antibiotic therapy is appropriate [20, 21].

Chen S. *et al.* studied a panel using three miRNAs as a prostate cancer diagnostic biomarker. One of the most common cancers afflicting men at any stage of their lives is Prostate Cancer (PCa). Prostate cancer mortality and incidence are rising annually as well. One extremely promising non-invasive method for diagnosing and assessing the prognosis of prostate cancer is the detection of microRNA expression in serum. The study included 224 trial participants in all, comprising 112 patients with prostate cancer and 112 healthy persons. There were three primary stages to the experiment: screening, testing, and validation. Quantitative reverse transcription-polymerase chain reaction was used to determine the expression levels of serum miRNAs in patients and healthy persons. The findings show that the diagnostic panel utilizing these three miRNAs has outstanding diagnostic performance, responsiveness, and selectivity [22, 23].

Imaging Biomarkers

A unique type of biomarker known as Imaging Biomarkers (IBs) is an indicator that is obtained from *in vivo* medical images. Several disorders can be diagnosed, prognosed, and responses can be evaluated using these IBs. They are also frequently utilized in patient care pathways. There are currently very few IBs directing clinical decision-making processes in hospitals, even though some are heavily used and others have great clinical potential. The presence of anomalies can be detected by qualitative medical imaging, but not their size. The size of an anomaly, or at least its relative size, can also be displayed via quantitative medical imaging. A variety of imaging techniques, including nonionizing radiation like MRI and radiant energy like Computed Tomography (CT), PET, and X-rays, can be used to identify quantitative imaging biomarkers [24] (QIBs) (Fig. **3**).

PET Scan as Imaging Biomarker

Positron Emission Tomography (PET) enables quantitative evaluation and non-invasive visualization of the body's physiological and metabolic processes. Positron-emitting radioisotopes of natural elements, such as oxygen-15, carbon-11, nitrogen-13, and fluorine-18, are employed in PET. These radioisotopes have been used to synthesize a variety of positron-emitting radiopharmaceuticals while maintaining their normal biological function. The clinical effectiveness of PET imaging can be attributed to three main elements. First, it has been demonstrated that the majority of human malignancies exhibit a notable accumulation of the glucose analog Fluorodeoxyglucose (FDG). Second, whole-body FDG-PET scans may now be completed in a clinically acceptable amount of time due to advancements in PET detector technology. Third, the accuracy of FDG-PET for tumor staging has been further enhanced by integrated PET/Computed Tomography (CT) equipment, which provides precise anatomic localization of focally increased FDG uptake. The most distinctive feature of PET-based biomarkers is that they typically measure a biologic process rather than the quantity of a specific cellular protein [25, 26].

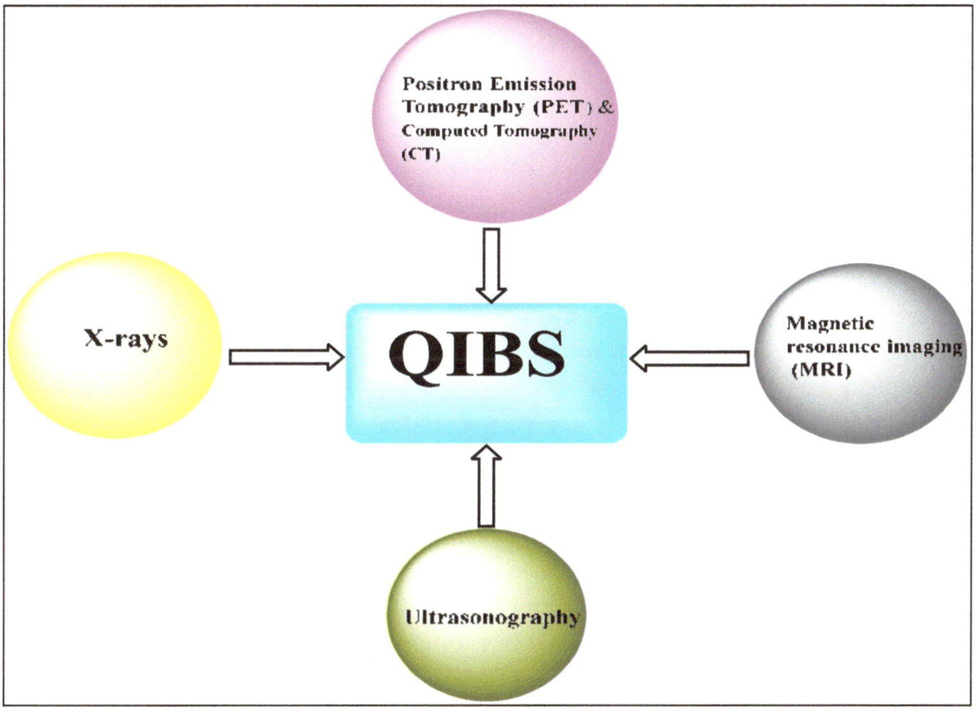

Fig. (3). Methods for imaging that can be applied to determine Quantitative Imaging Biomarkers (QIBs).

CT Scan as Imaging Biomarker

A computed tomography scan, or CT scan for short, is a kind of imaging that builds detailed images of the body using X-ray techniques. The only biophysical characteristic that CT biomarkers rely on is the differential absorption of X-rays due to variations in tissue density. This can occur during unenhanced scans or after the injection of an iodine-based contrast agent, which increases X-ray absorption in highly perfused tissues [27, 28].

Magnetic Resonance Imaging (MRI) as an Imaging Biomarker

Utilizing a magnetic field and computer-generated radio waves, Magnetic Resonance Imaging (MRI) is a medical imaging procedure that produces finely detailed images of your body's organs and tissues. The first magnetic resonance image appeared in 1973. Moreover, optical imaging is widely employed in preclinical animal research and medication development, and it is becoming more and more common in human clinics, such as with optical CT scanning. In MRI, radio waves and magnetic fields are used to create images on a computer. No ionizing radiation is used during an MRI. Organs, bones, muscles, and blood arteries can all be seen in the images obtained from an MRI scan [29].

BIOCHEMICAL AND PHYSIOLOGICAL BIOMARKERS

Blood and Urine Biomarkers

The most commonly utilized biomarkers for diagnosis support in clinical research are PET and CSF, yet their usage in clinical practice is limited due to cost and accessibility issues. New blood-based markers may be useful for general clinical usage and could speed up diagnosis because they are accessible, affordable, and accurate. Blood tests provide a wealth of data supported by reliable studies, such as the state of red blood cell populations, the detection of oxidative stress and inflammation, and the diagnosis of iron, vitamin, or energy deficiencies [30].

Elevated blood lead levels may suggest the need to test for neurological and cognitive disorders, particularly in children.

Hemoglobin A1c, or HbA1c, can be used as a blood marker to indicate the presence and severity of hyperglycemia, which suggests the presence of diabetes or prediabetes [31].

Urine is becoming a valuable biological fluid for studying illness biomarkers, as it is readily available and non-invasive. However, due to its complex composition and inter-individual variability, using urine as a source of biomarkers presents several difficulties. In order to overcome these obstacles, ultra-sensitive biosensing and microfluidic technologies reduce the number of sample handling and processing stages, which lowers the possibility of contamination and biomarker loss while enhancing detection accuracy and precision. By evaluating global metabolic profiles in readily available biofluids and identifying biomarkers to differentiate between individuals with and without disease, metabolomics is a potent new tool. The variety of metabolic effects that a disease condition can produce can be shown by detecting variations in metabolite concentrations. It has been possible to identify potential early diagnostic and prognostic indicators of disease through the use of urinary metabolomic techniques [32].

The most common cause of dementia, Alzheimer's Disease (AD), is a progressive neurological illness whose incidence is rising quickly. The development of intracellular neurofibrillary tangles, resulting from the aggregation of tau proteins and extracellular amyloid plaques, which impair neuronal signaling and cause cell death, is the hallmark of Alzheimer's Disease (AD). It has been demonstrated that acrolein, a lipid peroxidation byproduct, plays a role in the degenerative process of AD. Alzheimer's disease is caused in part by changes in the metabolism of acrolein. Reduced levels of 3-hydroxypropyl mercapturic acid (3-HPMA) metabolism, a urine acrolein metabolite, may be associated with neuronal impairment in AD patients [33].

Hormonal and Metabolic Biomarkers

A group of risk factors unique to cardiovascular disease, including high blood pressure, impaired fasting glucose, low HDL cholesterol, high triglyceride levels, and abdominal obesity, is collectively referred to as metabolic syndrome. Metabolic syndrome can be diagnosed and managed with the aid of biomarkers, and lifestyle modifications like exercise and diet can successfully lower risk and raise biomarker levels.

Hyperinsulinemia may be a much earlier indicator of the risk of metabolic diseases, the onset of metabolic syndrome, and the development of diabetes mellitus when poor glucose tolerance and a normal HbA1c are absent [34, 35].

The hormone leptin is generated by adipose tissue and is mostly secreted by white adipose tissue, although it is also produced by non-adipose tissues, including the stomach, placenta, immune system, and mammary gland. Elevated levels of leptin have been linked to metabolic, inflammatory, and homeostatic elements that are implicated in the pathophysiological mechanisms of Metabolic Syndrome (MetS), Hypertension (HTN), and other Cardiovascular Disorders (CVD). It also controls calcium homeostasis and the electrical characteristics of the left atrium. An inflammatory cytokine called tumor necrosis factor-alpha is mostly produced by macrophage cells, which are also a subset of other inflammatory cells. It is an acute inflammatory response protein that interacts with insulin receptors to control insulin resistance and raises the concentration of C-reactive proteins. In a similar vein, it influenced the control of cholesterol metabolism, adipokine production, and lipid metabolism. Tumor necrosis factor-alpha has been linked to all the metabolic syndrome components that are involved in various metabolic syndrome pathways and alterations, including systemic inflammation and activation of protein C kinase and mTOR in IR, according to a number of studies [36, 37].

VALIDATION AND STANDARDIZATION OF BIOMARKERS

Clinical validation and establishment of a unique quantitative multiplex protein panel were conducted to facilitate enhanced precision medicine in the field of Cardiovascular Disease (CVD). With only a single, small blood sample, low- and high-abundance proteins may be assessed concurrently in a single step thanks to the CVD-21 quantitative Proximity Extension Assay (PEA) tool's high specificity and sensitivity. With the exception of troponin I, whose sensitivity was too low, the tests in the CVD-21 tool accurately quantified 20 out of 21 proteins. The CVD-21 tool, in conjunction with high-sensitivity troponin T (hs-TnT), demonstrated a comparable improvement in biomarker-based risk stratification when compared to clinical variables, hs-TnT alone, and multiple individual biomarkers examined by traditional immunoassays in the nested case-cohort design, which comprised 4224 patients with Chronic Coronary Syndrome (CCS) [18, 38].

Proximity Extension Assay (PEA)

According to the study's findings, hyperhomocysteinemia raises the risk of coronary artery disease in obese people between the ages of 40 and 65. Transcriptomics and metabolomics combined have shed light on the metabolic distinctions between CAD obese and non-CAD obese people. This method presents a viable technique for the identification of obese patients at high risk. To confirm these results and investigate focused therapies, further study is necessary [39, 40].

Multivariable Cox Regression and Machine Learning Techniques

A reference model with understood biomarkers and a model with just clinical factors were some of the seven multi-marker models developed. Biomarker-based models excelled in models that only relied on clinical factors and offered comparable selectivity for all outcomes [41, 42].

Comparison with Gold Standard Immunoassays

NT-proBNP, hs-TnT, Cystatin C, and Growth Differentiating Factor-15 (GDF-15) concentrations have been detected using Roche electrochemiluminescence immunoassays. The Quantikine 1HS Human IL-6 Immunoassay was utilized for the analysis of interleukin-6. For validation, the results of the CVD-21 PEA instrument were compared to those of these standard tests.

Multiplex Proteomic Analysis

New DNA amplicons were produced by the proximity binding of particular antigens and DNA polymerase-aided joining; these were identified and measured it microfluidic qPCR. The analyses were carried out at Uppsala University's Science for Life Laboratory's Clinical Biomarkers Facility.

The CVD-21 tool represents the first instrument using PEA to simultaneously quantify 21 proteins associated with cardiovascular diseases. This study successfully identified novel pathophysiologic and prognostic information beyond established biomarkers, demonstrating the tool's potential in improving precision medicine for cardiovascular risk assessment [43 - 47]

Cell Type / Tissue Deconvolution

Tools: deconvolute function of deconvR, NNLS model, and BSmeth2Probe function. The objective is to map genomic loci to CpG probe IDs in order to deconvolve the WGBS ccfDNA samples into the cell types of origin.

Differential Methylation Analysis

Tools: SLIM technique and methylKit.

Objective: To induce differential methylation on sliding windows having a 500 bp gap between tiles.

DMR Annotation

Tools: Genomic Regions Enrichment of Annotations Tool (GREAT), rGREAT, AnnotateR package, chromHMM bed files.

Objective: The aim is to annotate DMRs with respect to enhancer, genomic, regulatory, and CpGareas [50, 51].

Linear Models of Methylation Levels on DMRs

Tools: R lm function.

Objective: To use linear regression models to obtain DMRs related to ACS types.

Machine Learning Models for Prediction of ACS Type

Tools: Random Forest (RF), Partial Least Squares Regression (PLS), Penalized Multinomial Regression (PMR), and caret package.

Purpose: To use ccfDNA methylation profiles to determine ACS type.

The research effectively shows the potential of applying methylation profiles of cell-free DNA (cfDNA) to identify between various forms of Acute Coronary Syndrome (ACS). The research demonstrates a unique differentiation between healthy individuals and ACS patients by the use of a combination of advanced machine learning models, differential methylation analysis, and cell type deconvolution. Although there is a noticeable difference between UA and healthy patients from NSTEMI and STEMI, further studies are needed to accurately classify all ACS subtypes. This non-invasive technique can enhance ACS diagnosis and patient description by providing an additional layer of information to established clinical biomarkers and ECG measurements.

This work shows the potential utility of a set of Differentially Methylated Regions (DMRs) in cell-free DNA (cfDNA) for patient stratification according to Acute Coronary Syndrome (ACS). Without the need for additional cardiac biomarkers, we were able to distinguish between STEMI and NSTEMI patients by using 254 DMRs. Targeted sequencing was used to validate our results, demonstrating the method's dependability and cost-effectiveness. Additionally, it offers repeatable code and an R package for comparable analyses in other disorders. Upcoming research will concentrate on increasing the number and variety of cohorts as well as refining the targeted sequencing technique for wider clinical use [52, 53].

ACS Classification

Classical biomarkers used for the classification of Acute Coronary Syndrome (ACS) include highly sensitive cardiac Troponin T (hs-cTnT), Left Ventricular Ejection Fraction (LVEF), and Creatine Kinases (CK, CK-MB). Troponin T levels are typically highest in STEMI and NSTEMI patients and lower in UA patients. LVEF is higher in healthy controls and lower in ACS patients, with UA patients showing a wide range of values. CK levels increase with the severity of ACS, with the highest levels observed in STEMI patients [54, 55].

ccfDNA Extraction

Using the Qiagen QIAamp Circulating Nucleic Acid Kit, newly thawed double-centrifuged citrate plasma was used to extract circulating cell-free DNA (ccfDNA). The DNA was extracted in 50 µl of ultrapure nuclease-free water, and its concentration was determined using a Qubit fluorometer and the Qubit dsDNA HS Assay kit. Using the ccfDNA or hsD1000 screentape and the Agilent TapeStation 4200, the fragment size was verified. The volume of plasma used to extract the ccfDNA was used to standardize the overall amount of the material [56, 57].

Whole-Genome Bisulfite Sequencing (WGBS)

Novogene received the separated circulating cell-free DNA (ccfDNA) and reviewed it for quality. Their low-input BS-seq (PBAT) approach was used for bisulfite conversion and sequencing. R software packages that were previously created internally were used to examine the sequencing data [58, 59].

Targeted Methylation Sequencing

Sheared genomic DNA was used to maximize targeted methylation sequencing, with 10 ng of separated cell-free DNA serving as the starting material per sample. The NEBNext Enzymatic Methyl-seq Module and the Nonacus Cell3TM Target: Library Preparation kit were used to carry out the enzymatic conversion. After libraries were combined, hybridized, and collected, quality assessments and sequencing were conducted using the Novaseq 6000 platform [60].

Cell Type / Tissue Deconvolution

For the omics-based deconvolution of cfDNA to origin cell types, the deconvR R package was utilized. Four models were available: Robust Linear Regression (RLM), Quadratic Programming (QP), Support Vector Regression (SVR), and Non-Negative Least Squares (NNLS). Among which, NNLS and RLM were the best models for methylation-based deconvolution. Three further cardiac tissues and tissue-specific CpG feature selection were added to Moss *et al.*'s extensive array-based human cell type methylation atlas [61, 62].

Validation and Standardization of Biomarkers

15% of people worldwide suffer from neurological illnesses, which are the primary cause of both cognitive and physical disability. The next 20 years will see a doubling of the prevalence of neurological disorders, especially neurodegenerative diseases, as a result of aging demographics. Regretfully, although current treatments alleviate the symptoms of cognitive and motor impairment, there is a pressing need to create disease-modifying treatments that reduce the rate of pathological advancement. Within that framework, biomarkers could detect prodromal and at-risk patients, track treatment responses, track the progression of the disease, and elucidate the causation of molecular processes to uncover new targets for additional clinical research. Finding biomarkers that distinguish different diseases and correspond to particular stages of pathology will therefore accelerate the identification and creation of treatment targets [63, 64].

Blood biomarkers play a crucial role in diagnosing and monitoring various diseases, particularly neurodegenerative disorders. Here are some notable types of blood biomarkers, their roles, and mechanisms as discussed in the provided document:

Calmodulin

Contributes to the survival and multiplication of cells. The abnormal regulation of calmodulin has been linked to neurodegenerative illnesses like Alzheimer's

Disease (AD) and cancer. Elevated levels of calmodulin in AD lymphoblasts suggest that this protein may be used as a biomarker to diagnose AD. Elevated levels of calmodulin in AD lymphoblasts suggest that this protein may be used as a biomarker to diagnose AD [65, 66].

miRNAs, or Micro RNAs

These short non-coding RNAs are essential for regulating gene expression as well as the development of neurons and apoptosis. Certain miRNAs have been linked to cancer and neurological diseases, including miR-107/miR-335-5p and miR-127-3p. Variations in miRNA expression can aid in the differentiation of AD from Frontotemporal Lobar Degeneration (FTLD), two kinds of dementia [67, 68].

Glial Fibrillary Acidic Protein (GFAP)

Glial Fibrillary Acidic Protein (GFAP) is a marker of reactive astrocytes, or astrogliosis. Neuroinflammation and gliosis surrounding amyloid plaques are reflected by increased GFAP expression in brain tissue and Cerebrospinal Fluid (CSF) in Alzheimer's Disease (AD).

Neurofilament Light Chain (NfL)

Neurofilament Light Chain (NfL) is a neurodegenerative marker. While elevated NfL levels have been linked to a number of neurodegenerative illnesses, their ability to distinguish AD from FTLD is not entirely reliable as a diagnostic tool.

β-Synuclein and Total Prion Protein (t-PrP)

The roles of total Prion Protein (t-PrP) and β-synuclein are linked to neurodegenerative illnesses. Elevated biofluid levels suggest neurodegenerative activity, but they do not provide enough diagnostic precision to differentiate between certain diseases like AD and FTLD.

α-Synuclein

α-synuclein is a possible indicator of dementia with Lewy bodies (DLB). The typical buildup of α-synuclein in the brain, known as Lewy bodies in DLB, is suggested by the lower serum levels of this protein in DLB when compared to AD and healthy individuals [69, 70].

KL-40 (Chitinase-3-like protein 1)

KL-40 acts as a neuroinflammation marker. Its importance in the etiology of AD and its potential as a diagnostic biomarker are highlighted by elevated levels in AD relative to other dementias.

Heart-type Fatty Acid-Binding Protein (H-FABP)

Heart-type Fatty Acid-Binding Protein (H-FABP) functions as a potential biomarker for tissue damage, being released into the bloodstream following cell injury. Elevated serum levels in DLB and Parkinson's Disease Dementia (PDD) suggest that it is useful as a biomarker. The intricacy and specificity needed to create diagnostic instruments for neurodegenerative illnesses are demonstrated by these biomarkers. The distinct function and mechanism of each biomarker aid in the identification and distinction of various illnesses. Cerebrospinal Fluid (CSF) is an essential component of neurodegenerative disease monitoring and diagnosis [71, 72].

Illness Biomarker Source

Because CSF comes into direct contact with the brain and spinal cord and is therefore largely free of peripheral protein contamination, it is a valuable biofluid that is used for identifying brain-derived markers. Because of this property, CSF is an essential tool for identifying biomarkers linked to neurodegenerative disorders.

Predictive and Diagnostic Value

The initiation and progress of disease can be predicted by CSF biomarkers. For example, increased concentrations of particular proteins, such as tau and chitinases, can indicate neurodegeneration, neuroinflammation, and the existence of illnesses like ALS and Alzheimer's. These biomarkers often rise before clinical symptoms appear, indicating an early illness [73].

TRACKING THE DEVELOPMENT OF NEURODEGENERATIVE DISEASES AND EVALUATING THERAPEUTIC RESPONSE

CSF biomarkers are also employed in the monitoring of therapeutic response and the tracking of illness development. Variations in biomarker levels can serve as an indicator of how well a treatment is working, providing a way to modify therapeutic approaches as necessary.

Chitinases

The CSF of ALS patients has higher levels of these enzymes, especially CHIT1, CHIT3L1, and CHIT3L2. Their levels are correlated with the aggressiveness of the disease and are linked to neuroinflammation and degeneration. Their continuous increase in CSF, but not in plasma, shows how useful CSF is for monitoring the course of the ALS disease [74].

Phosphorylated Tau (p-tau)

Total tau (t-tau) and p-tau levels in CSF and plasma rise as a result of hyperphosphorylated tau proteins being produced in response to amyloid-beta plaques in Alzheimer's disease. These alterations help in defining Alzheimer's apart from other neurological disorders. Certain variations of p-tau, such as p-tau181 and p-tau217, are especially good at predicting the pathology of Alzheimer's Disease and can detect patients years before symptoms appear [75].

Neurofilament Light (NfL)

NfL is a structural protein in neurons, especially concentrated in axons. It is released into CSF and plasma as neurons degenerate. Elevated NfL levels are observed in various neurodegenerative diseases and correlate with disease progression, although their specificity for Alzheimer's disease is limited. NfL is more useful for tracking responses to disease-modifying therapies across different neurodegenerative conditions.

CSF plays a vital role in the identification and tracking of biomarkers that offer important information about the initial phase, course, and response to treatment of neurodegenerative disorders. It is a vital tool in the clinical management of various disorders due to its capacity to reflect changes particular to the brain. The imaging method known as Single-Photon Emission Computed Tomography (SPECT) enables the observation of the body's functional network, especially in the brain [76].

Radiopharmaceutical Injection

The patient is injected with a radiopharmaceutical, which is a radioactive tracer, into the blood circulation. Technetium-99m (Tc-99m), iodine-123 (I-123), and thallium-201 (Tl-201) are examples of frequently used tracers. Based on the biological activity of the chemical, these tracers are compounds that include a radionuclide and are intended to target particular organs or cellular receptors. To image cerebral blood flow, for instance, Tc-99m HMPAO and Tc-99m ECD are frequently utilized [72, 77].

The radiopharmaceutical enters the bloodstream after injection and builds up in the intended tissues. The physiological and metabolic properties of the tissue determine where the tracer is distributed. Tracers can be used to see regions of the brain with high or low blood flow, metabolic activity, or the presence of particular receptors. The patient is positioned in front of a gamma camera, which can detect the gamma rays that the radionuclide emits. The components of a gamma camera include an array of detectors, photomultiplier tubes, a scintillation crystal (often constructed of sodium iodide), and a collimator to filter out scattered photons.

Getting Images

Multiple 2D images (also known as projections) of the patient are taken from various angles by a rotating gamma camera. To document the tracer's dispersion, these photos are acquired gradually. A three-dimensional representation of the tracer distribution inside the body is created by processing the gathered data. Algorithms like filtered back projection or iterative reconstruction are used to reconstruct the 2D projection images into 3D images. The resulting three-dimensional image highlights areas of aberrant function and displays the tracer's concentration across the body [78].

Image Interpretation

Radiologists or nuclear medicine experts interpret the reconstructed pictures. Tracer uptake patterns in certain areas can reveal a number of clinical diseases. In brain imaging, for instance, areas with decreased metabolic activity or blood flow may be indicative of disorders like Alzheimer's, stroke, tumors, or epilepsy.

SPECT Applications

Neurology: Measuring stroke risk, identifying epilepsy foci, as well as diagnosing and tracking neurodegenerative disorders (*e.g.*, Parkinson's, Alzheimer's).

Cardiology: Detecting coronary artery disease, assessing left ventricular function, and assessing myocardial viability and perfusion.

Oncology: Detecting metastases, tracking therapy response, and diagnosing and staging different forms of cancer. In orthopedics, bone metabolism is assessed, and fractures, infections, and osteomyelitis are identified.

In addition to supporting anatomical imaging methods like CT and MRI, SPECT offers useful functional information that helps with a variety of medical condition diagnoses and therapy [79].

ROLE AND MECHANISM OF MRI IN NEURODEGENERATIVE DISEASES

Magnetic Resonance Imaging (MRI) plays a crucial role in the diagnosis, monitoring, and research of neurodegenerative diseases. It offers high-resolution brain images that make it possible to identify structural alterations linked to a number of neurodegenerative diseases, including Parkinson's Disease (PD), Alzheimer's Disease (AD), Frontotemporal Dementia (FTD), Dementia with Lewy Bodies (DLB), and Vascular Dementia (VaD)

Diagnosis

Magnetic Resonance Imaging (MRI) facilitates the identification of distinctive patterns of brain atrophy linked to various neurodegenerative disorders. For instance, frontotemporal atrophy is suggestive of FTD, but medial temporal lobe atrophy is a hallmark of Alzheimer's disease [80].

Tracking the Progression of the Disease

Magnetic Resonance Imaging (MRI) can be utilized to monitor the rate of neurodegeneration by tracking the atrophy of the brain over time. This is very helpful when evaluating the effectiveness of treatment interventions in clinical trials.

Magnetic Resonance Imaging (MRI) is a useful tool for investigating the pathophysiology of neurodegenerative disorders. It makes it possible to examine the anatomy and function of the brain non-invasively, which advances our knowledge of how diseases are caused [81].

Mechanism of MRI

MRI uses powerful magnets and radio waves to generate detailed images of the brain. The process involves several steps:

Magnetization: To align the protons in the body's hydrogen atoms, a patient is positioned in a strong magnetic field.

Radiofrequency Pulses: To knock off this alignment, radiofrequency pulses are used. The protons realign themselves when the pulses are stopped, producing energy in the process.

Signal Detection: The MRI scanner's coils detect the energy produced when the protons realign. The brain can be observed in great detail because of these signals.

Image Reconstruction: A computer processes the signals to produce images that depict various brain structures. The production of distinct signal intensities by tissues with varying densities and compositions, such as cerebrospinal fluid, white matter, and gray matter, allows for the precise imaging of brain structures [82].

SPECIFIC APPLICATIONS IN NEURODEGENERATIVE DISEASES

MRI can measure the volume of specific brain regions. Atrophy or shrinkage of a brain region in neurodegenerative diseases is recognized. For example, hippocampal atrophy in Alzheimer's disease is identified. Cerebral cortex thickness is assessed through MRI, and this is a biological marker that indicates the neurodegeneration of neurons.

White matter hyperintensities, which are commonly observed in vascular dementia and linked to small vessel disease, are among the alterations in white matter that can be recognized by MRI.

Diffusion of water molecules provides information about the integrity of white matter in the brain through MRI techniques. It is useful for detecting microstructural changes in diseases like multiple sclerosis and other neurodegenerative conditions. fMRI measures brain activity by detecting changes in blood flow. It is used to study brain function and connectivity, which can be altered in neurodegenerative diseases [83].

CONCLUSION

MRI is a vital imaging tool in the field of neurodegenerative diseases, providing detailed structural and functional information that aids in diagnosis, monitoring, and research. Its ability to visualize brain atrophy, cortical thickness, and white matter changes, along with advanced techniques like DTI and fMRI, makes it indispensable in understanding and managing these complex conditions.

REFERENCES

[1] X. Dai, and L. Shen, "Advances and trends in omics technology development", *Front. Med. (Lausanne),* vol. 9, p. 911861, 2022. [http://dx.doi.org/10.3389/fmed.2022.911861] [PMID: 35860739]

[2] Z.N. Ling, Y.F. Jiang, J.N. Ru, J.H. Lu, B. Ding, and J. Wu, "Amino acid metabolism in health and disease", *Signal Transduct. Target. Ther.,* vol. 8, no. 1, p. 345, 2023. [http://dx.doi.org/10.1038/s41392-023-01569-3] [PMID: 37699892]

[3] D.C. Bou-Habib, G. Roderiquez, T. Oravecz, P.W. Berman, P. Lusso, and M.A. Norcross, "Cryptic nature of envelope V3 region epitopes protects primary monocytotropic human immunodeficiency virus type 1 from antibody neutralization", *J. Virol.,* vol. 68, no. 9, pp. 6006-6013, 1994. [http://dx.doi.org/10.1128/jvi.68.9.6006-6013.1994] [PMID: 8057475]

[4] G. Yu, H.C.H. Tam, C. Huang, M. Shi, C.K.P. Lim, J.C.N. Chan, and R.C.W. Ma, "Lessons and Applications of Omics Research in Diabetes Epidemiology", *Curr. Diab. Rep.,* vol. 24, no. 3, pp. 27-

44, 2024.
[http://dx.doi.org/10.1007/s11892-024-01533-7] [PMID: 38294727]

[5] K.A. Matthews, K. Räikkönen, L. Gallo, and L.H. Kuller, "Association between socioeconomic status and metabolic syndrome in women: Testing the reserve capacity model", *Health Psychol.,* vol. 27, no. 5, pp. 576-583, 2008.
[http://dx.doi.org/10.1037/0278-6133.27.5.576] [PMID: 18823184]

[6] D. L. Siehl, "The biosynthesis of tryptophan, tyrosine, and phenylalanine from chorismate," In: *Plant Amino Acids,* 1st ed., Boca Raton, FL: CRC Press, 1998, pp. 34.
[http://dx.doi.org/10.1201/9781482270068-6]

[7] M.A. Mallah, T. Soomro, M. Ali, S. Noreen, N. Khatoon, A. Kafle, F. Feng, W. Wang, M. Naveed, and Q. Zhang, "Cigarette smoking and air pollution exposure and their effects on cardiovascular diseases", *Front. Public Health,* vol. 11, p. 967047, 2023.
[http://dx.doi.org/10.3389/fpubh.2023.967047] [PMID: 38045957]

[8] P. Maechler, S. Carobbio, and B. Rubi, "In beta-cells, mitochondria integrate and generate metabolic signals controlling insulin secretion", *Int. J. Biochem. Cell Biol.,* vol. 38, no. 5-6, pp. 696-709, 2006.
[http://dx.doi.org/10.1016/j.biocel.2005.12.006] [PMID: 16443386]

[9] O.A. Almazroo, M.K. Miah, and R. Venkataramanan, "Drug Metabolism in the Liver", *Clin. Liver Dis.,* vol. 21, no. 1, pp. 1-20, 2017.
[http://dx.doi.org/10.1016/j.cld.2016.08.001] [PMID: 27842765]

[10] B. B. BRODIE, "Pathways of drug metabolism.," *J. Pharm. Pharmacol*., vol. 8, no. 1, pp. 1–17, Jan. 1956.
[http://dx.doi.org/10.1111/j.2042-7158.1956.tb12123.x]

[11] N. Isoherranen, and K.E. Thummel, "Drug metabolism and transport during pregnancy: how does drug disposition change during pregnancy and what are the mechanisms that cause such changes?", *Drug Metab. Dispos.,* vol. 41, no. 2, pp. 256-262, 2013.
[http://dx.doi.org/10.1124/dmd.112.050245] [PMID: 23328895]

[12] L.R. Engelking, "Gluconeogenesis," In: *Textbook of Veterinary Physiological Chemistry,* 2nd ed., 2015, pp. 225–230.
[http://dx.doi.org/10.1016/B978-0-12-391909-0.50037-2]

[13] S. Mathiarasan, and A. Hüls, "Impact of environmental injustice on children's health—interaction between air pollution and socioeconomic status", *Int. J. Environ. Res. Public Health,* vol. 18, no. 2, p. 795, 2021.
[http://dx.doi.org/10.3390/ijerph18020795] [PMID: 33477762]

[14] A. Singh, L.B. Prasad, K. Shiv, R. Kumar, and S. Garai, "Synthesis, characterization, and in vitro antibacterial and cytotoxic study of Co(II), Ni(II), Cu(II), and Zn(II) complexes of N-(--methoxybenzyl) N-(phenylethyl) dithiocarbamate ligand", *J. Mol. Struct.,* vol. 1288, no. May, p. 135835, 2023.
[http://dx.doi.org/10.1016/j.molstruc.2023.135835]

[15] M. Zeng, L. Yang, D. He, Y. Li, M. Shi, and J. Zhang, "Metabolic pathways and pharmacokinetics of natural medicines with low permeability", *Drug Metab. Rev.,* vol. 49, no. 4, pp. 464-476, 2017.
[http://dx.doi.org/10.1080/03602532.2017.1377222] [PMID: 28911247]

[16] A. Singh, R. Kumar, K. Shiv, S.K. Pandey, M.K. Bharty, R.J. Butcher, and L.B. Prasad, "Synthesis, crystal structure and screening for anticonvulsant and antianxiety activities of three new Ni(II), Cu(II), and Zn(II) dithiocarbamate complexes", *J. Mol. Struct.,* vol. 1298, no. P2, p. 137052, 2024.
[http://dx.doi.org/10.1016/j.molstruc.2023.137052]

[17] O. Robinson, A.R. Carter, M. Ala-Korpela, J.P. Casas, N. Chaturvedi, J. Engmann, L.D. Howe, A.D. Hughes, M.R. Järvelin, M. Kähönen, V. Karhunen, D. Kuh, T. Shah, Y. Ben-Shlomo, R. Sofat, C.H.E. Lau, T. Lehtimäki, U. Menon, O. Raitakari, A. Ryan, R. Providencia, S. Smith, J. Taylor, T. Tillin, J. Viikari, A. Wong, A.D. Hingorani, M. Kivimäki, and P. Vineis, "Metabolic profiles of socio-economic

position: a multi-cohort analysis", *Int. J. Epidemiol.,* vol. 50, no. 3, pp. 768-782, 2021. [http://dx.doi.org/10.1093/ije/dyaa188] [PMID: 33221853]

[18] F. Rajas, A. Gautier-Stein, and G. Mithieux, "Glucose-6 Phosphate, a Central Hub for Liver Carbohydrate Metabolism", *Metabolites,* vol. 9, no. 12, p. 282, 2019. [http://dx.doi.org/10.3390/metabo9120282] [PMID: 31756997]

[19] A. Singh, R. Kumar, R. Patel, Trishna, R. N. Gautam, M. K. Bharty, and L. B. Prasad, "Dithiocarbamate-based novel anti-histaminic agents: synthesis, characterization, crystal structure and thermal study," Dalton Trans., vol. 33, 2024. [http://dx.doi.org/10.1039/D4DT01777C]

[20] E. Cecchin and G. Stocco, "Pharmacogenomics and personalized medicine," *Genes (Basel),* vol. 11, no. 6, p. 679, 2020. [http://dx.doi.org/10.1007/978-1-59745-031-7_11]

[21] R. Garg, K. Dobhal, B. Kumar, A. K. Gupta, T. Saklani, R. Kumar, R. Yadav, and G. Kaur, "COVID-19: A challenge towards the sustainability of health in platform era," *NeuroQuantology*, vol. 20, no. 11, pp. 2503–2519, 2022. [http://dx.doi.org/10.47750/pnr.2022.13]

[22] S. Ekins, B.J. Ring, J. Grace, D.J. McRobie-Belle, and S.A. Wrighton, "Present and future in vitro approaches for drug metabolism", *J. Pharmacol. Toxicol. Methods,* vol. 44, no. 1, pp. 313-324, 2000. [http://dx.doi.org/10.1016/S1056-8719(00)00110-6] [PMID: 11274898]

[23] A. Kumar, B. Kumar, R. Kumar, A. Kumar, M. Singh, V. Tiwari, A. Trigunayat, P. Paul, and P. Singh, "Acute and subacute toxicity study of ethanolic extract of Calotropis procera (Aiton) Dryand flower in Swiss albino mice", *Phytomedicine Plus,* vol. 2, no. 2, p. 100224, 2022. [http://dx.doi.org/10.1016/j.phyplu.2022.100224]

[24] S. Gobena, B. Admassu, M.Z. Kinde, and A.T. Gessese, "Proteomics and Its Current Application in Biomedical Area: Concise Review", *Sci. World J.,,* vol. 2024, pp. 1-13, 2024. [http://dx.doi.org/10.1155/2024/4454744] [PMID: 38404932]

[25] B.A. Moffatt, and H. Ashihara, "Purine and pyrimidine nucleotide synthesis and metabolism", *Arabidopsis Book,* vol. 1, p. e0018, 2002. [http://dx.doi.org/10.1199/tab.0018] [PMID: 22303196]

[26] R. Dhiman, "Novel LC-MS/MS bioanalytical method validation for gabapentin in human plasma in accordance with current regulatory guidances", *Int J Pharm Chem Sci,* vol. 4, 2015no. 4, pp. 462-468. Available from: http://www.ijpcsonline.com/files/08-10-15/07-1035.pdf

[27] V.A. Ramchandani, W.F. Bosron, and T.K. Li, "Research advances in ethanol metabolism", *Pathol. Biol. (Paris),* vol. 49, no. 9, pp. 676-682, 2001. [http://dx.doi.org/10.1016/S0369-8114(01)00232-2] [PMID: 11762128]

[28] H. Pandey, S. Srivastava, R. Kumar, and Y. B. Tripathi, "Preclinical acute and repeated dose toxicity of Pueraria tuberosa (PTWE) on Charles Foster rats," Int. J. Pharm. Sci. Res., pp. 4572–4581, 2018. Available from: https://ijpsr.com/bft-article/preclinical-acute-and-repeated-dose-toxicity-of--ueraria-tuberosa-ptwe-on-charles-foster-rats/

[29] X. Clemente-Casares, and P. Santamaria, "Nanomedicine in autoimmunity", *Immunol. Lett.,* vol. 158, no. 1-2, pp. 167-174, 2014. [http://dx.doi.org/10.1016/j.imlet.2013.12.018] [PMID: 24406504]

[30] C. Wieder, R.P.J. Lai, and T.M.D. Ebbels, "Single sample pathway analysis in metabolomics: performance evaluation and application", *BMC Bioinformatics,* vol. 23, no. 1, p. 481, 2022. [http://dx.doi.org/10.1186/s12859-022-05005-1] [PMID: 36376837]

[31] B. Testa, and S.D. Krämer, "The biochemistry of drug metabolism--an introduction: part 1. Principles and overview", *Chem. Biodivers.,* vol. 3, no. 10, pp. 1053-1101, 2006. [http://dx.doi.org/10.1002/cbdv.200690111] [PMID: 17193224]

[32] I. Barroso, and M.I. McCarthy, "The Genetic Basis of Metabolic Disease", *Cell,* vol. 177, no. 1, pp. 146-161, 2019.
[http://dx.doi.org/10.1016/j.cell.2019.02.024] [PMID: 30901536]

[33] C.R. Stephens, J.F. Easton, A. Robles-Cabrera, R. Fossion, L. de la Cruz, R. Martínez-Tapia, A. Barajas-Martínez, A. Hernández-Chávez, J.A. López-Rivera, and A.L. Rivera, "The impact of education and age on metabolic disorders", *Front. Public Health,* vol. 8, p. 180, 2020.
[http://dx.doi.org/10.3389/fpubh.2020.00180] [PMID: 32671006]

[34] O.E. Owen, S.C. Kalhan, and R.W. Hanson, "The key role of anaplerosis and cataplerosis for citric acid cycle function", *J. Biol. Chem.,* vol. 277, no. 34, pp. 30409-30412, 2002.
[http://dx.doi.org/10.1074/jbc.R200006200] [PMID: 12087111]

[35] R. Dhiman, D. Kumar, B. Kumar, and B.L. Pandey, "Quantitative determination of atorvastatin, ortho-hydroxy atorvastatin, para-hydroxy atorvastatin in human plasma using rosuvastatin as internal standard by LC-MS / MS", *Int. J. Pharm. Chem. Sci.,* vol. 4, no. 4, pp. 487-500, 2016.

[36] A. El-Sayed, L. Aleya, and M. Kamel, "Microbiota's role in health and diseases", *Environ. Sci. Pollut. Res. Int.,* vol. 28, no. 28, pp. 36967-36983, 2021.
[http://dx.doi.org/10.1007/s11356-021-14593-z] [PMID: 34043164]

[37] M. Singh, B. Kumar, A. K Singh, C. Shekhar Azad, M. K Yadav, A. Kumar, R. Kumar, A. Kumar, and P. Chaudhary, "Pharmacokinetic screening to Estimate the drug likeliness characteristics of selected Herbal Anticancer Drugs", *Res. J. Pharm. Technol.,* vol. 16, no. 7, pp. 3422-3426, 2023.
[http://dx.doi.org/10.52711/0974-360X.2023.00566]

[38] B. Kumar, R. Kumar, H. Patel, and A. Kumar, "Development and validation of an RP-HPLC method for simultaneous estimation of amlodipine besylate and indapamide in bulk and tablet dosage form", *Indo Am. J. Pharm. Sci.,* vol. 4, no. 11, pp. 4278-4287, 2017.

[39] P. Adeqboyega, and L. Wang, "How does HIV cause depletion of CD4 lymphocytes? A mechanism involving virus signaling through its cellular receptors", *Curr. Mol. Med.,* vol. 1, no. 5, pp. 545-550, 2001.
[http://dx.doi.org/10.2174/1566524013363320] [PMID: 11899230]

[40] B. Kumar, R. Kumar, A. Kumar, and R. Patel, "Simultaneous RP-HPLC method development and validation of fenofibrate and rosuvastatin calcium in bulk and tablet dosage form," Zenodo, 2017. doi: 10.5281/zenodo.105788

[41] H. Maeda, and N. Dudareva, "The shikimate pathway and aromatic amino Acid biosynthesis in plants", *Annu. Rev. Plant Biol.,* vol. 63, no. 1, pp. 73-105, 2012.
[http://dx.doi.org/10.1146/annurev-arplant-042811-105439] [PMID: 22554242]

[42] K. Ajay, N. Manmath Kumar, K. Brijesh, K. Ashutosh, K. Rajesh, K. Vikas, and S. Ambrish Kumar, "Toxicity (acute and subacute) assessment and in-vivo antiurolithiatic activity of ethanolic extract of Caesalpinia bonducella seed in albino Wistar rat", *J. Appl. Pharm. Sci.,* vol. 12, no. 2, pp. 187-197, 2022.
[http://dx.doi.org/10.7324/JAPS.2022.120220]

[43] A. Casamassimi, A. Federico, M. Rienzo, S. Esposito, and A. Ciccodicola, "Transcriptome Profiling in Human Diseases: New Advances and Perspectives", *Int. J. Mol. Sci.,* vol. 18, no. 8, p. 1652, 2017.
[http://dx.doi.org/10.3390/ijms18081652] [PMID: 28758927]

[44] A. Siegbahn *et al.*, "Development and validation of a quantitative Proximity Extension Assay instrument with 21 proteins associated with cardiovascular risk (CVD-21).," *PLoS One*, vol. 18, no. 11, p. e0293465, 2023.

[45] R. Kumar, B. Kumar, and A. Kumar, "Pharmacokinetic profile of antitubercular drugs and their clinical applications in therapeutic drug monitoring", *Asian J Pharm,* vol. 3, 2017no. 5, pp. 145-154. Available from: http://ajpp.in/uploaded/p71.pdf

[46] A. Kumar, B. Kumar, R. Dhiman, A. Kumar, M. Singh, P. Singh, V. Tiwari, and A. Trigunayat,

"Assessment of pain-ameliorative potential of ethanolic extract of Calotropis procera and its active components characterization by HPTLC and GC-MS analysis," *Sys. Rev. Pharm.,* vol. 13, no. 9, pp. 581–586, 2022.

[47] R. Dhiman, V. Durga, and K. H. Meetei, "Determination of S-amlodipine in Human Plasma Using Tizanidine as Internal Standard by Lc/Ms/Ms Method," *J. Bioanal. Biomed.*, vol. 7, no. 3, pp. 91–96, 2015.
[http://dx.doi.org/10.4172/1948-593x.1000130]

[48] H. Li, Y. Feng, Q. Luo, Z. Li, X. Li, H. Gan, Z. Gu, Q. Gong, and K. Luo, "Stimuli-activatable nanomedicine meets cancer theranostics", *Theranostics,* vol. 13, no. 15, pp. 5386-5417, 2023.
[http://dx.doi.org/10.7150/thno.87854] [PMID: 37908735]

[49] A. Kumar, M.K. Nandi, B. Kumar, A. Kumar, R. Kumar, A.K. Singh, and M. Singh, "Antiurolithiasis, Antioxidant, Anti-inflammatory, Analgesic, and Diuretic Activity of Ethanolic Extract of Seeds of Caesalpinia bonducella", *Int. J. Pharm. Investig.,* vol. 11, no. 3, pp. 306-311, 2021.
[http://dx.doi.org/10.5530/ijpi.2021.3.54]

[50] F. Avila Cobos, J. Alquicira-Hernandez, J.E. Powell, P. Mestdagh, and K. De Preter, "Benchmarking of cell type deconvolution pipelines for transcriptomics data", *Nat. Commun.,* vol. 11, no. 1, p. 5650, 2020.
[http://dx.doi.org/10.1038/s41467-020-19015-1] [PMID: 33159064]

[51] R. Dhiman, V. Durga, K. Jayasankar, and S. Raj, "Solid phase extraction: Potential carbocisteine bioanalysis using carbocisteine 13c3 stable isotope: LC-MS/MS technique with pharmacokinetics on healthy human subjects", *Pharm. Lett.,* vol. 7, no. 8, pp. 126-134, 2015.

[52] J. Li, B. Gong, X. Chen, T. Liu, C. Wu, F. Zhang, C. Li, X. Li, S. Rao, and X. Li, "DOSim: An R package for similarity between diseases based on Disease Ontology", *BMC Bioinformatics,* vol. 12, no. 1, p. 266, 2011.
[http://dx.doi.org/10.1186/1471-2105-12-266] [PMID: 21714896]

[53] R. Dhiman, V. Durga, and A. J. Rio, "Bioanalytical method for lornoxicam determination in human plasma by using piroxicam as internal standard by LC-MS/MS", *J Chem Pharm Res,* vol. 7, no. 8, pp. 54-60, 2015.

[54] D. del Val Martin, M. Sanmartín Fernández, and J.L. Zamorano Gómez, "Biomarkers in acute coronary syndrome", *IJC Metab. Endocr.,* vol. 8, pp. 20-23, 2015.
[http://dx.doi.org/10.1016/j.ijcme.2015.04.003]

[55] R. Dhiman, B. Kumar, D. Kumar, and A. Kumar, "Liquid-liquid extraction method for ziprasidone (ZRS) bioanalysis by using ZRS-D8 (stable isotope) as internal standard", *J Chem Pharm Res,* vol. 8, no. 3, pp. 722-727, 2016.

[56] L. Kalaydjieva, "Congenital cataracts-facial dysmorphism-neuropathy", *Orphanet J. Rare Dis.,* vol. 1, no. 1, p. 32, 2006.
[http://dx.doi.org/10.1186/1750-1172-1-32] [PMID: 16939648]

[57] R. Kumar, B. Kumar, A. Kumar, A. Kumar, and M. Singh, "HPTLC and GC-MS analyses of biologically active extracts and fractions from Premna latifolia Roxb leaves", *Open Access J Neurol Disord,* vol. 13, no. 8, 2022.

[58] Y. Li, and T.O. Tollefsbol, "DNA methylation detection: bisulfite genomic sequencing analysis", *Methods Mol. Biol.,* vol. 791, pp. 11-21, 2011.
[http://dx.doi.org/10.1007/978-1-61779-316-5_2] [PMID: 21913068]

[59] R. Kumar, B. Kumar, A. Kumar, A. Kumar, and M. Singh, "GC-MS analysis of Phytocomponents in the Methanol Extract of Premna latifolia Roxb", *Pharmacognosy Res.,* vol. 14, no. 1, pp. 19-23, 2021.
[http://dx.doi.org/10.5530/pres.14.1.4]

[60] D.A. Moser, S. Müller, E.M. Hummel, A.S. Limberg, L. Dieckmann, L. Frach, J. Pakusch, V. Flasbeck, M. Brüne, J. Beygo, L. Klein-Hitpass, and R. Kumsta, "Targeted bisulfite sequencing: A

novel tool for the assessment of DNA methylation with high sensitivity and increased coverage", *Psychoneuroendocrinology,* vol. 120, p. 104784, 2020. [http://dx.doi.org/10.1016/j.psyneuen.2020.104784] [PMID: 32673938]

[61] X. Wang, J. Park, K. Susztak, N.R. Zhang, and M. Li, "Bulk tissue cell type deconvolution with multi-subject single-cell expression reference", *Nat. Commun.,* vol. 10, no. 1, p. 380, 2019. [http://dx.doi.org/10.1038/s41467-018-08023-x] [PMID: 30670690]

[62] H. Pandey, S. Srivastava, A. Kumar, R. Kumar, and Y. Bhusan Tripathi, "Pharmacokinetics Study Of Puerarin Absorption In Blood After Consumption Of Pueraria Tuberosa Water Extract (PTWE) By Rats", *Explor. Anim. Med. Res.,* vol. 9, no. 2, pp. 188-196, 2019.

[63] D.J. Hunter, E. Losina, A. Guermazi, D. Burstein, M.N. Lassere, and V. Kraus, "A pathway and approach to biomarker validation and qualification for osteoarthritis clinical trials", *Curr. Drug Targets,* vol. 11, no. 5, pp. 536-545, 2010. [http://dx.doi.org/10.2174/138945010791011947] [PMID: 20199395]

[64] R. Kumar, "HPTLC and GC-MS analyses of biologically active extracts and fractions from Premna latifolia Roxb leaves reveal anti-inflammatory and analgesic effects," Sys. Rev. Pharm., vol. 12, no. 10, pp. 811-818, 2021.

[65] C. Andrews, Y. Xu, M. Kirberger, and J.J. Yang, "Structural Aspects and Prediction of Calmodulin-Binding Proteins", *Int. J. Mol. Sci.,* vol. 22, no. 1, p. 308, 2020. [http://dx.doi.org/10.3390/ijms22010308] [PMID: 33396740]

[66] R. Kumar, Y. Singh, M. Piplani, P. Bhateja, R. Garg, and B. Kumar, *Chem. Insid. Spices Herbs Res. Dev.,* vol. 4, pp. 79-116, 2024. The Chemistry and Pharmacological Study of *Copaifera* spp. Oils. [http://dx.doi.org/10.2174/9789815196832124040006]

[67] J. O'Brien, H. Hayder, Y. Zayed, and C. Peng, "Overview of MicroRNA biogenesis, mechanisms of actions, and circulation", *Front. Endocrinol. (Lausanne),* vol. 9, p. 402, 2018. [http://dx.doi.org/10.3389/fendo.2018.00402] [PMID: 30123182]

[68] A. Kumar, P. Singh, R. Kumar, and S. Dutt, "Novel drug delivery systems for inflammatory bowel disease," In: *Nanobiotechnology and Artificial Intelligence in Gastrointestinal Diseases*, V. K. Chaturvedi, A. K. Singh, J. Singh, and D. P. Yadav, Eds. Bristol, UK: IOP Publishing, 2024, pp. 4-1, 4-24. [http://dx.doi.org/10.1080/19336896.2017.1423186] [PMID: 29308725]

[70] R. Kumar, S. Dutt, A. Goyal, A. Kumar, and B. Kumar, "Nanotechnology in gastrointestinal endoscopy," In: *Nanobiotechnology and Artificial Intelligence in Gastrointestinal Diseases*, V. K. Chaturvedi, A. K. Singh, J. Singh, and D. P. Yadav, Eds. Bristol, UK: IOP Publishing, 2024, pp. 5-1, 5-32. [http://dx.doi.org/10.1088/978-0-7503-6134-7ch5]

[71] R. Rezar, P. Jirak, M. Gschwandtner, R. Derler, T.K. Felder, M. Haslinger, K. Kopp, C. Seelmaier, C. Granitz, U.C. Hoppe, and M. Lichtenauer, "Heart-type fatty acid-binding protein (H-FABP) and its role as a biomarker in heart failure: What do we know so far?", *J. Clin. Med.,* vol. 9, no. 1, p. 164, 2020. [http://dx.doi.org/10.3390/jcm9010164] [PMID: 31936148]

[72] A. Hamid, R. Kumar, V. K. Chaturvedi, S. Dutt, G. Sulabh, V. Kumar, and D. P. Yadav, "Artificial intelligence (AI) based colonoscopy," In: *Nanobiotechnology and Artificial Intelligence in Gastrointestinal Diseases*, V. K. Chaturvedi, A. K. Singh, J. Singh, and D. P. Yadav, Eds. Bristol, UK: IOP Publishing, 2024, pp. 12-1, 12-23. [http://dx.doi.org/10.1088/978-0-7503-6134-7ch12]

[73] S. Safari, A. Baratloo, M. Elfil, and A. Negida, "Evidence Based Emergency Medicine Part 2: Positive and negative predictive values of diagnostic tests", *Emergency (Tehran),* vol. 3, no. 3, pp. 87-88, 2015. [PMID: 26495390]

[74] R.J. Martín-Palma, J. Hernández-Montelongo, V. Torres-Costa, M. Manso-Silván, and Á. Muñoz-

Noval, "Nanostructured porous silicon-mediated drug delivery", *Expert Opin. Drug Deliv.,* vol. 11, no. 8, pp. 1273-1283, 2014.
[http://dx.doi.org/10.1517/17425247.2014.919254] [PMID: 24941438]

[75] C. Wattmo, K. Blennow, and O. Hansson, "Cerebro-spinal fluid biomarker levels: phosphorylated tau (T) and total tau (N) as markers for rate of progression in Alzheimer's disease", *BMC Neurol.,* vol. 20, no. 1, p. 10, 2020.
[http://dx.doi.org/10.1186/s12883-019-1591-0] [PMID: 31918679]

[76] L. Gaetani, K. Blennow, P. Calabresi, M. Di Filippo, L. Parnetti, and H. Zetterberg, "Neurofilament light chain as a biomarker in neurological disorders", *J. Neurol. Neurosurg. Psychiatry,* vol. 90, no. 8, pp. 870-881, 2019.
[http://dx.doi.org/10.1136/jnnp-2018-320106] [PMID: 30967444]

[77] C.R. Santhosh, N.B. Shridhar, K. Narayana, S.G. Ramachandra, and S. Dinesh, "Studies on the luteolytic, oestrogenic and follicle-stimulating hormone like activity of plant Rhaphidophora pertusa (Roxb.)", *J. Ethnopharmacol.,* vol. 107, no. 3, pp. 365-369, 2006.
[http://dx.doi.org/10.1016/j.jep.2006.03.036] [PMID: 16713160]

[78] P.P. Bruyant, "Analytic and iterative reconstruction algorithms in SPECT", *J. Nucl. Med.,* vol. 43, no. 10, pp. 1343–1358, 2002.
Available from: https://jnm.snmjournals.org/content/43/10/1343

[79] C. McArthur, R. Jampana, J. Patterson, and D. Hadley, "Applications of cerebral SPECT", *Clin. Radiol.,* vol. 66, no. 7, pp. 651-661, 2011.
[http://dx.doi.org/10.1016/j.crad.2010.12.015] [PMID: 21371697]

[80] P. Huang, and M. Zhang, "Magnetic resonance imaging studies of neurodegenerative disease: From methods to translational research", *Neurosci. Bull.,* vol. 39, no. 1, pp. 99-112, 2023.
[http://dx.doi.org/10.1007/s12264-022-00905-x] [PMID: 35771383]

[81] R. Anand, and S. Chatterjee, "Tracking disease progression by searching paths in a temporal network of biological processes", *PLoS One,* vol. 12, no. 4, p. e0176172, 2017.
[http://dx.doi.org/10.1371/journal.pone.0176172] [PMID: 28448511]

[82] Y. Kumar, A. Koul, R. Singla, and M.F. Ijaz, "Artificial intelligence in disease diagnosis: a systematic literature review, synthesizing framework and future research agenda", *J. Ambient Intell. Humaniz. Comput.,* vol. 14, no. 7, pp. 8459-8486, 2023.
[http://dx.doi.org/10.1007/s12652-021-03612-z] [PMID: 35039756]

[83] C. Strafella, V. Caputo, M.R. Galota, S. Zampatti, G. Marella, S. Mauriello, R. Cascella, and E. Giardina, "Application of precision medicine in neurodegenerative diseases", *Front. Neurol.,* vol. 9, p. 701, 2018.
[http://dx.doi.org/10.3389/fneur.2018.00701] [PMID: 30190701]

CHAPTER 7

Treatment Strategies and Therapeutic Innovations

Saurav Kumar[1], **Jyoti**[2], **Parijat Pandey**[1] and **Dhirender Kaushik**[1,*]

[1] *Department of Pharmaceutical Sciences, Gurugram University, Gurugram, Haryana, India*

[2] *School of Pharmaceutical Sciences, MVN University, Palwal, Haryana, India*

Abstract: Incomplete metabolism, a condition characterized by the inability of cells to metabolize nutrients fully, has been linked to various health issues. This chapter provides an overview of the current treatment strategies and therapeutic innovations aimed at addressing the health impacts of incomplete metabolism. We discuss the traditional medical diet and cofactor treatments for inborn errors of metabolism, as well as enzyme replacement therapy and small molecule therapies as alternative approaches. Additionally, we explore the role of antioxidant defense, glutathione regeneration, and carnitine-mediated acyl group buffering in mitigating mitochondrial reductive stress and the potential benefits of intermittent fasting, increasing energy demand through strategies such as cold exposure and hypobaric hypoxia, and the use of SGLT2 inhibitors in promoting beiging of adipocytes. Finally, we highlight the future directions in gene therapy and cell reprogramming for inborn errors of metabolism, as well as the potential for repurposing existing drugs and the development of novel treatments through high-content screenings. This comprehensive review aims to provide a detailed understanding of the current treatment landscape and therapeutic innovations for incomplete metabolism, with the goal of improving patient outcomes and enhancing our understanding of this complex condition.

Keywords: Incomplete metabolism, Metabolism, Therapeutic innovations, Treatments.

INTRODUCTION

Incomplete metabolism refers to a situation where the body's metabolic processes are impaired, leading to an inability to fully break down, convert, or utilize nutrients and other biochemical substances. These include all the chemical reactions that comprise the living condition of cells and organisms, and thus the pathways involved in it are those such as glycolysis, citric acid cycle, and oxidative phosphorylation. These metabolic pathways are responsible for

* **Corresponding author Dhirender Kaushik:** Department of Pharmaceutical Sciences, Gurugram University, Gurugram, Haryana, India; E-mail: dkaushik.apti@gmail.com

Rajesh Kumar (Ed.)

converting food into energy, making up the building blocks that include proteins, lipids, and nucleic acids and ridding the body of waste products [1].

In normal metabolic conditions, the substrates are driven by the enzymes through these pathways to provide for the controlled processing of the metabolic by-products and their excretion. However, in conditions of incomplete metabolism, one or more steps of these pathways could get disrupted due to enzyme insufficiency or genetic disruption, among other factors that do not allow the complete processing of metabolic intermediates [2]. Consequently, few toxic substances may accumulate due to this process, or essential products may not get synthesized, thus leading to various metabolic disorders. Examples of incomplete metabolism include:

Phenylketonuria (PKU): The failure of the body to change phenylalanine into tyrosine due to the absence of an enzyme called phenylalanine hydroxylase; toxic accumulation of phenylalanine causes damage to the brain [3].

Maple Syrup Urine Disease (MSUD): This is a defect in the branched-chain alpha-keto acid dehydrogenase complex. It results in the accumulation of the branched-chain amino acids and their toxic by-products [4].

Lysosomal Storage Disorders: These include disorders like Gaucher's disease, where there is a deficiency of the enzyme glucocerebrosidase, which in turn causes intracellular accumulation of glucocerebroside [5].

IMPORTANCE OF ADDRESSING INCOMPLETE METABOLISM

Prevention of Complications

If incomplete metabolism is left untreated, it can become the only cause of serious complications like developmental delay, neurological damage, organ damage, and even death. For example, PKU, if allowed to remain untreated, could result in serious intellectual disability [6].

Improving Quality of Life

Management of metabolic disorders effectively can improve the quality of life among patients. For instance, dietary management in PKU allows patients to lead normal lives by preventing toxic effects caused by the buildup of phenylalanine [7].

Reducing Health Care Costs

Early diagnosis and early intervention reduce the management costs of the long-term complications associated with metabolic disorders. For example, newborn screening for metabolic disorders prevents hospitalization and other costly treatments resulting from the advanced stages of the disease [6].

Development of Research and Therapeutics

Knowledge of mechanisms underlying incomplete metabolism may provide the impetus needed in the development of new therapies. Gene therapies, emerging enzyme replacement therapies, and other novel approaches give promise for the cure or effective management of diseases caused by these disorders [8].

Public Health Impact

Metabolic disorders, although individually rare, collectively contribute significantly to public health burdens. Addressing these conditions through effective treatment strategies can reduce the overall burden on healthcare systems.

OVERVIEW OF TREATMENT STRATEGIES AND INNOVATIONS

The treatment strategies for incomplete metabolism are diverse and have evolved tremendously over the years, right from simple dietary intervention to sophisticated gene and stem cell therapies. A few of them are as follows:

Dietary Interventions: Dietary control is the fundamental approach for treating numerous metabolic illnesses, and it is strongly recommended in this case as well. The following are measures that follow the principle of restricting the intake of substances that an enzyme deficiency is unable to metabolize properly. For example:

Low-Phenylalanine Diet: In PKU, this diet is employed to prevent an accumulation of phenylalanine.

Low-Protein Diets: In urea cycle disorders, low-protein diets help promote the limitation of ammonia production.

Pharmacological Treatments: A prime significance is attached to drugs for symptom-free and disease-preventing strategies. These include:

<u>Substrate Reduction Therapy:</u> Drugs that decrease the formation of toxic substrates.

Cofactor Supplementation: Provision of required cofactors to potentiate residual enzyme activity (*e.g.*, tetrahydrobiopterin in the setting of PKU).

ERT: Replacement of deficient enzyme is performed by injecting the needed enzyme intravenously into the body. The injected enzyme would help degrade the accumulated substrate. It is applied to diseases such as Gaucher, Fabry, and Pompe disease. In some conditions, ERT has been lifesaving to those patients, and it does come with lifelong treatments and monitor sessions.

Gene Therapy: Gene therapy corrects the basic genetic cause of metabolic disorders. With the help of recent techniques using gene editing or replacement by CRISPR-Cas9, there is even a possibility of a single-dose cure. Various preclinical trials are in action, and several clinical trials have already given promising results in conditions like Leber's congenital amaurosis.

Stem Cell Therapy: This is a line of treatment currently under study, devised to replace or repair cells that are damaged or deficient in metabolic disorders. This line of therapy is promising in disorders for which cells producing the enzymes are nonfunctional and thereby has the possibility of offering the resumption of physiological and metabolic functions.

Pharmacological Chaperones: These small molecules protect the misfolded protein and make it functional. This is specifically useful in diseases where misfolded enzymes can be partially restored to function in the presence of chaperones, as in Fabry disease.

Personalized Medicine: With advances in molecular profiling, the treatment can be personalized based on individual needs. For instance, in personalized mitochondrial disorders, the regimen involves specific mutations for better outcomes.

Combination Therapies: ERT can be combined with various modules of treatment to increase the effectiveness of the therapy. For instance, co-treatment of ERT with substrate reduction therapy in lysosomal storage disorders is more effective in improving the outcome of various diseases. Clinical research is being conducted in search of the best combinations in order to get maximal therapeutic benefits.

Preventive Measures and Early Intervention: Early detection in the management of metabolic conditions is the key before issues become damaging in the long term. Public health efforts directed toward early detection along with lifestyle changes are very important preventive issues.

Innovations and Future Directions: The management of metabolic disorders is a continuously evolving specialty. Newer technologies, including RNA-based therapies, CRISPR gene editing, and synthetic biology, are likely to bring newer, more effective, and accessible treatments in the future. Interdisciplinary collaboration involving researchers in the fields of genetics, biochemistry, and clinicians is key to translating these innovations into clinical practice (Table **1**).

CURRENT TREATMENT STRATEGIES

Role of Diet in Managing Metabolic Disorders

Dietary interventions are at the heart of controlling most of the metabolic disorders, especially the ones in which the specific nutrients or metabolites function abnormally. In such conditions, the symptoms can be managed, and possible complications, including the accumulation of toxic substances, can be prevented by carefully controlling the intake of specific foods or nutrients that one excessively consumes or, *vice versa*, supplies in supplementary amounts to compensate for deficiencies in others. For example, in PKU, which is the inability to metabolize phenylalanine, a diet low in phenylalanine is crucial to protect against the toxic accumulation of this amino acid, which can cause severe neurological damage. In MSUD, dietary limitation of the branched-chain amino acids leucine, isoleucine, and valine is required to prevent harmful metabolites from accumulating [29].

Dietary management should avoid excessive food restrictions but rather supplementation of food on the other hand. For instance, arginine or citrulline may be supplemented for patients with urea cycle disorders in order to facilitate the excretion of nitrogen and prevent the occurrence of hyperammonemia [30].

Specific Dietary Strategies

Ketogenic Diets

The ketogenic diet is recognized as a therapeutic diet in the management of many diseases involving incomplete metabolism and is, therefore, a fat-enriched, less carbohydrate, moderate protein diet. This diet helps to change the primary energy substrate of the body from glucose to fat and raises the levels of fats, *i.e.*, ketone bodies, in the blood by the process of ketogenesis. A ketogenic diet helps to meet the energy requirement of the body, especially of the brain, which is able to utilize ketones when glucose cannot be metabolized. This is advantageous, especially in patients suffering from certain inborn errors of metabolism in which carbohydrates or glucose metabolism is not normal or completely defective. For instance, in glucose transporter 1 deficiency syndrome (GLUT1-DS), the blood-brain barrier (BBB) is not able to transport glucose, and the ketogenic diet provides a fuel source for the brain's ketone energy, alleviating epileptic symptoms. Also, low-carbohydrate high fat diets have been effective in treating some mitochondrial diseases caused by deficiency of nuclear genes related to mitochondrial function affecting cellular bioenergy. A ketogenic diet, through its provision of ketones, may be useful in pathological conditions where the body's mechanisms of energy production are defective since it may omit harmful metabolic pathways that depend on glucose usage and excess toxic waste accumulation [31].

Table 1. Diseases caused by incomplete metabolism (often due to inborn errors of metabolism) along with their corresponding treatments.

Category	Metabolic Disorder	Cause	Treatment	References
Amino Acid Metabolism Disorders	Phenylketonuria (PKU)	Deficiency in phenylalanine hydroxylase	Avoidance of high-protein foods, use of Sapropterin dihydrochloride	[6]
	Maple Syrup Urine Disease (MSUD)	Lack in the branched-chain alpha-keto acid dehydrogenase complex	Low-protein, low-BCAA diet	[9]
	Homocystinuria	Deficiency in cystathionine beta-synthase	Low-methionine diet, Vitamin B6, B12, and folate supplementation	[10]
	Tyrosinemia Type I	Deficiency in fumarylacetoacetate hydrolase	Low-tyrosine and low-phenylalanine diet. Liver transplantation in severe cases	[11]
	Alkaptonuria	Deficiency in homogentisate 1,2-dioxygenase	Low-protein diet	[12]

(Table 1) cont.....

Category	Metabolic Disorder	Cause	Treatment	References
Carbohydrate Metabolism Disorders	Galactosemia	Deficiency in galactose- 1-phosphate uridylyltransferase	Strict avoidance of lactose and galactose, Use of soy-based or lactose-free formulas for infants	[13]
	Hereditary Fructose Intolerance (HFI)	Deficiency in aldolase B	Avoidance of fructose, sucrose, and sorbitol in the diet	[14]
	Glycogen Storage Diseases (GSDs) Von Gierke's Disease (GSD I)	Lack of glucose-- -phosphatase,	Frequent small meals high in complex carbohydrates	[15]
	Glycogen Storage Diseases (GSDs) Pompe Disease (GSD II)	Deficiency in acid alpha-glucosidase	Enzyme replacement therapy with alglucosidase alfa	[16]
	Glycogen Storage Diseases (GSDs) McArdle Disease (GSD V)	Deficiency in muscle glycogen phosphorylase	Regular moderate exercise, High-carbohydrate diet	[17]
Lipid Metabolism Disorders	Familial Hypercholesterolemia	Mutations in the LDL receptor gene	Statins to lower cholesterol levels.	[18]
	Medium-Chain Acyl-CoA Dehydrogenase Deficiency (MCADD)	Deficiency in MCADD	Avoidance of fasting. High-carbohydrate diet	[19]
	Carnitine Palmitoyltransferase Deficiency (CPT I & II)	Defects in enzymes that transport long-chain fatty acids into the mitochondria	High-carbohydrate, low-fat diet along with medium-chain triglyceride (MCT) supplementation	[20]
Organic Acid Metabolism Disorders	Methylmalonic Acidemia (MMA)	Deficiency in methyl malonyl-CoA mutase	Low-protein diet with restrictions on isoleucine, valine, methionine, and threonine	[21]
	Propionic Acidemia	Deficiency in propionyl-CoA carboxylase	Low-protein diet with restriction on isoleucine, valine, methionine, and threonine	[21]

(Table 1) cont.....

Category	Metabolic Disorder	Cause	Treatment	References
Urea Cycle Disorders	Ornithine Transcarbamylase Deficiency (OTC)	X-linked lack in ornithine transcarbamylase	Low-protein diet to reduce ammonia production	[22]
	Arginase Deficiency	Deficiency in arginase,	Low-protein diet, Medications to lower ammonia levels	[23]
Lysosomal Storage Disorders	Gaucher Disease	Deficiency in glucocerebrosidase	ERT with imiglucerase, velaglucerase alfa, or taliglucerase alfa	[24]
	Tay-Sachs Disease	Deficiency in hexosaminidase A	No cure; supportive care and symptomatic management	[25]
Mitochondrial Disorders	Leigh Syndrome	Mutations affecting mitochondrial energy production	Supportive care, Coenzyme Q10, creatine, and other mitochondrial support supplements	[26]
	Mitochondrial Myopathy	Mutations in mitochondrial DNA affecting muscle function	Supportive therapy	[27]
Purine and Pyrimidine Metabolism Disorders	Lesch-Nyhan Syndrome	Deficiency in hypoxanthine-guanine phosphoribosyltransferase (HGPRT)	Allopurinol to reduce uric acid levels	[28]
	Adenosine Deaminase Deficiency	Deficiency in adenosine deaminase	Bone marrow transplantation	[28]

Moreover, the ketogenic diet has been found to have neuroprotective and anti-inflammatory effects, which might be beneficial in managing oxidative stress and cellular damage associated with various metabolic disorders. It helps stabilize blood sugar levels and reduces insulin resistance, which can be particularly advantageous in metabolic conditions like fatty acid oxidation disorders. However, the ketogenic diet requires careful medical supervision and individualized planning to ensure nutritional adequacy and prevent potential side effects such as nutrient deficiencies, dehydration, and metabolic imbalances [32]. It requires that a patient on a ketogenic diet have regular testing of their ketone levels having a tight balance of macronutrients to keep them in ketosis but not into acidosis. With these considered challenges, the ketogenic diet, nonetheless, is a very useful treatment for select metabolic disorders characterized by incomplete metabolism. The dietary intervention may be targeted to improve the clinical outcome and quality of life greatly.

Low in Protein Diet

Low-protein diet is one of the closely monitored dietary interventions in the management of various metabolic disorders characterized by incomplete metabolism, where the main ones involve amino acid metabolism. In this regimen, dietary restrictions are limited to those proteins and amino acids that, due to enzyme deficiencies or other metabolic defects, cannot be adequately metabolized by the body. Conditions like PKU, MSUD, and homocystinuria, which involve the toxic accumulation of certain amino acids, phenylalanine, leucine, valine, isoleucine, or methionine, require a low-protein diet to minimize toxic accumulation of the amino acids and thereby reduce symptoms. In PKU, for example, the deficiency of the enzyme phenylalanine hydroxylase mediates the inability to convert phenylalanine into tyrosine, leading to toxic levels of phenylalanine in the blood that, if not treated, lead to severe intellectual disability and other neurological complications. A diet low in protein, together with special medical foods low in phenylalanine, effectively keeps the levels of phenylalanine low to avoid damage to the development and functions of the brain [33].

In addition, low-protein intake is very important in treating urea cycle disorders of the body when it cannot clear ammonia from the bloodstream for one reason or another. A low-protein diet reduces the quantity of ammonia resulting from protein metabolism, hence reducing the threat of hyperammonemia. Hyperammonemia is a condition that is toxic due to an excessive level of ammonia in the blood, leading to neurological damage and metabolic crises. The diet is often supplemented with all essential amino acids, with the exception of the affected metabolic pathways, for proper nutrition of the patient. In other disorders, such as MSUD-a disorder arising due to an accumulation of branched, chain amino acids resulting from enzyme deficiencies in their catabolism, the diet needs to be specifically tailored to restrict these amino acids while still providing necessary growth requirements. The low-protein diet thus needs close planning and monitoring to ensure that patients receive the proper balance of essential nutrients & amino acids to neglect deficiencies and for general health. Undeniably, despite the many challenges, the low-protein diet remains one of the cornerstone therapies for many inborn errors of metabolism (Fig. **1**). It provides a selective strategy toward the biochemical imbalances of these diseases, significantly improving the outcome and quality of life of patients with such diseases [34].

Low Carbohydrate Diets

The low carbohydrate diet is a nutritional intervention that has widely gained use in the management of selected metabolic disorders characterized by incomplete carbohydrate digestion. This dietary approach tends to severely limit dietary consumption of carbohydrates, typically metabolized into glucose, the principal source of energy for most cells. Low-carbohydrate diets in individuals with disorders where the metabolism of carbohydrates is impaired prevent the accumulation of toxic intermediates and reduce metabolic stress in diseases like glycogen storage diseases, HFI, and PDCD. For example, in hereditary fructose intolerance, the lack of aldolase B prevents proper breakdown and usage of fructose upon ingestion and thus creates toxic accumulations of fructose- 1-phosphate. Patients can avoid metabolic crises and liver damage through dieting to restrict or eliminate fructose, sucrose, and sorbitol from intake completely.

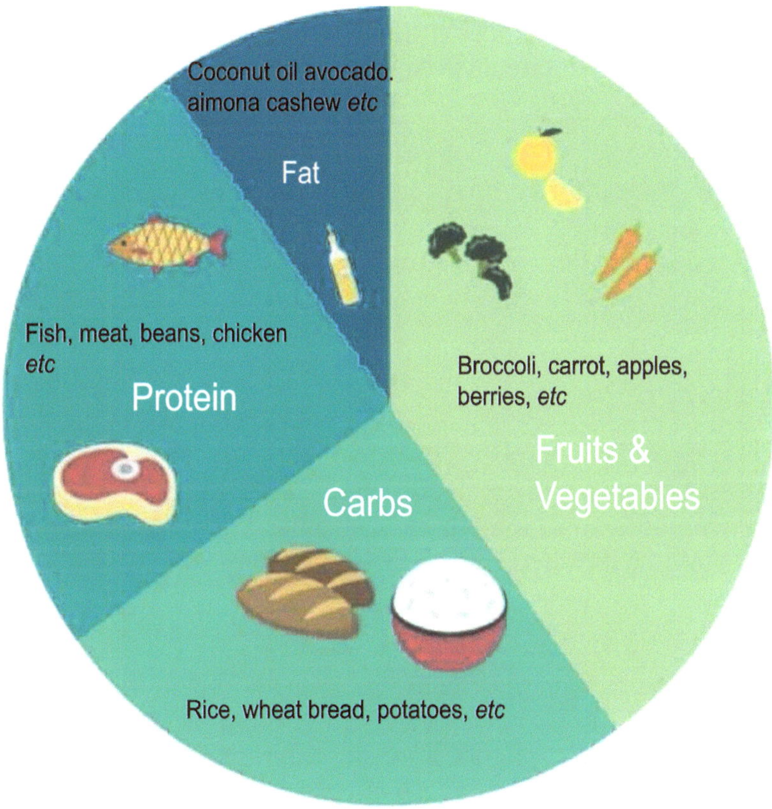

Fig. (1). Balanced diet chart.

On the other hand, diseases of glycogen storage, as in type I (also termed von Gierke disease), apt to glucose-6-phosphatase deficiency that impairs gluconeogenesis and glycogenolysis, low carbohydrate intake, avoids hypoglycemia through stimulation of other energy sources including frequent feeding with cornstarch or complex carbohydrates that provide a more sustained release of glucose. Deficiency of pyruvate dehydrogenase complex renders pyruvate nonconvertible to acetyl-CoA, leading to cellular energy deficiency. A low-carbohydrate diet, usually in combination with a ketogenic diet, shifts metabolism away from glycolysis and dependence upon glucose. This dietary intervention encourages the use of fats and ketones as energy sources, bypassing the metabolic blockade due to enzyme deficiency. A diet low in carbohydrates diminishes reliance on impaired metabolic pathways through judicious use of carbohydrate intake and diminishes the production of lactic acid and other potentially toxic metabolites, thereby improving overall metabolic balance. Although a low-carbohydrate diet requires very cautious planning and continuous follow-up to avoid nutritional deficiencies and ensure adequate caloric intake, its use is of paramount necessity in the therapeutic management of various metabolic disorders, as it plays a key role in stabilizing metabolic function, preventing acute complications, and improving long-term health outcomes [35].

Low-Fat Diets

The diet for individuals affected by disorders of fat metabolism, such as MCADD, is a low-fat diet that restricts the absorption of long-chain fatty acids. The dietary pattern in disorders of fat metabolism minimizes the accumulation of toxic intermediates and supplies a readily available source of energy by emphasizing MCTs [36].

Pharmacological Treatments

Most metabolic disorders are treated pharmacologically by the alleviation of their symptoms or the correction of certain biochemical imbalances. Examples of such drugs are enzyme cofactors, inhibitors, and other agents that may be useful in modifying metabolic pathways.

Enzyme Cofactor

A complete course of treatment of incomplete metabolism utilizes cofactors that enhance the activity of enzymes, catalyzing the reaction of substances in metabolic pathways. The cofactors may be vitamins, minerals, or other small molecules necessary for the proper functioning of the enzymes. This section presents the function of enzyme cofactors in metabolic processes and their therapeutic implications. In PKU, sapropterin dihydrochloride (Kuvan) is a coenzyme for phenylalanine hydroxylase. It helps lower blood phenylalanine levels in responsive patients [37].

Mechanism: The main mechanism of enzyme cofactor treatment is based on supplementation with certain cofactors that enhance the residual activity of partially functional enzymes or compensate for deficiencies in some cofactors.

With the increase in availability, the catalytic efficiency of enzymes improves by reducing toxic metabolite accumulations and restoring normal metabolic pathways. This method especially works when the enzyme is not completely inactivated but only has reduced activity due to a genetic alteration that reduces its affinity for the cofactor [37].

Substrate Reduction Therapy (SRT)

Substrate Reduction Therapy is a therapeutic approach to metabolic disorders aimed at the reduction in production or accumulation of toxic substrates. Many inborn errors of metabolism involve genetic mutations leading to enzyme deficiencies or dysfunctions, which impede the regular breakdown of a particular substrate, thus resulting in a toxic accumulation of the same. The SRT tries to decrease the formation of these toxic substrates by reducing production or increasing other pathways to eliminate them. It thereby reduces symptoms and prevents complications in those cases of disorders that affect complete metabolism. Miglustat is also used in substrate reduction for Niemann-Pick Disease Type C & Gaucher's Disease. They unload the burden from the impaired metabolic pathways by limiting the synthesis of problematic substrates [38].

Enzyme Inhibitors

The inhibitors to enzymes are a class of drugs that decrease the activity of some enzymes associated with metabolic pathways. They have critical importance in controlling incomplete metabolisms, especially in metabolic disorders where the overactivity of a certain enzyme leads to toxic metabolites accumulating or essential substrates being exhausted. This is the way that enzyme inhibitors, through the selective inhibition of these enzymes, help in the restoration of metabolic balance and avoid the accumulation of harmful compounds, as well as the relief of symptoms in various metabolic disorders. The use of allopurinol for the management of hyperuricemia represents its ability to inhibit xanthine oxidase, an enzyme embroiled in the metabolism of purine. Inhibitors block the activity of specific enzymes, preventing the formation of toxic metabolites [39].

Ammonia Scavengers

Ammonia scavengers include the therapeutic agents that reduce the level of ammonia elevation in the body, particularly in conditions where the metabolism of ammonia is defective. In relation to that fact, an excess amount of ammonia accumulation in the blood, commonly known as hyperammonemia, can lead to neurotoxicity and other fatal complications, especially in patients suffering from UCDs or advanced liver diseases. Ammonia-scavenging agents promote the excretion of ammonia or its conversion into less injurious-to-health substances that can safely leave the body. Drugs like sodium phenylbutyrate and sodium benzoate are two types of medications used in urea cycle disorders to facilitate ammonia removal from the body. These drugs convert ammonia into compounds that can be more easily excreted, thus averting hyperammonemia [40].

ENZYME REPLACEMENT THERAPY (ERT)

The activity of ERT has been developed to replace enzyme deficiencies due to genetic mutations in a series of metabolic disorders. Such disorders are very often due to toxic substrate accumulation because of the shortage of enzymes responsible for metabolizing a substance of a particular type [41]. Thus, ERT offers the deficient enzyme to restore normal metabolic function and minimize the toxic effects of accumulated metabolites. This section provides a comprehensive overview of the mode of action, clinical applications, advantages, pitfalls, and future perspectives in ERT [42].

Mechanism of Action

ERT involves the intravenous administration of recombinant or purified enzymes to patients who lack or have deficient activity of specific enzymes. The replacement enzymes are designed to mimic the activity of the endogenous enzymes that are defective due to genetic mutations [43].

Delivery and Targeting: The exogenous enzyme is delivered into the bloodstream, where it circulates and is taken up by target cells. Many lysosomal storage disorders (LSDs) are treated using ERT, as the infused enzymes are typically directed to lysosomes, the cellular compartments where these enzymes normally function. Enzymes are often tagged with mannose-6-phosphate (M6P), a signal that directs them to lysosomes *via* cellular receptors [44].

Function Restoration: Once inside the lysosome, the replacement enzyme degrades accumulated substrates, thus alleviating the cellular dysfunction and associated symptoms. For example, in Gaucher disease, the infused enzyme glucocerebrosidase breaks down glucocerebroside, a substrate that accumulates due to the deficient enzyme activity [44].

Clinical Applications

ERT is used to treat various lysosomal storage disorders and other conditions characterized by enzyme deficiencies. Key examples include:

Gaucher Disease

It is caused by a lack of glucocerebrosidase, and glucocerebroside accumulates in macrophages. ERT with recombinant glucocerebrosidase (*e.g.*, imiglucerase, velaglucerase alfa) reduces spleen and liver enlargement and improves hematologic and skeletal symptoms [45].

Fabry Disease

It is caused by a lack of alpha-galactosidase A, resulting in globotriaosylceramide accumulation. ERT with recombinant alpha-galactosidase A-agalsidase alfa, for example-reduces Gb3 levels and relieves symptoms like pain and renal dysfunction [46].

Pompe Disease

Caused by lack of acid alpha-glucosidase accumulation of glycogen in muscle tissues. ERT using recombinant alpha-galactosidase A, such as agalsidase alfa, leads to improved muscle strength and function and decreased levels of glycogen accumulation [47].

Mucopolysaccharidosis (MPS) Disorders

Diseases include but are not limited to MPS I, II, and VI, caused by a deficiency in various lysosomal enzymes that catalyze the breakdown of glycosaminoglycans. ERT substituting with recombinant laronidase in MPS I, idursulfase in MPS II, and galsulfase substitution in MPS VI have had the prime role in symptomatic relief and the reduction of GAG accumulation [48].

Benefits of Enzyme Replacement Therapy

Symptom Management

ERT may improve or stabilize clinical symptoms related to toxic metabolite accumulation, which includes but is not limited to organ enlargement, pain, and neurologic deficits. In this respect, one would expect ERT to enhance the quality of life and general individual health with these disorders through the restoration of enzyme function [49].

Disease Progression

ERT has the potential to slow or completely halt disease progression in selected LSDs, preventing further deterioration in the status of affected organs and tissues. The earlier the therapy is instituted, the better the optimization of outcomes and prevention of irreversible damage can be achieved [50].

The Patient Outcomes

ERT has proven to improve many characteristics of patients' health, including hematologic, skeletal, and visceral symptoms, leading to improved functional capacity and reduced disability in many conditions [51].

Limitations and Challenges

Despite its benefits, ERT has limitations and challenges:

Limited Efficacy in Neurological Symptoms: The main problem of crossing through the BBB makes most of the ERTs have limited efficacy in neurological symptoms. Certain diseases like Gaucher disease and Fabry disease have neurological manifestations not taken care of by the current ERTs effectively [52].

Immune Reactions: A variety of immune reactions includes the most common adverse events, for example, those resulting from immune responses against the infused enzyme, which induce antibodies that can neutralize the activity of the enzyme or lead to allergic reactions. This may require pre-medication supplementation with antihistamines and corticosteroids to manage these reactions [53].

Frequent Infusions: Treatment with ERT is usually invasive, as it requires periodic intravenous infusions. These procedures could be potentially very burdensome for patients. The actual frequency of such procedures would vary based on the disorder and enzyme used [54].

High Cost: Generally, ERT is expensive; hence, the cost becomes an inhibiting factor in accessing this treatment by some patients. The high cost of therapy is attributed to the complex production and purification process of recombinant enzymes [54].

Combination Therapies

In other words, combination therapies make use of two or more therapeutic approaches for the treatment of a certain disease or condition. In metabolic disorders, combination therapy can improve the efficacy of treatment and reduce the risk of drug resistance by bringing about multifaceted aspects that are targeted in disease pathology [55]. The rationale for using combination therapies includes:

Synergistic Effects

The use of multiple treatments can cause a synergistic effect, whereby the outcome of combined treatment is higher compared to that achieved by separate treatments. This will be notably important in complex metabolic disorders in which a single form of therapy cannot address all conditions associated with a disease. For instance, in Cancer, this combination of chemotherapy and targeted therapies has the potential to increase tumor cell death while reducing the possibility of resistance to drugs [56].

Reducing Drug Resistance

One of the strategies to avoid the development of resistance in chronic diseases or infections is combination therapies. Targeting through different pathways or mechanisms, combination therapies reduce the likelihood that a disease will evolve resistance to a single agent. Example: For instance, in HIV, combination antiretroviral therapy targeting drugs against different stages of the life cycle of the virus helps in preventing the emergence of drug-resistant strains [57].

Addressing Multiple Disease Mechanisms

Many metabolic disorders involve multiple biochemical pathways or cellular processes. Combination therapies can simultaneously target various mechanisms, leading to a more comprehensive approach to treatment. Example: In Type 2 Diabetes, combining medications that enhance insulin sensitivity with those that promote insulin secretion can provide better glycemic control than using a single agent alone [58].

Enhancing Patient Adherence

Combination therapies can sometimes simplify treatment regimens by reducing the number of medications or the frequency of administration. This can improve patient adherence and overall treatment outcomes (Table **2**). Example: In Hypertension, fixed-dose combination pills that include an antihypertensive drug along with a diuretic can simplify the treatment regimen and improve adherence [59].

INNOVATIVE THERAPEUTIC APPROACHES

Gene Therapy

Gene therapy is a very new approach to the treatment of incomplete metabolisms, particularly for IEMs, due to genetic mutations that result in either the absence or dysfunction of particular enzymes [61]. Unlike conventional therapies that aim to manage symptoms or replace missing enzymes, gene therapy targets the root cause of metabolic disorders by correcting or compensating for the defective gene. This approach offers the potential for a long-lasting, possibly curative solution. Gene therapy involves various methods to deliver therapeutic genes into patients' cells to restore normal metabolic function [62].

Table 2. Gene therapy comparison with enzyme replacement treatment and hematopoietic stem cell transplantation [60].

Features	Gene Therapy	Enzyme Replacement Therapy (ERT)	Hematopoietic Stem Cell Transplantation (HSCT)
The essence of the method	Vector-based delivery of the proper gene to cells	Administration of the appropriate enzyme into the organism	Donor cells transplanted with the appropriate gene
Number of therapeutic interventions	Single	Multiple	Single
The main advantages of the method	Consistent gene expression, a single treatment that has a long-lasting impact, and enhancements to the CNS and other organs	Intravenous administration is a relatively safe mode of delivery and the only FDA-approved approach to date.	The capacity to use monocytes to transport the enzyme to the brain in a single application
The main disadvantages of The method	The possibility of mutagenesis in the case of viral vectors, as well as the danger of mounting an immune reaction against the vector and some transgene components	The risks of building neutralising antibodies against the enzyme, the necessity of multiple enzyme administrations, and the enzyme's inability to pass the blood-brain barrier	Immunosuppression, graft versus host disease (GvHD) risk, conditioning prior to HSCT, and restricted gene expression in the central nervous system

Methods of Gene Therapy for Incomplete Metabolism

There are several methods used in gene therapy to treat metabolic disorders, each with distinct mechanisms and applications:

Viral Vector-Mediated Gene Therapy

The high efficiency of transduction and the possibility of long-lasting expression of the introduced genes make viral vectors the most common vehicles for therapeutic gene delivery into cells.

AAV(Adeno Associated Virus)

Vectors are the most common viral vectors for gene therapy due to the fact that they can infect both dividing and non-dividing cells and have a low risk for integration into the host genome, reducing insertional mutagenesis [63]. AAVs have entered clinical trials for the treatment of genetic metabolic disorders such as SMA and Hemophilia. Specifically, AAV-mediated gene therapy has shown potential in providing active copies of genes to treat conditions such as GSDIa and MMA [64].

Lentiviral Vectors

Because they may incorporate their genetic material into the host genome, lentiviruses—a kind of retrovirus—are useful in treating conditions that require long-term gene expression. For *ex vivo* gene therapy, in which patient cells are altered outside of the body and subsequently reinfused, lentiviral vectors are frequently utilised. As demonstrated by the success of this strategy in the treatment of haematopoietic stem cell gene therapy for disorders like X-linked adrenoleukodystrophy (X-ALD) and metachromatic leukodystrophy (MLD) [65].

Adenovirus Vectors

Although adenoviruses allow the expression of large genes and achieve high levels of transgene expression, their potential to provoke strong immune responses makes them less used. However, adenovirus vectors have been used in preclinical models for disorders requiring high gene expression levels over a short period, such as urea cycle disorders [66, 67].

Non-Viral Gene Delivery Methods

Non-viral gene delivery methods offer an alternative to viral vectors, providing benefits such as reduced immunogenicity and the ability to deliver larger genetic payloads.

Plasmid DNA Delivery

Plasmid DNA is a circular DNA molecule used to deliver therapeutic genes into the cells. Methods such as electroporation or the use of liposome-lipid-based carriers facilitate the uptake of plasmid DNA into cells. However, this is less efficient compared to viral vectors and much safer because it does not integrate into the host genome, reducing the risk of insertional mutagenesis. This strategy is under investigation for diseases such as phenylketonuria and lysosomal storage diseases [68].

Gene Editing With CRISPR-Cas9

The CRISPR-Cas9 gene-modifying technology enables precise modification of the genome. The system works by correcting specific mutations in defective genes directly in metabolic disorders [69]. The potential of this technique for permanently curing disorders by correcting the genetic defect at its origin is thus bright. Examples include the correction of mutations in the PAH gene for PKU using CRISPR-Cas9 and the *OTC* gene for ornithine transcarbamylase deficiency using CRISPR-Cas9 [70, 71].

RNA-Based Therapies (siRNA and mRNA)

Under development are also RNA-based therapies, including small interfering RNA (siRNA) and messenger RNA (mRNA), which can be used to modulate gene expression or replace defective proteins. mRNA therapy entails the administration of synthetic mRNA, encoding the missing or dysfunctional protein, thus enabling cells to produce the protein transiently. This strategy has been investigated for diseases such as methylmalonic acidemia (MMA) and propionic acidemia (PA), where mRNA treatment aids in the replacement of the missing enzymes (Fig. **2**) [72, 73].

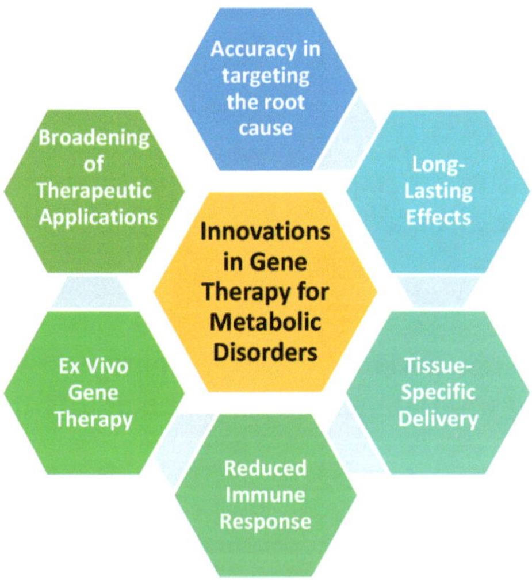

Fig. (2). Innovations in gene therapy for metabolic disorders.

Approaches For Gene Therapy in Incomplete Metabolism

Gene therapy is aimed at correcting or compensating for the defect caused by a metabolic disorder through the introduction of a functional gene copy-editing of the genome to correct mutations or modulation of gene expression. There is not just one method for gene therapy; rather, there are several approaches, each with its singular mechanisms, applications, and challenges. Approaches have included gene augmentation, gene editing, RNA-based therapies, and *ex vivo* gene therapy.

Gene Augmentation Therapy

Gene augmentation therapy involves the introduction of a functional copy of a defective or missing gene into a patient's cells. This is quite an effective approach, especially for disorders caused by loss-of-function mutations where the expression of the defective gene results in no production of a non-functional protein [74].

Mechanism: A functional gene is delivered into the patient's cells through viral or non-viral vectors. This introduced gene will then produce the missing or defective protein by itself and hence restore the normal function of the protein. This approach does not remove or modify the existing defective gene but rather complements it with a functional one [75].

These vectors are widely employed owing to their high efficiency in gene delivery and their stable expression; AAV, lentivirus, and adenovirus. AAV vectors are preferred due to low immunogenicity and infection capabilities regarding dividing and non-dividing cells. Examples of non-viral vectors include plasmid DNA, lipid nanoparticles, and polymeric carriers, among others, which have some advantages in their safety profile and the ability to carry larger genetic material. Gene augmentation represents one of the most common applications in the treatment of LSDs, hemophilia, and other disorders in which an inactivating mutation of a single gene results in a loss of enzyme or protein activity. Examples include highly successful treatments of SMA using AAV vectors for delivery of the SMN1 gene [76].

Gene Editing

Gene editing techniques enable targeted alteration of the genome so as to fix abnormal genes that cause diseases of metabolism. This strategy tries to fix the genetic defect at its origin permanently which can enable a cure of the disease in one attempt.

CRISPR-Cas9

In the hands of a skilled scientist, the powerful CRISPR-Cas9 system allows the erasure and replacement of the genetic information contained in a molecule of DNA with extreme accuracy. A highly diverse assembly of CRISPR proteins, using a short synthetic molecule of RNA as a guide, is able to bind to a predetermined region of DNA and cut it within the two strands. The repair processes of the cell then utilize a template provided to correct the errors or add new DNA [77].

Other Gene Editing Tools

Advanced techniques include the gene-editing techniques: TALENS, standing for transcription-activator-like effector nuclease, and ZFNS, standing for zinc-finger nuclease. These are in-organism gene-modification techniques without necessarily having to use CRISPR. These techniques also allow targeted alteration of DNA; however, they are less used as compared to CRISPR because of the complications and the somewhat lower success rate of the methods [77].

Applications

There are records to suggest that gene editing techniques have been used to manage metabolic diseases like phenylketonuria (PKU), a condition where gene mutation of the PAH gene can be reinstated to normal and sickle cell anemia where alteration of the diseased HBB gene is performed. This technique may also work for other diseases, such as cystic fibrosis and muscular dystrophy [78].

RNA-Based Therapies

Genetic modification as RNA therapies: Modulation of gene expression at the RNA level, other than by directly altering DNA itself [79]. These can further degrade faulty RNA transcripts, replace defective RNA, or modify RNA splicing to make functional proteins.

Antisense Oligonucleotides (ASOs)

They are small synthetic DNA or RNA strands designed to bind with a particular mRNA molecule. Binding to their target is how the ASOs inhibit. This may prevent faulty mRNA from translating into protein or may initiate degradation of the mRNA, thereby reducing the level of toxic protein. They can also be used to modify splicing patterns to create a functional protein [80].

Small Interfering RNA (siRNA)

siRNA therapies are based on the principle of small RNA molecules for induction of RNA interference-a natural cellular process wherein specific mRNA molecules are degraded with no translation into proteins. siRNA could be used either for downregulation of the expression of a pathogenic protein or for modulation of gene expression [81].

mRNA Therapy

mRNA therapy involves delivering synthetic mRNA encoding a functional protein to cells, allowing them to produce the missing or dysfunctional protein transiently. Unlike DNA-based gene therapies, mRNA therapies do not integrate into the genome, reducing the risk of insertional mutagenesis. RNA-based therapies are particularly useful for disorders where reducing the levels of a toxic protein is beneficial, such as in Huntington's disease or amyloidosis. mRNA therapy has been explored for metabolic disorders like propionic acidemia and methylmalonic acidemia to provide transient enzyme replacement [82, 83].

Ex Vivo Gene Therapy

It defines *ex vivo* gene therapy as the technique in which cells taken directly from the patient, after some manipulative and modifying procedures outside the body, are then reinfused into the same patient. It can also be applied to diseases affecting the blood cells, such as specific metabolic and immunodeficiency disorders. In the process, patient cells are usually HSCs or T cells. They are usually collected and then changed in the laboratory by either viral or non-viral vectors to deliver the therapeutic gene.

These modified cells are then expanded and subsequently reinfused into the patient, in whom they *in vivo* engraft and expand, creating a therapeutic effect. This is very effective in *ex vivo* gene therapy applications for hematopoietic disorders like beta-thalassemia, sickle cell disease, and SCID. Applications also occur in cancer treatments through CAR-T cell therapies that genetically modify T cells to target certain cancer cells [84, 85].

In vivo Gene Therapy

The approach of *in vivo* gene therapy involves the administration of the corrective gene of interest directly into the patient's body, where it will be taken up by the target cells. This happens to be convenient for those disorders in which separation and culture of some cell types outside the body is difficult. Most *in vivo* gene therapies make use of viral vectors for delivering the corrective gene directly to the diseased tissue or organ. Vectors are most often administered intravenously or directly injected into the target site, such as the liver, muscle, or brain, depending on the disorder being treated [86]. *In vivo* gene therapy is applied to diseases of specific tissues or organs, such as those of the retina, for example, Leberongenital amaurosis-liver-based metabolic diseases-for example, ornithine transcarbamylase deficiency-and neuromuscular diseases, for example, Duchenne muscular dystrophy.

Clinical Applications and Success Stories

Several gene therapy applications have shown success in clinical settings for metabolic disorders:

Spinal Muscular Atrophy (SMA)

Gene therapy against Spinal Muscular Atrophy using a vector AAV that expresses the SMN1 gene has proved extremely effective in the treatment of this genetic motor neuron disorder. The treatment is known as onasemnogene abeparvovec or, by its trade name, Zolgensma, which has an outstanding improvement in motor functions and survival in infants [87].

Haemophilia

The available AAV-mediated gene therapies for both hemophilia A and B have indeed proven their potential in reducing bleeding incidences and, generally, the need for regular factor replacement. All these therapies aim at the restoration of the production of clotting factors VIII or IX by introducing functional copies of the genes into liver cells [88].

Aromatic L-Amino Acid Decarboxylase (AADC) Deficiency

Gene therapy for aromatic L-amino acid decarboxylase deficiency, a rare disorder of neurotransmitter synthesis, using AAV vectors has shown remarkable gains in motor function and symptomatic reductions in treated patients. This gene therapy uses the AAV vector to introduce the gene DDC into the brain [89].

Leber's Congenital Amaurosis (LCA)

Genetic LCA, or Leber's Congenital Amaurosis, is a genetic disorder characterized by progressive blindness. This condition is allowed to be cured using Luxturna, a gene therapy. This works by delivering a healthy copy of the *RPE65* gene to the retinal cells to revive their outstanding photo-response properties (Fig. **3**) [90].

Stem Cell Therapy

General applications for stem cell therapy involve the use of pluripotent or multipotent stem cells to replace injured tissues and restore their normal function. This may also include replacing defective cells with healthy ones in metabolic disorders capable of producing the needed enzymes or proteins.

A. Applications in Metabolic Disorders

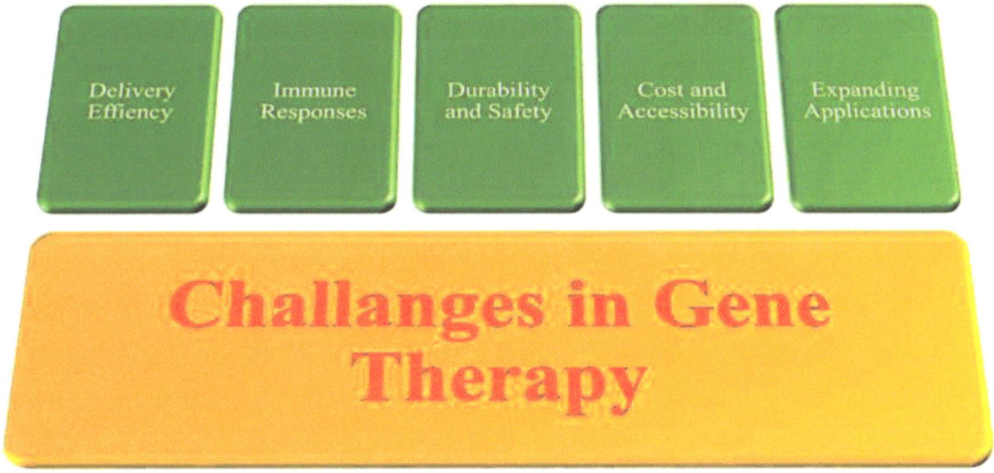

Fig. (3). Challenges in gene therapy.

Hematopoietic Stem Cell Transplantation (HSCT)

HSCT has applications for several diseases, including LSDs: Hurler Syndrome (MPS I) and Adrenoleukodystrophy (ALD). In such cases, hematopoietic stem cell transplantation from a healthy donor may provide cells capable of synthesizing the enzyme in question, thereby preventing the accumulation of toxic end products. HSCT has been demonstrated in the case of MPS I to improve outcomes through the provision of a constant source of alpha-L-iduronidase enzyme to the individual and facilitating a reduction in the accumulation of glycosaminoglycans in tissues [91].

Induced Pluripotent Stem Cells (iPSCs)

iPSCs are adult cells that have been reprogrammed back to a state of pluripotency and can thus differentiate into any cell type. These iPSCs hold great promise for developing patient-specific therapies in which the generation of healthy tissue for transplantation is performed using cells from the same patient. This includes, for Gaucher's Disease, the use of iPSCs for the generation of neurons that were genetically corrected, producing glucocerebrosidase capable of potentially addressing neurological symptoms [92].

Mesenchymal Stem Cells (MSCs)

MSCs have anti-inflammatory and immunomodulatory properties, making them useful in treating metabolic disorders with an inflammatory component. MSCs can be derived from bone marrow, adipose tissue, or umbilical cord blood [93].

Example: MSCs are being investigated for their potential to treat Mitochondrial Disorders, where they may help improve mitochondrial function and reduce oxidative stress.

CHALLENGES AND FUTURE PERSPECTIVES

Stem cell therapy is still in its early stages for many metabolic disorders, with several challenges to overcome:

Graft-Versus-Host Disease: GVHD is a complication of HSCT wherein the immune cells of the donor may attack the tissues of the recipient. Better matching of donor and recipient, on the one hand, and strategies aimed at dampening the immune response, such as immunosuppressive therapies, are being developed or utilized to minimize this risk.

Tumorigenicity: iPSCs have the potential to form tumors (teratomas) if not fully differentiated before transplantation. Ensuring the complete and safe differentiation of iPSCs is a key focus of ongoing research.

Ethical Issues: The use of embryonic stem cells raises ethical concerns, leading to a preference for iPSCs, which do not involve the destruction of embryos. However, the long-term safety and efficacy of iPSC-based therapies are still under investigation.

Pharmacological Chaperones

Pharmacological chaperones represent small molecules that bind to and stabilize their misfolded cognate proteins, facilitating the attainment of their appropriate conformation and restoration of activity. This approach is particularly useful in disorders caused by mutations that result in the production of misfolded but potentially functional enzymes [94].

Stabilization of Misfolded Proteins: Many metabolic disorders are caused by mutations that result in the production of enzymes with reduced stability or folding efficiency. Pharmacological chaperones bind to these misfolded enzymes, stabilizing them and promoting proper folding, which allows the enzyme to escape degradation and perform its normal function. In Fabry Disease, the pharmacological chaperone migalastat binds to and stabilizes the alpha-galactosidase A enzyme, enhancing its activity in patients with amenable mutations [95].

Enhancement of Enzyme Activity: By stabilizing the enzyme, pharmacological chaperones can increase the amount of active enzyme in the cell, improving the breakdown of substrates that would otherwise accumulate to toxic levels. In

Gaucher's Disease, chaperones are being explored to enhance the activity of the glucocerebrosidase enzyme, potentially providing an oral alternative to enzyme replacement therapy [96].

Methods for Managing Incomplete Metabolism

Incomplete metabolism often results in the accumulation of toxic metabolites and oxidative stress, which can lead to cellular damage and various metabolic disorders. Innovative therapeutic approaches aim to enhance the body's natural defenses, improve metabolic flexibility, and promote cellular health. Below are several novel strategies for managing incomplete metabolism.

Antioxidant Defence and Glutathione Regeneration

One of the important consequences of incomplete metabolism deals with oxidative stress. This could be defined as a kind of imbalance between the generation of ROS and the antioxidant defenses. In situations involving an excess of ROS, this will lead to oxidative damage of cellular components; lipids, proteins, and DNA. At the end of it all, this leads to a dysfunction of cells and, ultimately, death [97].

Glutathione is a significant intracellular antioxidant that scavenges ROS and maintains cellular redox balance. It acts as the first line of defence against oxidative stress by donating electrons to neutralize reactive radicals. In incomplete metabolism, GSH levels are decreased due to the overwhelming production of ROS. Therefore, strategies to enhance GSH regeneration and antioxidant defence become imperative. One of the most promising therapeutic strategies is supplementation with precursors, including N-acetylcysteine, which provides cysteine for GSH synthesis. Supplementation with NAC restores GSH levels, diminishes oxidative damage, and may improve overall cellular health. Vitamin E, vitamin C, and alpha-lipoic acid supplementation also seem to support cellular antioxidant capacity and protect against oxidative damage in metabolic disorders [98, 99].

Carnitine-Mediated Acyl Group Buffering

It is thus of vital importance that carnitine works well in mitochondrial fatty acid oxidation and metabolic homeostasis. Energy production, on the other hand, should be subjected to the transportation of long-chain fatty acids into mitochondria for beta-oxidation. Besides that, carnitine helps in buffering and removal when the accumulation of toxic acyl-CoA intermediates takes place in disorders. In cases of primary carnitine deficiencies or secondary deficiencies due to a mitochondrial fatty acid oxidation defect, the accumulation of toxic acyl may occur owing to a lack of carnitine. This results in an impairment in cellular metabolism, leading to an energy deficit with possible metabolic crises. L-carnitine supplementation is, therefore standard therapy to reach metabolic balance in cases of carnitine deficiencies. This therapy repletes carnitine stores, facilitates the transport of fatty acids into mitochondria, and thus promotes energy production and prevents the toxic accumulation of acyl-CoA intermediates. L-carnitine supplementation may improve clinical outcomes by minimizing metabolic decompensation and decreasing the severity of symptoms related to carnitine deficiency [100, 101].

Mitochondrial Reductive Stress and its Management

Mitochondrial reductive stress is a result of the buildup of reducing equivalents from NADH and $FADH_2$. This results in an imbalance in redox status, which in turn impairs mitochondrial function, lessens the efficiency of oxidative phosphorylation and enhances the production of ROS.

Consequences of Reductive Stress: Reductive stress compromises cellular energy production and augments the formation of ROS, causing further deterioration in mitochondrial function. Chronically, this form of stress results in cellular damage, inflammation, and the development of metabolic disorders [102].

Management Strategies: Improvement in mitochondrial function and reduction of reductive stress can be achieved through various strategies such as exercise, caloric restriction, and the use of pharmacological agents. Regular exercise stimulates mitochondrial biogenesis and elevates oxidative metabolism, improving mitochondrial efficiency. The caloric restriction seems to reduce oxidative stress and enhance mitochondrial function by increasing autophagy and reducing mitochondrial injury. Pharmacological agents, including metformin and mitoQ (a mitochondrially targeted antioxidant), may be used to optimize oxidative phosphorylation and reduce mitochondrial ROS production [103].

Intermittent Fasting and Increasing Energy Demand

Intermittent fasting is a form of dietary intervention that involves periodic alternation between times of feeding and fasting. Indeed, such an intervention might improve metabolic flexibility, promote mitochondrial function, and stimulate cellular mechanisms of repair.

Metabolic Flexibility: IF forces the body to switch from glucose to fat and *vice versa* as a primary energy source. This forces the body to acquire greater adaptability to meet changing energy demands. For such reasons, metabolic flexibility may have widespread applications in disease states where metabolism is defective or impaired or in diseases where glucose utilization is impaired. Other means of increasing energy demand include cold exposure and hypobaric hypoxia. While cold exposure promotes non-shivering thermogenesis and increases energy expenditure, hypobaric hypoxia enhances the use of oxygen by the mitochondria and cellular adaptation to metabolic stress [104 - 106].

SGLT2 Inhibitors

SGLT2 inhibitors belong to a class of drugs majorly utilized in the treatment of diabetes through the inhibition of glucose reabsorption along the renal nephrons and inducing glycosuria. Some recent literature has identified the potential of these drugs in inducing beiging in white adipocytes, hence offering a new perspective on therapeutic intervention in metabolic disorders [107].

Beiging of Adipocytes: Beiging is the process by which white adipocytes are trans-differentiated to beige adipocytes, featuring a higher amount of mitochondria and increased thermogenic capacity. Because of their capability to dissipate energy as heat through the action of uncoupling protein 1, beige adipocytes may increase energy expenditure, enhance insulin sensitivity, and promote metabolic health [108].

Potential Therapeutic Benefits: SGLT2 inhibitors contribute to enhanced metabolic flexibility and mitochondrial function in incompletely metabolizing individuals by facilitating the beiging of adipocytes, which helps to counteract metabolic imbalances and reduce fat deposition and insulin resistance (Fig. **4**) [109].

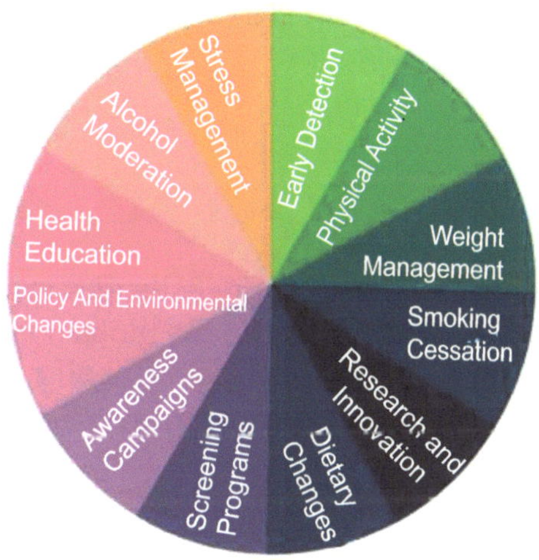

Fig. (4). Preventive strategies and early intervention.

FUTURE DIRECTIONS IN RESEARCH AND INNOVATION

Novel technologies will, of course, continue to revolutionize metabolic studies by bringing in new concepts of understanding, treatment of metabolic disorders, and diagnosis, making a real difference in patient care and outcomes. Genomic and epigenomic facilities, like next-generation sequencing, have helped us identify most of the genetic and epigenetic factors associated with these diseases. All this deeper understanding paves the way for precision medicine, in which treatments are tailored to the genetic makeup, lifestyle, and environment of each individual. Such a personal approach not only heightens effectiveness in treatment but also minimizes side effects. Another important area of interest has been the gut microbiome. Various metabolic problems are associated with gut flora disturbances. Future treatments may include the administration of probiotics, prebiotics, and even fecal microbiota transplantation as ways to hopefully better control gut bacteria and enhance metabolic health.

Advanced neuroimaging like functional MRI and PET scans offer real-time images of metabolic activity, while AI and machine learning make complex data analyses more available to clinicians for the identification of patterns and prediction of disease risk and treatment response. Exuberant potential also lies in gene-editing technologies like CRISPR-Cas9, which can correct genetic defects, and those areas of regenerative medicine and stem cell therapy that are able to repair damaged tissues. New biomarker discoveries make early detection possible and build personalized strategies for treatment. In a similar vein, interdisciplinary collaboration-developing a comprehensive solution requires the ability to integrate genetics, bioinformatics, pharmacology, and clinical medicine. Clinical trials and collaboration with industry players put the fruits of this research into practical application. Most importantly, patients' involvement and healthcare professionals' education are crucial for the assimilation of such advances, ultimately aiming at improving patient care.

CONCLUSION

Incomplete metabolism is one of the greatest challenges in medical science, where the body fails to metabolize certain important biochemicals due to enzymatic deficiency, genetic mutation, or a disruption of metabolic routes. This leads to the accumulation of toxic intermediates; the inability to synthesize an important compound may have severe consequences that will result in defective development, neurological damage, organ failures, and even sudden death. The incompletely metabolized conditions are of utmost importance in the context of preventing complications, improvement of quality of life, cost reduction of healthcare, and development of better therapeutic approaches.

Current treatments are diverse and targeted toward biochemical defects, including dietary interventions that regulate nutrient intake to prevent the accumulation of toxic metabolites and pharmacological interventions that replace enzyme cofactors and reduce substrate accumulation. ERT supplies the missing enzyme to reconstitute a metabolic pathway, though it too has its drawbacks in terms of limited efficiency and enormous expense. Combined therapies now represent an integrated approach whereby enhanced effectiveness of treatment addresses multiple disease mechanisms. Long-term symptomatic management therapies include innovative approaches, targeting the genetic mutation itself through gene therapy, replacement of defective cells by stem cell therapy, and pharmacological chaperones that stabilize misfolded proteins. The novel strategies aimed at symptom management encompass the improvement of antioxidant defenses, optimization of mitochondrial function, and development of new pharmacological agents to improve metabolic health. New treatments and therapies are developed and researched in continuous efforts to make the treatments of metabolic disorders more effective, personalized, and curative. Translating research and technology into real-world benefits will take continued collaboration across disciplines with the engagement of patients and ultimately improve the lives of individuals affected by incomplete metabolism.

REFERENCES

[1] A. W. El-Hattab, "Inborn errors of metabolism.," *Clin. Perinatol.*, vol. 42, no. 2, pp. 413–39, x, Jun. 2015.
[http://dx.doi.org/10.1016/j.clp.2015.02.010] [PMID: 26042912]

[2] F. Ezgu, "Inborn Errors of Metabolism.," *Adv. Clin. Chem.*, vol. 73, pp. 195–250, 2016
[http://dx.doi.org/10.1016/bs.acc.2015.12.001]

[3] J. Vockley, H.C. Andersson, K.M. Antshel, N.E. Braverman, B.K. Burton, D.M. Frazier, J. Mitchell, W.E. Smith, B.H. Thompson, and S.A. Berry, "Phenylalanine hydroxylase deficiency: diagnosis and management guideline", *Genet. Med.,* vol. 16, no. 2, pp. 188-200, 2014.
[http://dx.doi.org/10.1038/gim.2013.157] [PMID: 24385074]

[4] P. R. Blackburn, J. M. Gass, F. P. Vairo, K. M. Farnham, H. K. Atwal, S. Macklin, E. W. Klee, and P. S. Atwal, "Maple syrup urine disease: mechanisms and management", *Appl. Clin. Genet.,* vol. 10, pp. 57-66, 2017.
[http://dx.doi.org/10.2147/TACG.S125962]

[5] M. Siebert, E. Sidransky, and W. Westbroek, "Glucocerebrosidase is shaking up the synucleinopathies", *Brain,* vol. 137, no. 5, pp. 1304-1322, 2014.
[http://dx.doi.org/10.1093/brain/awu002] [PMID: 24531622]

[6] N. McWhorter, M.K. Ndugga-Kabuye, M. Puurunen, and S.L. Ernst, "Complications of the low phenylalanine diet for patients with phenylketonuria and the benefits of increased natural protein", *Nutrients,* vol. 14, no. 23, p. 4960, 2022.
[http://dx.doi.org/10.3390/nu14234960] [PMID: 36500989]

[7] J.C. Rocha, and A. MacDonald, "Dietary intervention in the management of phenylketonuria: current perspectives", *Pediatric Health Med. Ther.,* vol. 7, pp. 155-163, 2016.
[http://dx.doi.org/10.2147/PHMT.S49329] [PMID: 29388626]

[8] A.C. Puhl, and S. Ekins, "Advancing the research and development of enzyme replacement therapies

for lysosomal storage diseases", *GEN Biotechnol.,* vol. 1, no. 2, pp. 156-162, 2022.
[http://dx.doi.org/10.1089/genbio.2021.0013] [PMID: 35706761]

[9] P. Blackburn, J. Gass, F. Pinto e Vairo, K. Farnham, H. Atwal, S. Macklin, E. Klee, and P. Atwal, "Maple syrup urine disease: mechanisms and management", *Appl. Clin. Genet.,* vol. 10, pp. 57-66, 2017.
[http://dx.doi.org/10.2147/TACG.S125962] [PMID: 28919799]

[10] S. J. Sacharow, J. D. Picker, and H. L. Levy, "Homocystinuria caused by cystathionine beta-synthase deficiency," In:GeneReviews, 2017.

[11] J.M. Chinsky, R. Singh, C. Ficicioglu, C.D.M. van Karnebeek, M. Grompe, G. Mitchell, S.E. Waisbren, M. Gucsavas-Calikoglu, M.P. Wasserstein, K. Coakley, and C.R. Scott, "Diagnosis and treatment of tyrosinemia type I: a US and Canadian consensus group review and recommendations", *Genet. Med.,* vol. 19, no. 12, pp. 1380-1395, 2017.
[http://dx.doi.org/10.1038/gim.2017.101] [PMID: 28771246]

[12] S. Judd, M. Khedr, A.M. Milan, A.S. Davison, A.T. Hughes, A. Needham, E.E. Psarelli, A. Shenkin, and L.R. Ranganath, "The nutritional status of people with alkaptonuria: An exploratory analysis suggests a protein/energy dilemma", *JIMD Rep.,* vol. 53, no. 1, pp. 45-60, 2020.
[http://dx.doi.org/10.1002/jmd2.12084] [PMID: 32395409]

[13] I. Badiu Tişa, A.C. Achim, and A. Cozma-Petruţ, "The importance of neonatal screening for galactosemia", *Nutrients,* vol. 15, no. 1, p. 10, 2022.
[http://dx.doi.org/10.3390/nu15010010] [PMID: 36615667]

[14] J. K. Aronson, Ed., *Meyler's Side Effects of Drugs: The International Encyclopedia of Adverse Drug Reactions and Interactions,* 16th ed. Amsterdam, Netherlands: Elsevier Science, Oct. 2015.

[15] C.J. Hendriksz, and P. Gissen, "Glycogen storage disease", *Paediatr. Child Health (Oxford),* vol. 21, no. 2, pp. 84-89, 2011.
[http://dx.doi.org/10.1016/j.paed.2010.09.013]

[16] N. A. van der Beek, M. L. Hagemans, A. T. van der Ploeg, A. J. Reuser, and P. A. van Doorn, "Pompe disease (glycogen storage disease type II): clinical features and enzyme replacement therapy", Acta Neurol. Belg., vol. 106, no. 2, pp. 82-86, 2006.
[PMID: 16898258]

[17] J. Walter, P.A. Labrune, and P. Laforet, "The glycogen storage diseases and related disorders", In: *Inborn Metabolic Diseases.* Springer Berlin Heidelberg: Berlin, Heidelberg, 2016, pp. 121-137.
[http://dx.doi.org/10.1007/978-3-662-49771-5_5]

[18] A. Benito-Vicente, K.B. Uribe, S. Jebari, U. Galicia-Garcia, H. Ostolaza, and C. Martin, "Familial hypercholesterolemia: The most frequent cholesterol metabolism disorder caused disease", *Int. J. Mol. Sci.,* vol. 19, no. 11, p. 3426, 2018.
[http://dx.doi.org/10.3390/ijms19113426] [PMID: 30388787]

[19] D. Sulistya Ekaputri, I. Gusti Lanang Sidiartha, and I. Gusti Ayu Putu Eka Pratiwi, "Medium-chain acyl-CoA dehydrogenase deficiency (MCADD): A case of recurrent vomiting and diarrhea", *Int. J. Genet. Genom.,* vol. 8, no. 4, p. 127, 2020.
[http://dx.doi.org/10.11648/j.ijgg.20200804.12]

[20] W. L. Nyhan and G. F. Hoffmann, *Atlas of Inherited Metabolic Diseases,* 4th ed. Boca Raton, FL, USA: CRC Press, 2020.
[http://dx.doi.org/10.1201/9781315114033]

[21] M. Sowa, "Nutrition Management of Propionic Acidemia and Methylmalonic Acidemia", In: *Nutrition Management of Inherited Metabolic Diseases.* , L. E. Bernstein, F. Rohr, and S. van Calcar, Eds. Cham, Switzerland: Springer, 2022, pp. 295-305.
[http://dx.doi.org/10.1007/978-3-030-94510-7_21]

[22] V. Walker, "Ammonia toxicity and its prevention in inherited defects of the urea cycle," *Diabetes*

Obes. Metab., vol. 11, no. 9, pp. 823–835, 2009.
[http://dx.doi.org/10.1111/j.1463-1326.2009.01054.x]

[23] E. A. Crombez and S. D. Cederbaum, "Hyperargininemia due to liver arginase deficiency," *Mol. Genet. Metab.,* vol. 84, no. 3, pp. 243–251, Mar. 2005.
[http://dx.doi.org/10.1016/j.ymgme.2004.11.004]

[24] T. Dinur, U. Grittner, S. Revel-Vilk, M. Becker-Cohen, M. Istaiti, C. Cozma, A. Rolfs, and A. Zimran, "Impact of long-term enzyme replacement therapy on glucosylsphingosine (Lyso-gb1) values in patients with type 1 gaucher disease: Statistical models for comparing three enzymatic formulations", *Int. J. Mol. Sci.,* vol. 22, no. 14, p. 7699, 2021.
[http://dx.doi.org/10.3390/ijms22147699] [PMID: 34299318]

[25] V. V. Solovyeva, A. A. Shaimardanova, D. S. Chulpanova, K. V. Kitaeva, L. Chakrabarti, and A. A. Rizvanov, "New approaches to Tay-Sachs disease therapy", *Front. Physiol.,* 2018.
[http://dx.doi.org/10.3389/fphys.2018.01663]

[26] D. Orsucci, E. Caldarazzo Ienco, G. Siciliano, and M. Mancuso, "Mitochondrial disorders and drugs: what every physician should know," *Drugs Context,* vol. 8, pp. 212588, 2019.
[http://dx.doi.org/10.7573/dic.212588]

[27] A.S. Ahuja, "Understanding mitochondrial myopathies: a review", *PeerJ,* vol. 6, no. 5, p. e4790, 2018.
[http://dx.doi.org/10.7717/peerj.4790] [PMID: 29844960]

[28] R.E. Kelley, and H.C. Andersson, "Disorders of purines and pyrimidines", In: *Handbook of Clinical Neurology.* vol. 120. Elsevier B.V., 2014, pp. 827-838.
[http://dx.doi.org/10.1016/B978-0-7020-4087-0.00055-3]

[29] J.H. Walter, P. Labrune, and P. Laforêt, "The Glycogen Storage Diseases and Related Disorders", In: *Inborn Metabolic Diseases.* Springer Berlin Heidelberg: Berlin, Heidelberg, 2022, pp. 179-200.
[http://dx.doi.org/10.1007/978-3-662-63123-2_5]

[30] J. de Toro-Martín, B. J. Arsenault, J. P. Després, and M. C. Vohl, "Precision nutrition: a review of personalized nutritional approaches for the prevention and management of metabolic syndrome", *Nutrients,* vol. 9, no. 8, article 913, 2017.
[http://dx.doi.org/10.3390/nu9080913]

[31] A. Christ, M. Lauterbach, and E. Latz, "Western diet and the immune system: an inflammatory connection", *Immunity,* vol. 51, no. 5, pp. 794□811, 2019.
[http://dx.doi.org/10.1016/j.immuni.2019.09.020]

[32] L. Guariguata, D.R. Whiting, I. Hambleton, J. Beagley, U. Linnenkamp, and J.E. Shaw, "Global estimates of diabetes prevalence for 2013 and projections for 2035", *Diabetes Res. Clin. Pract.,* vol. 103, no. 2, pp. 137-149, 2014.
[http://dx.doi.org/10.1016/j.diabres.2013.11.002] [PMID: 24630390]

[33] W. van den Brink, J. van Bilsen, K. Salic, F. P. M. Hoevenaars, L. Verschuren, R. Kleemann, J. Bouwman, G. V. Ronnett, B. van Ommen, and S. Wopereis, "Current and Future Nutritional Strategies to Modulate Inflammatory Dynamics in Metabolic Disorders," *Front. Nutr.,* vol. 6, no. 129, 2019.
[http://dx.doi.org/10.3389/fnut.2019.00129]

[34] C. R. Balistreri, C. Caruso, and G. Candore, "The role of adipose tissue and adipokines in obesityrelated inflammatory diseases", *Mediators Inflamm.,* vol. 2010.
[http://dx.doi.org/10.1155/2010/802078]

[35] S. Espina, D. Casas-Deza, V. Bernal-Monterde, M. J. Domper-Arnal, S. García-Mateo, and A. Lué, "Evaluation and management of nutritional consequences of chronic liver diseases", *Nutrients,* vol. 15, no. 15, Art. no. 3487, 2023.
[http://dx.doi.org/10.3390/nu15153487]

[36] N. Di Daniele, A. Noce, M. Vidiri, E. Moriconi, G. Marrone, M. Annicchiarico-Petruzzelli, G. D'Urso, M. Tesauro, V. Rovella, and A. De Lorenzo, "Impact of Mediterranean diet on metabolic

syndrome, cancer and longevity," Oncotarget, vol. 8, pp. 8947–8979, 2017. Available from: www.impactjournals.com/oncotarget/ [http://dx.doi.org/10.18632/oncotarget.13553]

[37] Y. Sun, T. Zhang, B. Lu, X. Li, and L. Jiang, "Application of cofactors in the regulation of microbial metabolism: A state of the art review," *Front. Microbiol.,* vol. 14, 2023. [http://dx.doi.org/10.3389/fmicb.2023.1145784]

[38] W. W. Yue, S. Mackinnon, and G. A. Bezerra, "Substrate reduction therapy for inborn errors of metabolism", *Emerg. Top. Life Sci.,* vol. 3, no. 1, pp. 63–73, 2019. [http://dx.doi.org/10.1042/ETLS20180058]

[39] R. R. Ramsay, and K. F. Tipton, "Assessment of enzyme inhibition: a review with examples from the development of monoamine oxidase and cholinesterase inhibitory drugs", *Molecules*, vol. 22, no. 8, 2017. [http://dx.doi.org/10.3390/molecules22071192]

[40] V. Walker, "Ammonia metabolism and hyperammonemic disorders", In: *Adv. Clin. Chem.,* vol. 67, G. S. Makowski, Ed. Elsevier, 2014, pp. 73–150. [http://dx.doi.org/10.1016/bs.acc.2014.09.002]

[41] K. F. Tjhung, M. N. Shokhirev, D. P. Horning, and G. F. Joyce, "An RNA polymerase ribozyme that synthesizes its own ancestor," *Proc. Natl. Acad. Sci. U.S.A.,* vol. 117, no. 6, pp. 2906–2913, 2020. [http://dx.doi.org/10.1073/pnas.1914282117]

[42] C. Adamson, and M. Kanai, "Integrating abiotic chemical catalysis and enzymatic catalysis in living cells", *Org. Biomol. Chem.,* vol. 19, no. 1, 2021. [http://dx.doi.org/10.1039/D0OB01898H]

[43] R. El Dib, H. Gomaa, R. P. Carvalho, S. E. Camargo, R. Bazan, P. Barretti, and F. C. Barreto, "Enzyme replacement therapy for Anderson-Fabry disease," *Cochrane Database Syst. Rev.,* vol. 2016, no. 7, 2016. [http://dx.doi.org/10.1002/14651858.CD006663.pub4]

[44] L. López de Frutos, E. García-González, B. García-Rodríguez, Y. González-Irazabal, C. Lahoz, P. Irún, J.J. Cebolla, and P. Giraldo, "Serum protein profile analysis in lysosomal storage disorders patients", *Clin. Chim. Acta,* vol. 510, pp. 430-436, 2020. [http://dx.doi.org/10.1016/j.cca.2020.07.056] [PMID: 32745579]

[45] E. Shemesh, L. Deroma, B. Bembi, P. Deegan, C. Hollak, N. J. Weinreb, and T. M. Cox, "Enzyme replacement and substrate reduction therapy for Gaucher disease," Cochrane Database Syst. Rev., vol. 2015, no. 3, 2015. [http://dx.doi.org/10.1002/14651858.CD010324.pub2]

[46] B. Chan, and D.N. Adam, "A Review of Fabry Disease", *Skin Therapy Lett.,* vol. 23, no. 2, pp. 4-6, 2018. [PMID: 29562089]

[47] A. Toscano, and B. Schoser, "Enzyme replacement therapy in late-onset Pompe disease: a systematic literature review," *J. Neurol.,* vol. 260, pp. 951–959, 2013. [http://dx.doi.org/10.1007/s00415-012-6636-x]

[48] E. Jameson, S. Jones, and T. Remmington, "Enzyme replacement therapy with laronidase (Aldurazyme®) for treating mucopolysaccharidosis type I," *Cochrane Database Syst. Rev.,* vol. 2019, no. 6, 2019. [http://dx.doi.org/10.1002/14651858.CD009354.pub5]

[49] E. Kuperus, M.E. Kruijshaar, S.C.A. Wens, J.M. de Vries, M.M. Favejee, J.C. van der Meijden, D. Rizopoulos, E. Brusse, P.A. van Doorn, A.T. van der Ploeg, and N.A.M.E. van der Beek, "Long-term benefit of enzyme replacement therapy in Pompe disease", *Neurology,* vol. 89, no. 23, pp. 2365-2373, 2017. [http://dx.doi.org/10.1212/WNL.0000000000004711] [PMID: 29117951]

[50] J. Muenzer, “Early initiation of enzyme replacement therapy for the mucopolysaccharidoses,” *Mol. Genet. Metab.,* vol. 111, no. 2, pp. 63–72, 2014.
[http://dx.doi.org/10.1016/j.ymgme.2013.11.015]

[51] D.P. Germain, J. Charrow, R.J. Desnick, N. Guffon, J. Kempf, R.H. Lachmann, R. Lemay, G.E. Linthorst, S. Packman, C.R. Scott, S. Waldek, D.G. Warnock, N.J. Weinreb, and W.R. Wilcox, "Ten-year outcome of enzyme replacement therapy with agalsidase beta in patients with Fabry disease", *J. Med. Genet.,* vol. 52, no. 5, pp. 353-358, 2015.
[http://dx.doi.org/10.1136/jmedgenet-2014-102797] [PMID: 25795794]

[52] D. Concolino, F. Deodato, and R. Parini, "Enzyme replacement therapy: efficacy and limitations", *Ital. J. Pediatr.,* vol. 44, suppl. 2, p. 120, 2018.
[http://dx.doi.org/10.1186/s13052-018-0562-1] [PMID: 30442189]

[53] L. Zhang, W. Wang, and S. Wang, “Effect of vaccine administration modality on immunogenicity and efficacy,” *Expert Rev. Vaccines,* vol. 14, no. 11, pp. 1509–1523, 2015.
[http://dx.doi.org/10.1586/14760584.2015.1081067]

[54] E. Fuentes, M. Fuentes, M. Alarcón, and I. Palomo, "Immune system dysfunction in the elderly", An Acad Bras Cienc, vol. 89, no. 1, pp. 285–299, 2017.

[55] V. J. Clemente-Suárez, A. Martín-Rodríguez, L. Redondo-Flórez, C. López-Mora, R. Yáñez-Sepúlveda, and J. F. Tornero-Aguilera, "New Insights and Potential Therapeutic Interventions in Metabolic Diseases", *Int. J. Mol. Sci.,* vol. 24, no. 13, 2023.
[http://dx.doi.org/10.3390/ijms241310672]

[56] J. M. A. Delou, A. S. O. Souza, L. C. M. Souza, and H. L. Borges, "Highlights in resistance mechanism pathways for combination therapy", *Cells,* vol. 8, no. 9, 2019.
[http://dx.doi.org/10.3390/cells8091013]

[57] M. Pi, H. Kuang, C. Yue, Q. Yang, A. Wu, Y. Li, Y.G. Assaraf, D.H. Yang, and S. Wu, "Targeting metabolism to overcome cancer drug resistance: A promising therapeutic strategy for diffuse large B cell lymphoma", *Drug Resist. Updat.,* vol. 61, p. 100822, 2022.
[http://dx.doi.org/10.1016/j.drup.2022.100822] [PMID: 35257981]

[58] N.E. Sunny, F. Bril, and K. Cusi, "Mitochondrial adaptation in nonalcoholic fatty liver disease: Novel mechanisms and treatment strategies", *Trends Endocrinol. Metab.,* vol. 28, no. 4, pp. 250-260, 2017.
[http://dx.doi.org/10.1016/j.tem.2016.11.006] [PMID: 27986466]

[59] P.K. Prabhakar, "Combination therapy: A new tool for the management of obesity", *Endocr. Metab. Immune Disord. Drug Targets,* vol. 24, no. 4, pp. 402-417, 2024.
[http://dx.doi.org/10.2174/1871530323666230825140808] [PMID: 37641995]

[60] P. Zapolnik, and A. Pyrkosz, "Gene therapy for mucopolysaccharidosis type ii—a review of the current possibilities", *Int. J. Mol. Sci.,* vol. 22, no. 11, 2021.
[http://dx.doi.org/10.3390/ijms22115490]

[61] A. M. Keeler, M. K. ElMallah, and T. R. Flotte, "Gene Therapy 2017: Progress and Future Directions", Clin. Transl. Sci., vol. 10, no. 4, pp. 242–248, 2017.
[http://dx.doi.org/10.1111/cts.12466]

[62] G.A.R. Gonçalves, and R.M.A. Paiva, "Gene therapy: advances, challenges and perspectives", *Einstein (Sao Paulo),* vol. 15, no. 3, pp. 369-375, 2017.
[http://dx.doi.org/10.1590/s1679-45082017rb4024] [PMID: 29091160]

[63] G. Gao, M. R. Alvira, S. Somanathan, Y. Lu, L. H. Vandenberghe, J. J. Rux, R. Calcedo, J. Sanmiguel, Z. Abbas, and J. M. Wilson, “Adeno-associated viruses undergo substantial evolution in primates during natural infections,” *Proc. Natl. Acad. Sci. U.S.A.,* vol. 100, no. 10, pp. 6081–6086, 2003.
Available from: https://www.pnas.org
[http://dx.doi.org/10.1073/pnas.0937739100]

[64] G. Gao, M. R. Alvira, L. Wang, R. Calcedo, J. Johnston, and J. M. Wilson, "Novel adeno-associated viruses from rhesus monkeys as vectors for human gene therapy," *Proc. Natl. Acad. Sci. U.S.A.*, vol. 99, no. 18, pp. 11854–11859, 2002. [http://dx.doi.org/10.1073/pnas.182412299]

[65] O. Negre, A.-V. Eggimann, Y. Beuzard, J.-A. Ribeil, P. Bourget, S. Borwornpinyo, S. Hongeng, S. Hacein-Bey, M. Cavazzana, P. Leboulch, and E. Payen, "Gene therapy of the β-hemoglobinopathies by lentiviral transfer of the βA(T87Q)-globin gene," *Hum. Gene Ther.*, vol. 27, no. 2, 2016. [http://dx.doi.org/10.1089/hum.2016.007]

[66] S. Shen, E.D. Horowitz, A.N. Troupes, S.M. Brown, N. Pulicherla, R.J. Samulski, M. Agbandje-McKenna, and A. Asokan, "Engraftment of a galactose receptor footprint onto adeno-associated viral capsids improves transduction efficiency", *J. Biol. Chem.*, vol. 288, no. 40, pp. 28814-28823, 2013. [http://dx.doi.org/10.1074/jbc.M113.482380] [PMID: 23940044]

[67] B.E. Deverman, P.L. Pravdo, B.P. Simpson, S.R. Kumar, K.Y. Chan, A. Banerjee, W.L. Wu, B. Yang, N. Huber, S.P. Pasca, and V. Gradinaru, "Cre-dependent selection yields AAV variants for widespread gene transfer to the adult brain", *Nat. Biotechnol.*, vol. 34, no. 2, pp. 204-209, 2016. [http://dx.doi.org/10.1038/nbt.3440] [PMID: 26829320]

[68] M. Wallen, F. Aqil, W. Spencer, and R. C. Gupta, "Exosomes as an Emerging Plasmid Delivery Vehicle for Gene Therapy", "Exosomes as an emerging plasmid delivery vehicle for gene therapy," *Pharmaceutics*, vol. 15, no. 7, 2023. [http://dx.doi.org/10.3390/pharmaceutics15071832]

[69] E. J. Sontheimer, and R. Barrangou, "The Bacterial Origins of the CRISPR Genome-Editing Revolution", *Hum. Gene Ther.*, vol. 26, no. 7, 2015. [http://dx.doi.org/10.1089/hum.2015.091]

[70] P. Mali, L. Yang, K. M. Esvelt, J. Aach, M. Guell, J. E. DiCarlo, J. E. Norville, and G. M. Church, "RNA-guided human genome engineering via Cas9," *Science*, vol. 339, pp. 823–828, 2013. [http://dx.doi.org/10.1126/science.1232033]

[71] L. Cong, F.A. Ran, D. Cox, S. Lin, R. Barretto, N. Habib, P.D. Hsu, X. Wu, W. Jiang, L.A. Marraffini, and F. Zhang, "Multiplex genome engineering using CRISPR/Cas systems", *Science*, vol. 339, no. 6121, pp. 819-823, 2013. [http://dx.doi.org/10.1126/science.1231143] [PMID: 23287718]

[72] J.K. Nair, J.L.S. Willoughby, A. Chan, K. Charisse, M.R. Alam, Q. Wang, M. Hoekstra, P. Kandasamy, A.V. Kel'in, S. Milstein, N. Taneja, J. O'Shea, S. Shaikh, L. Zhang, R.J. van der Sluis, M.E. Jung, A. Akinc, R. Hutabarat, S. Kuchimanchi, K. Fitzgerald, T. Zimmermann, T.J.C. van Berkel, M.A. Maier, K.G. Rajeev, and M. Manoharan, "Multivalent N-acetylgalactosamine-conjugated siRNA localizes in hepatocytes and elicits robust RNAi-mediated gene silencing", *J. Am. Chem. Soc.*, vol. 136, no. 49, pp. 16958-16961, 2014. [http://dx.doi.org/10.1021/ja505986a] [PMID: 25434769]

[73] J.S. Butler, A. Chan, S. Costelha, S. Fishman, J.L.S. Willoughby, T.D. Borland, S. Milstein, D.J. Foster, P. Gonçalves, Q. Chen, J. Qin, B.R. Bettencourt, D.W. Sah, R. Alvarez, K.G. Rajeev, M. Manoharan, K. Fitzgerald, R.E. Meyers, S.V. Nochur, M.J. Saraiva, and T.S. Zimmermann, "Preclinical evaluation of RNAi as a treatment for transthyretin-mediated amyloidosis", *Amyloid*, vol. 23, no. 2, pp. 109-118, 2016. [http://dx.doi.org/10.3109/13506129.2016.1160882] [PMID: 27033334]

[74] E.M. Brydon, R. Bronstein, A. Buskin, M. Lako, E.A. Pierce, and R. Fernandez-Godino, "AAV-mediated gene augmentation therapy restores critical functions in mutant PRPF31+/− iPSC-derived RPE cells", *Mol. Ther. Methods Clin. Dev.*, vol. 15, pp. 392-402, 2019. [http://dx.doi.org/10.1016/j.omtm.2019.10.014] [PMID: 31890732]

[75] E. Banin, E. Gootwine, A. Obolensky, R. Ezra-Elia, A. Ejzenberg, L. Zelinger, H. Honig, A. Rosov, E. Yamin, D. Sharon, E. Averbukh, W.W. Hauswirth, and R. Ofri, "Gene augmentation therapy restores

retinal function and visual behavior in a sheep model of CNGA3 achromatopsia", *Mol. Ther.,* vol. 23, no. 9, pp. 1423-1433, 2015.
[http://dx.doi.org/10.1038/mt.2015.114] [PMID: 26087757]

[76] A. De Caneva, F. Porro, G. Bortolussi, R. Sola, M. Lisjak, A. Barzel, M. Giacca, M.A. Kay, K. Vlahoviček, L. Zentilin, and A.F. Muro, "Coupling AAV-mediated promoterless gene targeting to SaCas9 nuclease to efficiently correct liver metabolic diseases", *JCI Insight,* vol. 4, no. 15, p. e128863, 2019.
[http://dx.doi.org/10.1172/jci.insight.128863] [PMID: 31211694]

[77] N. Zabaleta, M. Barberia, C. Martin-Higueras, N. Zapata-Linares, I. Betancor, S. Rodriguez, R. Martinez-Turrillas, L. Torella, A. Vales, C. Olagüe, A. Vilas-Zornoza, L. Castro-Labrador, D. Lara-Astiaso, F. Prosper, E. Salido, G. Gonzalez-Aseguinolaza, and J. R. Rodriguez-Madoz, "CRISPR/Cas9-mediated glycolate oxidase disruption is an efficacious and safe treatment for primary hyperoxaluria type I," *Nat. Commun.,* vol. 9, 2018.
[http://dx.doi.org/10.1038/s41467-018-07827-1]

[78] F.P. Pankowicz, M. Barzi, X. Legras, L. Hubert, T. Mi, J.A. Tomolonis, M. Ravishankar, Q. Sun, D. Yang, M. Borowiak, P. Sumazin, S.H. Elsea, B. Bissig-Choisat, and K.D. Bissig, "Reprogramming metabolic pathways in vivo with CRISPR/Cas9 genome editing to treat hereditary tyrosinaemia", *Nat. Commun.,* vol. 7, no. 1, p. 12642, 2016.
[http://dx.doi.org/10.1038/ncomms12642] [PMID: 27572891]

[79] A. Sparmann, and J. Vogel, "RNA-based medicine: from molecular mechanisms to therapy", *EMBO J.,* vol. 42, no. 21, p. e114760, 2023.
[http://dx.doi.org/10.15252/embj.2023114760] [PMID: 37728251]

[80] M. Gagliardi, and A. T. Ashizawa, "The challenges and strategies of antisense oligonucleotide drug delivery", *Biomedicines,* vol. 9, no. 4, 2021.
[http://dx.doi.org/10.3390/biomedicines9040433]

[81] S. Dyawanapelly, S.B. Ghodke, R. Vishwanathan, P. Dandekar, and R. Jain, "RNA interference-based therapeutics: Molecular platforms for infectious diseases," *J. Biomed. Nanotechnol.,* vol. 10, no. 9, pp. 1998–2037, 2014.
[http://dx.doi.org/10.1166/jbn.2014.1929]

[82] X. Liang, D. Li, S. Leng, and X. Zhu, "RNA-based pharmacotherapy for tumors: From bench to clinic and back", *Biomed. Pharmacother.,* vol. 125, p. 109997, 2020.
[http://dx.doi.org/10.1016/j.biopha.2020.109997] [PMID: 32062550]

[83] Y. Zhou and C. Chen, "From bench to bedside: Current developments in RNA-based therapies for treatment of hyperlipidemia," Int. J. Drug Discov. Pharmacology, vol. 1, no. 1, p. 7, 2022.
Doi: 10.53941/ijddp.v1i1.141.

[84] J. Estève, J.M. Blouin, M. Lalanne, L. Azzi-Martin, P. Dubus, A. Bidet, J. Harambat, B. Llanas, I. Moranvillier, A. Bedel, F. Moreau-Gaudry, and E. Richard, "Generation of induced pluripotent stem cells-derived hepatocyte-like cells for ex vivo gene therapy of primary hyperoxaluria type 1", *Stem Cell Res. (Amst.),* vol. 38, p. 101467, 2019.
[http://dx.doi.org/10.1016/j.scr.2019.101467] [PMID: 31151050]

[85] D.A. Carbonaro-Sarracino, K. Chun, D.N. Clark, M.L. Kaufman, X. Jin, X. Wang, and D.B. Kohn, "Gene delivery using AAV8 in vivo for disease stabilization in a bimodal gene therapy approach for the treatment of ADA-deficient SCID", *Mol. Ther. Methods Clin. Dev.,* vol. 20, pp. 765-778, 2021.
[http://dx.doi.org/10.1016/j.omtm.2021.02.007] [PMID: 33738330]

[86] S. Salabarria, B.J. Byrne, C. Liberati, and M. Corti, "Preclinical gene therapy studies for metabolic myopathy", In: *Muscle Gene Ther* Springer International Publishing, 2019, pp. 579-596.
[http://dx.doi.org/10.1007/978-3-030-03095-7_33]

[87] M. Bowerman, C. G. Becker, R. J. Yáñez-Muñoz, K. Ning, M. J. A. Wood, T. H. Gillingwater, K. Talbot, and The UK SMA Research Consortium, "Therapeutic strategies for spinal muscular atrophy:

SMN and beyond," *Dis. Model. Mech.,* vol. 10, no. 8, pp. 943–954, 2017.
[http://dx.doi.org/10.1242/dmm.030148]

[88] G. Spadarella, A. Di Minno, N. Brunetti-Pierri, J. Mahlangu, and G. Di Minno, "The evolving landscape of gene therapy for congenital haemophilia: An unprecedented, problematic but promising opportunity for worldwide clinical studies", *Blood Rev.,* vol. 46, 2021.
[http://dx.doi.org/10.1016/j.blre.2020.100737]

[89] M.J. Malaquias, F. Magrinelli, and K.P. Bhatia, "Breakthrough news in adenoviral vector-mediated AADC gene therapy: Lessons from the success in AADC deficiency and possible future applications", *Mov. Disord. Clin. Pract. (Hoboken),* vol. 9, no. 6, pp. 737-738, 2022.
[http://dx.doi.org/10.1002/mdc3.13503] [PMID: 35937482]

[90] M. Daich Varela, T. A. Cabral de Guimaraes, M. Georgiou, and M. Michaelides, "Leber congenital amaurosis/early-onset severe retinal dystrophy: current management and clinical trials," *Br. J. Ophthalmol.,* vol. 106, pp. 445–451, 2022.
[http://dx.doi.org/10.1136/bjophthalmol-2020-318483]

[91] E.A. Kulagin, A.G. Volkova, I.Y. Nikolaev, O.V. Goloshchapov, A.G. Smirnova, T.A. Rudakova, E.I. Darskaya, E.V. Morozova, N.B. Mikhaylova, J.D. Rabik, V.G. Timchik, T.I. Shchemelinina, R.D. Skvortsova, T.S. Razumovskaya, S.N. Bondarenko, I.S. Moiseev, V.N. Marchenko, V.I. Trofimov, and B.V. Afanasyev, "Clinical features and prognostic value of functional disorders and pulmonary comorbidity in adult patients prior to allogeneic hematopoietic stem cell transplantation", *Cell. Ther. Transplant.,* vol. 8, no. 2, pp. 26-37, 2019.
[http://dx.doi.org/10.18620/ctt-1866-8836-2019-8-2-26-37]

[92] V. Singh, N. Kumar, M. Kalsan, and A. Saini, "Mechanism of induction: Induced pluripotent stem cells (iPSCs)," *J. Stem Cells,* vol. 10, no. 1, 2015.
Available from: https://www.researchgate.net/publication/277575563

[93] J.Y. Chen, X.Z. Mou, X.C. Du, and C. Xiang, "Comparative analysis of biological characteristics of adult mesenchymal stem cells with different tissue origins", *Asian Pac. J. Trop. Med.,* vol. 8, no. 9, pp. 739-746, 2015.
[http://dx.doi.org/10.1016/j.apjtm.2015.07.022] [PMID: 26433660]

[94] L. Liguori, M. Monticelli, M. Allocca, B. Hay Mele, J. Lukas, M.V. Cubellis, and G. Andreotti, "Pharmacological chaperones: A therapeutic approach for diseases caused by destabilizing missense mutations", *Int. J. Mol. Sci.,* vol. 21, no. 2, p. 489, 2020.
[http://dx.doi.org/10.3390/ijms21020489] [PMID: 31940970]

[95] Y.X. Tao, and P.M. Conn, "Pharmacoperones as novel therapeutics for diverse protein conformational diseases", *Physiol. Rev.,* vol. 98, no. 2, pp. 697-725, 2018.
[http://dx.doi.org/10.1152/physrev.00029.2016] [PMID: 29442594]

[96] P. Beerepoot, R. Nazari, and A. Salahpour, "Pharmacological chaperone approaches for rescuing GPCR mutants: Current state, challenges, and screening strategies", *Pharmacol. Res.,* vol. 117, pp. 242–251, 2017.
[http://dx.doi.org/10.1016/j.phrs.2016.12.036]

[97] K. Ulrich, and U. Jakob, "The role of thiols in antioxidant systems", *Free Radic. Biol. Med.,* vol. 140, pp. 14–27, Aug. 2019.
[http://dx.doi.org/10.1016/j.freeradbiomed.2019.05.035]

[98] B. Kaptaner, A. Doğan, C. Yılmaz, H. Aykut, E. Doğan, C.F. Babat, and F. Dönmez, "Histopathological and biochemical alterations in the adrenal gland of male rats exposed to bisphenol F", *Biol. Bull. Russ. Acad. Sci.,* vol. 51, no. 4, pp. 873-882, 2024.
[http://dx.doi.org/10.1134/S1062359023606717]

[99] J. Dumanović, E. Nepovimova, M. Natić, K. Kuča, and V. Jaćević, "The Significance of Reactive Oxygen Species and Antioxidant Defense System in Plants: A Concise Overview", *Front. Plant Sci.,* vol. 11, p. 552969, 2021.

[http://dx.doi.org/10.3389/fpls.2020.552969]

[100] M.N. Davies, L. Kjalarsdottir, J.W. Thompson, L.G. Dubois, R.D. Stevens, O.R. Ilkayeva, M.J. Brosnan, T.P. Rolph, P.A. Grimsrud, and D.M. Muoio, "The acetyl group buffering action of carnitine acetyltransferase offsets macronutrient-induced lysine acetylation of mitochondrial proteins", *Cell Rep.,* vol. 14, no. 2, pp. 243-254, 2016.
[http://dx.doi.org/10.1016/j.celrep.2015.12.030] [PMID: 26748706]

[101] Y.M.H. Bruls, M. de Ligt, L. Lindeboom, E. Phielix, B. Havekes, G. Schaart, E. Kornips, J.E. Wildberger, M.K.C. Hesselink, D. Muoio, P. Schrauwen, and V.B. Schrauwen-Hinderling, "Carnitine supplementation improves metabolic flexibility and skeletal muscle acetylcarnitine formation in volunteers with impaired glucose tolerance: A randomised controlled trial", *EBioMedicine,* vol. 49, pp. 318-330, 2019.
[http://dx.doi.org/10.1016/j.ebiom.2019.10.017] [PMID: 31676389]

[102] L. Zhang, and K.D. Tew, "Reductive stress in cancer", In: *Advances in Cancer Research.* vol. 152. Academic Press Inc., 2021, pp. 383-413.
[http://dx.doi.org/10.1016/bs.acr.2021.03.009]

[103] R. Sharma, B. Reinstadler, K. Engelstad, O.S. Skinner, E. Stackowitz, R.G. Haller, C.B. Clish, K. Pierce, M.A. Walker, R. Fryer, D. Oglesbee, X. Mao, D.C. Shungu, A. Khatri, M. Hirano, D.C. De Vivo, and V.K. Mootha, "Circulating markers of NADH-reductive stress correlate with mitochondrial disease severity", *J. Clin. Invest.,* vol. 131, no. 2, p. e136055, 2021.
[http://dx.doi.org/10.1172/JCI136055] [PMID: 33463549]

[104] R. Antoni, K.L. Johnston, A.L. Collins, and M.D. Robertson, "Effects of intermittent fasting on glucose and lipid metabolism", *Proceedings of the Nutrition Society*, pp. 361-368, vol. 76, no. 3, 2017.
[http://dx.doi.org/10.1017/S0029665116002986]

[105] B. D. Horne, J. B. Muhlestein, and J. L. Anderson, "Health effects of intermittent fasting: Hormesis or harm? A systematic review", *Am. J. Clin. Nutr.,* vol. 102, no. 2, pp. 464–470, Aug. 2015.
[http://dx.doi.org/10.3945/ajcn.115.109553]

[106] S. Brogi, R. Tabanelli, S. Puca, and V. Calderone, “Intermittent fasting: Myths, fakes and truth on this dietary regimen approach,” *Foods,* vol. 13, p. 1960, 2024.
[http://dx.doi.org/10.3390/foods13131960]

[107] X. Cai, W. Yang, X. Gao, Y. Chen, L. Zhou, S. Zhang, X. Han, and L. Ji, "The association between the dosage of SGLT2 inhibitor and weight reduction in type 2 diabetes patients: A meta-analysis", *Obesity (Silver Spring),* vol. 26, no. 1, pp. 70-80, 2018.
[http://dx.doi.org/10.1002/oby.22066] [PMID: 29165885]

[108] Y. J. M. op den Kamp, A. Gemmink, M. de Ligt, B. Dautzenberg, E. Kornips, J. A. Jorgensen, G. Schaart, R. Esterline, D. A. Pava, J. Hoeks, V. B. Schrauwen-Hinderling, S. Kersten, B. Havekes, T. R. Koves, D. M. Muoio, M. K. C. Hesselink, J. Oscarsson, E. Phielix, and P. Schrauwen, Effects of SGLT2 inhibitor dapagliflozin in patients with type 2 diabetes on skeletal muscle cellular metabolism", *Mol. Metab., vol. 66, p. 101620, 2022.*
[http://dx.doi.org/10.1016/j.molmet.2022.101620]

[109] J. Singh, “A Comprehensive Analysis of SGLT2-inhibition in Type 2 Diabetes and Heart Failure,” *Doctor of Medicine*, 2019.

CHAPTER 8

Nutritional and Lifestyle Interventions

Arvinder Pal Singh[1], **Ankush Goyal**[1,*], **Amit Aggarwal**[1,*] and **Navjot Singh Sethi**[2]

[1] *Maharaja Agrasen School of Pharmacy, Maharaja Agrasen University, Solan, Himachal Pradesh, India*

[2] *IEC University, Solan, Himachal Pradesh, India*

Abstract: Nutritional and lifestyle interventions are crucial in shaping humans mental and physical health and well-being over time. Based on the most recent research and evidence-based practices, the diet and lifestyle choices we make have an impact on health outcomes. It delves into the mechanisms through which specific interventions can improve health and lower the risk of chronic diseases. The effect of dietary patterns on health is a key focus. Numerous studies demonstrate the significant influence that dietary decisions have on the development and course of chronic illnesses, such as obesity, diabetes, and cardiovascular disease. A diet rich in fruits, vegetables, whole grains, and lean proteins and low in processed foods, sugar, and saturated fats can lower the risk of developing disease and raise the probability of staying healthy. The general state of health and metabolic processes are also significantly impacted by micronutrients. Several health problems, including a weakened immune system, cognitive decline, and weakened bone health, are linked to deficiencies in micronutrients. When needed, adequate intake of vitamins, minerals, and antioxidants, through diet or supplements, is vital for maintaining health and energy. Lifestyle decisions, including physical activity, stress management, and sleep quality, also significantly influence health. Frequent exercise improves muscular, cardiovascular, and mental health by reducing stress and anxiety. Emotional resilience and mental clarity can be achieved by incorporating stress-reduction practices like yoga, meditation, and deep breathing. Sustaining a healthy sleep pattern is essential for better general health and higher energy levels.

Keywords: Cardiovascular diseases, Diabetes, Lifestyle interventions, Metabolic disorders, Metabolic health, Metabolism, Nutritional interventions, Obesity.

INTRODUCTION

Nutritional and lifestyle interventions are crucial in shaping humans' mental and physical health and well-being over time. Understanding overall metabolic health

* **Corresponding authors Amit Aggarwal and Ankush Goyal:** Maharaja Agrasen School of Pharmacy, Maharaja Agrasen University, Solan, Himachal Pradesh, India;
E-mails: goyal.ankush35@gmail.com; amitaggarwal78@yahoo.co.in

Rajesh Kumar (Ed.)

considering various factors such as lifestyle interventions, metabolic responses to stress, and the impact of different oncogenes on metabolic pathways is also unnecessary. These lifestyle modifications include regular physical activity, quality sleep, and healthy eating habits (Fig. **1**) [1 - 4].

Nutritional interventions aid in improving various aspects of human health across different age groups. Based on research-based studies, the diet and lifestyle choices we make have an impact on health outcomes. Research studies on nutritional interventions have shown positive outcomes [5].

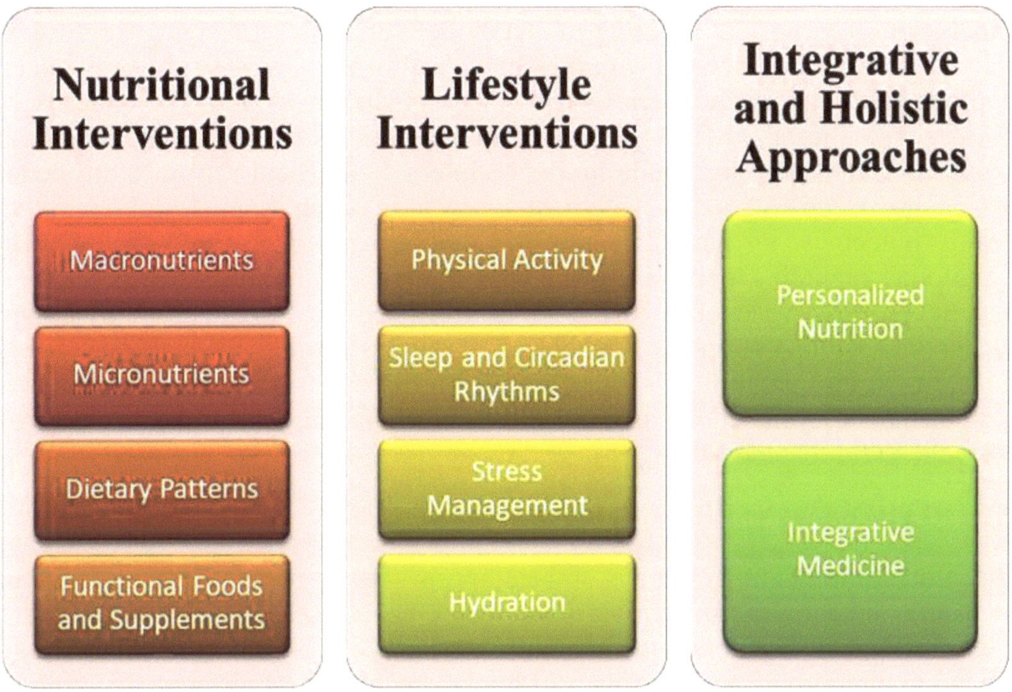

Fig. (1). Nutritional and lifestyle interventions.

Several health problems such as immunological disorders, bone-related issues, and cognitive disorders are associated with nutritional deficiency. In addition to this, the metabolic processes are also significantly affected by nutrients. These nutrients are generally classified as micronutrients and macronutrients [6]. The effect of dietary patterns on health is a key focus. These patterns can be more useful in providing a better way to understand how diet influences metabolic health. When examining the overall impact of food and nutrients, these dietary patterns reflect real eating habits, *i.e.*, consumption of foods or meals rather than just individual dietary components. This is crucial for tailoring effective nutritional interventions to enhance metabolic health [7]. Functional foods and supplements have also increased in this era to reduce the complex interactions between the risk factors produced by metabolic syndrome (MetS) [8].

Lifestyle interventions, including physical activity, smoking cessation, stress management, and quality sleep, significantly influence metabolic health in individuals with various conditions like metabolic syndrome, prediabetes, and Type 2 diabetes [9]. These interventions, can reduce the risk of cardiovascular diseases, and improve clinical parameters such as blood glucose level and blood pressure [3, 5, 7, 10 - 71]. Further, sleep deprivation is closely related to metabolic dysfunction. It is also responsible for the chronic misalignment of endogenous rhythms contributing to metabolic issues [72]. Engaging in physical activity is directly associated with metabolic health. Regular physical activity can improve sleep quality and manage sleep disorders like insomnia. The intensity of physical activities also affects sleep quality. High-intensity physical activities can cause difficulty sleeping, while moderate-intensity physical activities are better for quality sleep. Stress management interventions can significantly improve health outcomes in patients with metabolic syndrome and psychological symptoms, including depression, anxiety, and schizophrenia. Stress reduction techniques improve overall metabolic health [44, 73, 74]. Further, hydration (water intake) also supports metabolism through different mechanisms. Some intervention and preclinical studies point to a direct association of dehydration with metabolic dysfunction and elevated arginine vasopressin (AVP) levels, a key hormone in fluid homeostasis [75].

NUTRITIONAL INTERVENTIONS

Nutritional interventions are customized approaches to manage a wide range of metabolic disorders including obesity, diabetes, heart disease, digestive problems, allergies, and deficiencies in essential nutrients [45]. These interventions are crucial in improving various aspects of human health across different age groups. Studies have shown that implementing these interventions can lead to positive outcomes such as enhanced physical development in children under 8 years old, with longer intervention periods significantly impacting weight-for-height and weight-for-age scores [76]. In physically active women aged 50 to 70 years, combining nutritional interventions with different training models resulted in enhanced physical capacities, quality of life, and level of physical activity, regardless of the type of training used [46].

Generally, macronutrients and micronutrients are classified as nutritional strategies to overcome health issues related to incomplete metabolism. Additionally, in school settings, nutritional interventions were found to potentially reduce the prevalence of overweight and obesity among students, with multicomponent interventions ranking as the most effective for reducing body mass index (BMI) and improving the intake of fruit and vegetables. These findings underscore the importance of tailored nutritional interventions in promoting overall health and well-being across different populations (Fig. **2**) [78].

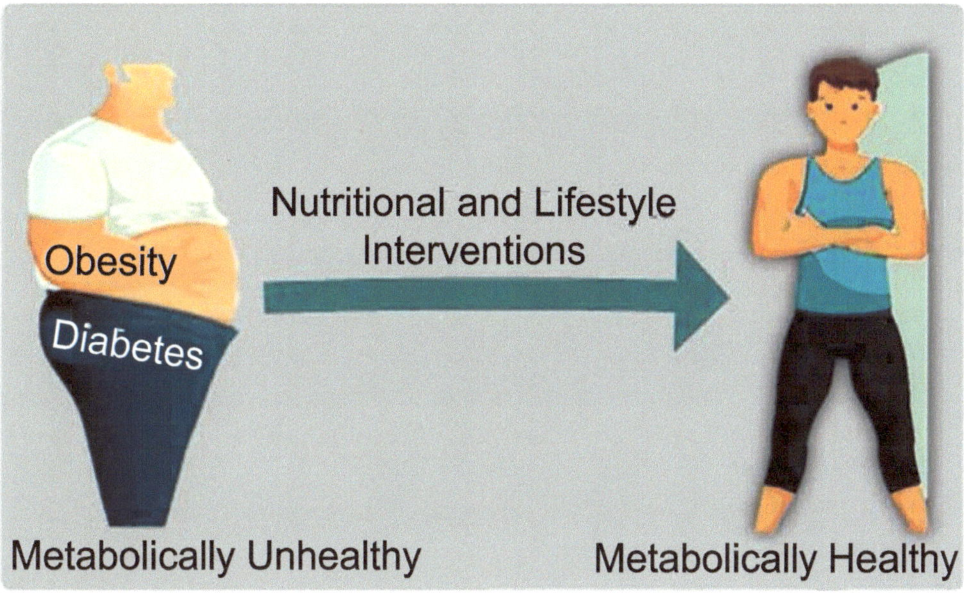

Fig. (2). Diagram showing the impact of nutritional and lifestyle interventions.

Macronutrients

Macronutrient (carbohydrates, lipids, and proteins) interventions are crucial in various health aspects. Macronutrients are the largest part of nutrition. Carbohydrates and lipids are the major energy sources while proteins are considered building blocks of the human body. In addition to this, macronutrients also serve as substrates for many metabolic pathways necessary for growth and development, regulatory processes, and adaptation. In various illnesses, all macronutrients protect the human body from injury or microbial invasion, and proper wound healing. Further, these nutrients successfully recover the person into normal conditions from a diseased state [79]. Studies have also shown that manipulating carbohydrate intake for at least 24 hours can impact sleep outcomes among healthy individuals, with higher carbohydrate diets positively influencing rapid eye movement sleep but potentially negatively affecting non-rapid eye movement sleep [80]. Furthermore, dietary macronutrient modifications have been linked to changes in body composition and metabolic markers in children and adolescents, highlighting the importance of considering macronutrient composition in obesity management [81]. These findings emphasize the significance of macronutrients in promoting overall metabolic health and addressing specific health conditions across different age groups and populations (Fig. **3**).

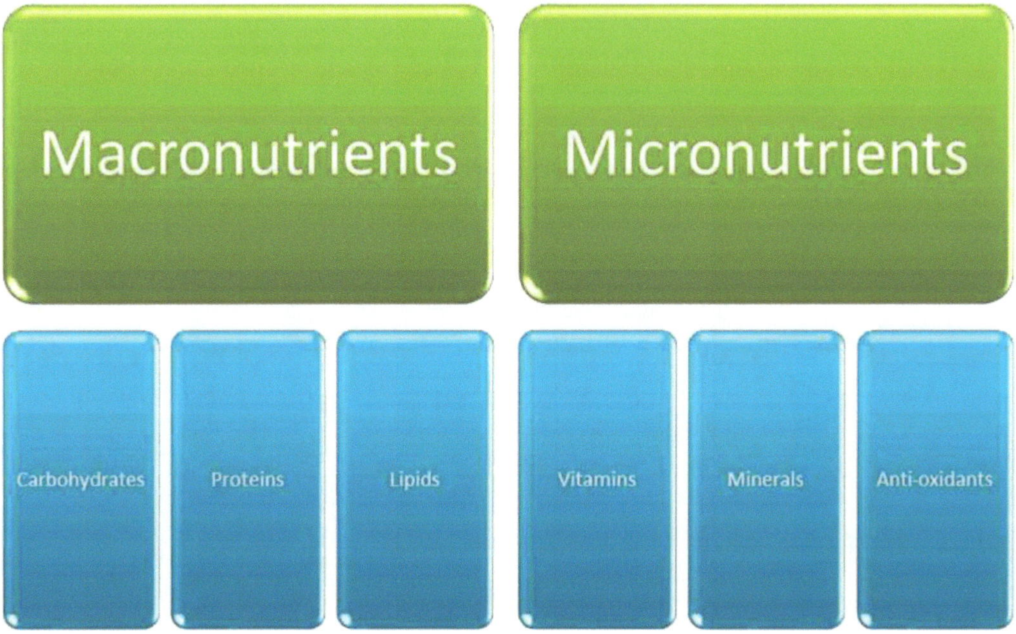

Fig. (3). General components of macronutrients and micronutrients.

Role of Carbohydrates, Proteins, And Fats in Metabolism

Quantitatively, carbohydrates are the largest energy source in a normal diet; these macronutrients meet 40-55% of the daily energy intake. The impact of carbohydrate consumption on both physical and mental health is significant, with evidence suggesting that excessive intake of simple and refined carbohydrates can lead to metabolic diseases and negatively affect mental well-being. In carbohydrates, glucose is the major carbohydrate involved in several metabolic functions [79, 82, 83]. Therefore, implementing interventions on carbohydrate metabolism is crucial for promoting overall health and well-being. In blood sugar regulation, controlling the type and amount of carbohydrates consumed can help manage blood sugar levels, mainly important for individuals with diabetes or insulin resistance. Complex sugars, like those found in whole grains and vegetables, play a key role in maintaining adequate glycemic control in individuals, especially those with diabetes. This includes choosing complex carbohydrates that digest more slowly and have a gentle impact on blood sugar levels [84]. In managing chronic diseases, carbohydrate intake can be crucial. For example, reducing the intake of refined carbohydrates (sugars) while increasing the consumption of whole grains and fiber-rich foods can improve metabolic health markers [85].

Protein metabolism is crucial in various diseases, impacting disease progression and treatment outcomes. Research indicates that muscle wasting, a result of deranged protein metabolism, is associated with increased complications in acute and chronic diseases, such as cancer treatment toxicity [86]. Furthermore, changes in plasma branched-chain amino acid levels are closely linked to the risk of developing diabetes, showcasing the significance of protein metabolism in disease development [87, 88]. Additionally, rare genetic diseases like familial hypercholesterolemia have been pivotal in advancing our understanding of lipid metabolism, leading to the development of statins for managing lipid levels and reducing cardiovascular disease risks. Moreover, a common metabolic pathway has been identified linking obesity, type-2 diabetes, hypertension, and cardiovascular diseases, emphasizing the interconnectedness of these conditions through protein-protein interactions and gene networks. These findings indicate the critical role of protein metabolism in the pathophysiology and management of various diseases [88, 89].

Lipid metabolism interventions play a crucial role in human health and disease management. Research indicates that perturbations in cholesterol and fatty acid metabolism contribute significantly to the pathophysiology of various diseases, including cardiovascular disorders, neurodegenerative diseases, and chronic pulmonary conditions [90 - 92]. Moreover, lipid-related interventions have been found to extend lifespan in various organisms through genetic, dietary, and pharmacological means, highlighting the potential for lipid metabolism modulation to improve human health span and serve as biomarkers for ageing [93]. These results are important to understanding the metabolism. Therefore, these findings target lipid metabolism to prevent and treat several human diseases.

Optimal Macronutrient Ratios for Supporting Complete Metabolism

To improve metabolism, at least 80% of energy needs with a balanced intake of macronutrients is required. Specifically, 70% of energy should come from carbohydrates and no more than 30% from fats, ensuring the inclusion of essential fatty acids [94]. Protein intake should be about 1.5 grams per kilogram of body weight per day, with higher amounts needed for patients experiencing severe stress or trauma, such as those with burn injuries. During stress states, sparing protein mass is crucial, with glucose primarily serving as a precursor for essential cellular functions, while fatty acids and ketone bodies cover energy needs, especially for the central nervous system [95]. For critically ill patients, adjusting nutrient intake based on their metabolic and physical condition is essential to prevent rapid muscle loss and support recovery. The ODMDC (Optimal Dietary Macronutrient Distribution in China) trial in China is exploring how different fat and carbohydrate intake patterns affect weight management and cardiometabolic health, underscoring the importance of a well-balanced macronutrient distribution for overall health in non-obese individuals [96, 97].

Micronutrients

Micronutrients are vital for various metabolic functions and can significantly influence health conditions such as obesity, diabetes, heart arrhythmias, and heart failure. Research studies indicate that improving micronutrient intake can enhance metabolic health by improving insulin resistance, glucose metabolism, and heart function [20]. Deficiencies in key micronutrients like magnesium, potassium, B vitamins, coenzyme Q10, calcium, zinc, iron, thiamine, vitamins D, E, and K, and folate are linked to heart failure and metabolic syndrome, emphasizing the need for adequate micronutrient consumption [98]. Additionally, studies suggest that micronutrient supplementation can affect gene expression and DNA methylation, influencing lipid metabolism and growth, as seen in organisms like Atlantic salmon. These findings emphasize the crucial role of micronutrients in metabolic health and the potential benefits of targeted micronutrient interventions for improving overall health [30].

ESSENTIAL VITAMINS AND MINERALS INVOLVED IN METABOLIC PROCESSES

Essential vitamins and minerals are fundamental to metabolic processes, supporting critical cellular functions such as energy production, DNA synthesis, oxygen transport, and neuronal activity, all of which are necessary for brain and muscle function. B vitamins (B_1, B_2, B_3, B_5, B_8, B_9, and B_{12}) and vitamin C, iron, magnesium, and zinc are particularly significant in these processes. Their deficiency can affect cognitive and psychological well-being and lead to fatigue [23]. Furthermore, minerals are vital for plant growth and metabolism, with nutrients such as nitrogen, phosphorus, potassium, iron, and zinc playing key roles in synthesizing essential metabolites for plant development and environmental adaptation. The relationship between vitamins, minerals, and metabolic pathways highlights their importance in sustaining overall health and functionality in both humans and plants [99].

Micronutrient deficiencies, such as iron, vitamin A, iodine, and zinc, are widespread globally, particularly in developing countries like India. These deficiencies contribute to serious health issues, including high rates of morbidity and mortality among children, with vitamin A deficiency alone responsible for about 0.33 million child deaths each year [100]. Although micronutrient supplementation is vital during pregnancy to support optimal fetal development, excessive supplementation can lead to negative outcomes such as spontaneous abortion and gestational diabetes mellitus [34]. In developed countries, inadequate intake of specific micronutrients has been associated with chronic diseases like cardiovascular disease and cancer. This has led to efforts to enhance micronutrient levels through dietary changes, fortification, and supplements. However, clinical trials have not consistently demonstrated benefits, suggesting that a balanced diet rich in micronutrients from whole foods may be more effective in lowering disease risk [12].

Deficiency and Supplementation Considerations

Further, in patients with Enteropathic Spondyloarthritis (ESpA), deficiencies in micronutrients like folic acid and vitamin D are prevalent, especially in those with Crohn's Disease, impacting disease phenotype and treatment choices. In heart failure patients, up to 50% are deficient in micronutrients like coenzyme Q10, zinc, and iron, affecting mitochondrial energy production and exacerbating the disease progression [101]. Moreover, obesity is associated with deficiencies in micronutrients such as iron, zinc, and vitamins A, C, D, and E, due to poor-quality diets and altered metabolism, highlighting the importance of addressing these deficiencies in obesity management [35]. Additionally, micronutrient deficiencies like iodine, iron, zinc, and selenium can worsen thyroid dysfunction, emphasizing the critical role of micronutrients in thyroid hormone synthesis and metabolism· In rheumatoid arthritis (RA), poor nutrient status and the impact of drug therapies on nutrient absorption underscore the significance of micronutrients in managing disease symptoms and progression.

Dietary Patterns

Dietary patterns are fundamental to nutritional interventions, influencing various health outcomes. Research reveals that distinct macronutrient compositions can produce different health effects, with high carbohydrate plus high fat (HCHF) diets associated with obesity and inflammatory issues [102]. Moreover, micronutrients such as iron, zinc, and multi-micronutrient supplements have proven effective in improving nutritional status, preventing anaemia, and boosting immune response in adolescents [47]. Additionally, the research underscores the role of specific nutrients like vitamins D, E, A, iron, zinc, selenium, magnesium, omega-3 fatty acids, and fiber in reducing respiratory infections, including COVID-19, and further studies are needed to establish optimal doses and mechanisms [103]. Dietary patterns, incorporating both macronutrients and micronutrients, are crucial for crafting effective nutritional strategies to enhance health and prevent conditions such as obesity, diabetes, and cardiovascular diseases [17].

Benefits of Balanced and Varied Diets

Balanced and diverse dietary patterns, such as Mediterranean and plant-based diets, offer numerous health advantages, as evidenced by extensive research. The Mediterranean diet, which emphasizes plant-based foods, olive oil, and moderate amounts of dairy and fish, is associated with a lower risk of Type 2 diabetes, obesity, cardiovascular diseases, and enhanced cognitive function [36, 104, 105]. Conversely, plant-based diets, which focus on whole grains, fruits, vegetables, and plant proteins while minimizing meat consumption, are linked to a 23% reduction in the risk of Type 2 diabetes and beneficial impacts on body weight and HbA1c levels [36]. Both dietary approaches positively affect gut microbiota, subsequently improving inflammatory responses, better metabolic health, and lower cancer risk. Therefore, it is valuable to adopt well-rounded and varied dietary practices for overall health and well-being [105].

Role of Fiber and Antioxidants

Fiber and antioxidants are essential as both macronutrient and micronutrient elements in fostering human health. Fibers including β-glucans and specific phenolics found in oats, enhance immune function and prevent several conditions such as cancer and obesity [106]. Antioxidants, which are abundant in fruits and vegetables, mitigate oxidative stress by neutralizing free radicals and promoting overall health [18, 106]. These dietary elements, abundant in bioactive compounds and vitamins, significantly impact human well-being by improving metabolic processes, decreasing inflammation, and warding off degenerative diseases. This indicates the importance of integrating fiber and antioxidants into daily diets for optimal health and disease prevention [48].

Functional Foods and Supplements

Functional foods and supplements are essential for enhancing public health outcomes and preventing diseases by providing added health benefits and reducing the risk of chronic conditions. The research underscores the value of functional foods with bioactive compounds in improving health outcomes, preventing chronic diseases, and fostering overall wellness [107, 108]. For instance, fermented milk products have demonstrated their ability to balance gut microbiota, boost immune function, and offer substantial nutritional benefits, leading to better health and well-being [109]. Additionally, functional foods are associated with lower blood pressure and cholesterol levels and reduced platelet aggregation, playing a crucial role in the primary prevention of cardiovascular diseases and other diet-related disorders [110]. Integrating functional foods and supplements into the diet, can address nutritional deficiencies, manage disease processes, and enhance protective health mechanisms, subsequently influencing metabolic health with positive outcomes [108].

PROBIOTICS AND PREBIOTICS

Probiotics and prebiotics are pivotal in boosting human health by modifying the gut microbiota. Probiotics are living microorganisms that aid the host by improving the balance of intestinal flora, whereas prebiotics are non-digestible dietary fibers that promote the growth of beneficial gut bacteria. Evidence suggests that the combined use of probiotics and prebiotics, known as synbiotics, can have both additive and synergistic health benefits, enhancing the survival and establishment of live microbial supplements and optimizing the microbial equilibrium in the gastrointestinal tract [111, 112]. Moreover, research indicates that diets rich in prebiotic fibers like inulin can alleviate the adverse effects of environmental pollutants on gut microbial health and metabolic stability, thereby lowering the risk of diseases related to toxins such as polychlorinated biphenyls (PCBs). Probiotics and prebiotics in dietary plans support overall health and disease prevention [111].

Omega-3 and Omega-6 Fatty Acids

Omega-3 fatty acids, such as alpha-linolenic acid (ALA), eicosapentaenoic acid (EPA), and docosahexaenoic acid (DHA), are commonly found in marine and plant sources, fortified foods, and dietary supplements, protecting against cardiovascular diseases, cancer, hypertension, and arthritis and omega-6 fatty acids are abundant in sources like sunflower oil but need to be balanced with omega-3 intake for optimal health benefits [37]. Research suggests that a higher prevalence of omega-3 over omega-6 intake is associated with benefits in hypertension, dyslipidemia, cardiovascular risk, and neuropathic pain, while also potentially impacting neurodegenerative and mental disorders. Combining omega-3 and omega-6 fatty acids with supplements like spirulina algae, chitosan, probiotics, and vitamin D can yield intriguing health outcomes [113]. Therefore, maintaining a balanced intake of omega-3 and omega-6 fatty acids is essential for promoting overall human health and preventing various diseases.

Herbal Supplements and Adaptogens

Herbal supplements and adaptogens are beneficial nutritional interventions due to their potential, to address stress-related issues and boost overall health. These natural substances, derived from medicinal plants, are noted for their safety, effectiveness, and affordability in treating various conditions [19, 31].

Adaptogens like Ashwagandha, *Rhodiola rosea*, and Tulsi have been well-researched for their ability to enhance resilience to stress and support physical performance [19, 38]. Moreover, traditional herbs such as *Panax ginseng* and *Withania somnifera* have long been used in Chinese, Japanese, and Indian medicine to improve athletic performance and general well-being. These herbal remedies not only mitigate the impact of stress but also bolster resistance to challenging conditions, potentially enhancing life quality and safeguarding against stress-related illnesses [19, 31].

LIFESTYLE INTERVENTIONS

Lifestyle interventions refer to comprehensive programs that incorporate dietary modifications and exercise regimens to improve metabolic health in individuals with conditions like metabolic syndrome (MetS), prediabetes, and Type 2 diabetes (T2D). These interventions are intended to reduce obesity, improve metabolism, and lower the risk of cardiovascular diseases and diabetes-related mortality. Lifestyle modifications, like diet and exercise, can significantly improve clinical parameters such as glucose levels, systolic blood pressure, weight, and waistline [114, 115]. The metabolic parameters have substantial progress due to these interventions that resulted in remission of prediabetes and even type 2 diabetes [49]. Moreover, researchers are investigating the potential of using epigenetic signatures as biomarkers to predict metabolic outcomes and develop personalized nutritional strategies for treating metabolic alterations [116]. Implementing standardized lifestyle intervention programs in primary care clinics will impact metabolic syndrome through early diagnosis and consistent education, highlighting the importance of holistic approaches to metabolic health management [114].

Physical Activity

Physical activity is any bodily movement that results in energy expenditure, encompassing activities like walking, housework, structured exercise, sports, and occupational tasks. Physical activity mainly targets metabolic conditions like MetS and T2D lifestyle interventions. Various studies have shown that interventions promoting physical activity greatly affect metabolic parameters. For instance, lifestyle interventions incorporating physical activity have been associated with reductions in waist circumference, blood pressure, fasting plasma glucose, and improvements in high-density lipoproteins (HDL) levels [24, 25]. Engaging in physical activity has been linked to enhanced physical fitness, including aerobic capacity, muscular fitness, and flexibility, which are essential components of metabolic health [24].

Types of Exercises Beneficial for Metabolic Health

Aerobic Exercise

Aerobic exercise has been extensively studied for its positive effects on metabolic health. Research indicates that aerobic exercise significantly improves glucose levels, blood pressure, glycosylated haemoglobin, HDL levels, and triglyceride levels in diabetic older adults without significant adverse effects [62]. Additionally, regular aerobic exercise has been shown to improve lipid profiles, reduce the risk of cardiovascular diseases, and positively impact physical and mental well-being, making it a valuable non-pharmacological treatment option for various diseases [63]. Aerobic exercise training raises antiatherogenic apolipoproteins and lipoprotein sub-fractions while lowering atherogenic ones, ultimately reducing cardiovascular disease risk when prescribed as treatment or prevention therapy [117]. In patients with insulin resistance, aerobic exercise improves vasodilation, increasing blood flow to engaged skeletal muscles with better glucose absorption [118].

Resistance Training

Regular participation in well-designed resistance training programs enhances muscular strength and power in children and adolescents. It also improves bone mineral density, cardiovascular risk factors, metabolic health, and psychosocial well-being, preparing them for physical activities and sports [119]. In adults, especially those who have undergone bariatric surgery, resistance training ameliorates body composition, muscular strength, and weight management, with a good quality of life [50]. Resistance exercise has been shown to have noteworthy positive effects on cardiovascular health, musculoskeletal health, neuropsychological well-being, and metabolic parameters, making it an effective intervention for disease prevention and management, including MetS [120, 121].

Flexibility and Balance Exercises

Flexibility and balance exercises promote metabolic health by enhancing flexibility and restoring energy balance across the mitochondria. Various investigations demonstrated that maneuvers like exercise, caloric restriction, and exposure to cold or hypobaric hypoxia can help maintain energy balance and activate pathways that repair damaged molecules, thus improving metabolic health [122]. High-intensity exercise can improve metabolic outcomes in individuals with non-alcoholic fatty liver disease (NAFLD) by enhancing fat oxidation at rest and carbohydrate oxidation during exercise, ultimately reducing hepatic fat content and improving metabolic profiles [64]. A novel method for assessing metabolic flexibility through indirect calorimetry has been proposed, linking metabolic flexibility to the balance of specific biomarkers and endocannabinoid systems, and highlighting the importance of exercise intensity in influencing metabolic health [65].

Sleep and Circadian Rhythms

Sleep and circadian rhythms promote various physiological functions and metabolic pathways. The circadian clock system, primarily controlled by the suprachiasmatic nucleus, orchestrates molecular clock rhythms in peripheral tissues, influencing metabolic homeostasis at a systemic level. Disruptions in circadian rhythms, such as those caused by shift work or jet lag, can lead to metabolic diseases like obesity and type 2 diabetes. Food intake acts as a powerful zeitgeber for entraining molecular clocks and regulating metabolic pathways, emphasizing the importance of the timing of food consumption for metabolic health. Sleep quality and quantity are closely linked to metabolic changes. Understanding the intricate interplay between sleep, circadian rhythms, and nutrition is essential for developing tailored strategies to improve metabolic health and prevent disease [66, 67, 123 - 125].

IMPORTANCE OF SLEEP IN METABOLIC REGULATION

Metabolic Processes During Sleep

Metabolic processes along with sleep disorders are associated with adverse metabolic effects such as obesity, insulin resistance, and type 2 diabetes mellitus [126]. Sleep induces a shift in oxidized substrates from carbohydrates to fats, leading to a decrease in respiratory quotient, indicating metabolic flexibility [127]. Studies on metabolomics reveal alterations in specific metabolic pathways during sleep-wake disorders, particularly affecting branched-chain amino acid, glucose, and lipid metabolism, potentially contributing to the association with psychiatric disorders and medical conditions [128]. Physiological regulation during non-rapid eye movement (NREM) and rapid eye movement (REM) sleep differs, with NREM sleep focused on maintaining body homeostasis at a low energy expenditure level, while REM sleep shows cardiorespiratory instability and thermoregulatory impairment [51]. Sleep is essential for anabolic activities, and clearance of waste products, and parallels ageing processes, with metabolomics offering insights into the interconnected effects of sleep, ageing, and related disorders at a molecular level [129].

Sleep Deprivation and Metabolism

Sleep deprivation is prevalent in modern society and is linked to neurodegenerative and cardiometabolic disorders, impacting learning, memory, and metabolic features across different tissues [52]. Inadequate sleep disrupts nutrient metabolism, appetite regulation, and energy balance, increasing the risk of obesity, diabetes, and heart disease [130]. Insufficient sleep induces a positive energy balance, leading to weight gain due to increased energy intake, while animal models show conflicting results regarding energy expenditure and weight changes [53]. Moreover, the relationship between sleep deficiency and obesity is well-established, with suboptimal sleep recognized as a determinant of metabolic dysfunction, emphasizing the importance of addressing sleep as a crucial lifestyle factor for metabolic health [131].

Strategies for Improving Sleep Quality

Quality sleep is enhanced by various strategies that have been identified through research. Behavioural approaches, such as Benson's relaxation technique, progressive muscle relaxation, education, aromatherapy, acupressure, massage, and eye masks, have shown efficacy in improving postoperative sleep quality [132]. Among adolescents, interventions like Surya Namaskar and deep breathing exercises proved effective for sleep quality and reducing stress levels [54]. Quality improvement projects (QIP) are essential in healthcare systems, including those related to sleep health, and involve common models, tools, and data analysis techniques [55]. Further, sleep health involves considering sleep quality, quantity, and timing of the sleep window, with a shift towards behavioural approaches like cognitive-behavioural therapy for insomnia management over hypnotic medications [133].

STRESS MANAGEMENT

Research has shown that stress management interventions can significantly improve health outcomes in patients with metabolic syndrome and psychological symptoms, including depression, anxiety, and schizophrenia [134]. Sestrins, a group of stress-responsive proteins, have been identified as key regulators of metabolic stress and redox homeostasis, offering potential preventive and therapeutic benefits in metabolic stress-related pathologies and ageing processes [135]. Financial stress has been linked to an increased risk of metabolic syndrome, with physical activity and optimal sleep duration acting as moderators that can attenuate this association, highlighting the importance of addressing behavioural adaptations in financially stressed individuals to reduce the metabolic risk [56]. Smoking cessation has been associated with positive changes in metabolic risk behaviours, with readiness to change behaviours, and stress management playing crucial roles in mediating and moderating these effects, emphasizing the interconnectedness of stress management and metabolic health [21].

Impact of Stress on Metabolism

Stress exerts a significant impact on metabolism across various organisms. Maternal stress during pregnancy can lead to metabolic programming effects in offspring, resulting in low birth weight and hyperglycemia [136]. In mice, acute stress induces the immediate production of lactate, a central metabolite in unstressed mammalian metabolism, and alters glucose flux, highlighting the role of lactate in stress response [137]. Environmental stressors like heat and drought affect grape berry metabolism, with heat stress significantly influencing organic acid content and gene expression in berries, indicating a distinct response to stress in different plant organs [138]. Chronic stress, particularly from factors like aircraft noise, can induce detrimental metabolic shifts in the heart within a short exposure time, resembling patterns seen in cardiovascular diseases, emphasizing the need to understand molecular consequences of environmental stressors for targeted interventions [68]. Stress, including that from wartime conditions, is associated with metabolic-associated fatty liver disease (MAFLD) development, involving disturbances in various processes and the activation of stress hormones and proinflammatory biomarkers, underscoring the relevance of investigating stress effects on MAFLD pathogenesis and potential therapeutic interventions [139].

Techniques for Stress Reduction

Stress reduction techniques help in improving overall metabolic health. Studies have shown that teaching adults stress management techniques, such as focused attention on breathing, muscular relaxation, guided imagery, and cognitive reorganization, can lead to a relaxation response and enhance the metabolic control of conditions like diabetes mellitus type 2 (DM2) [140]. Also, the B-R-E-A-T-H-E technique, a stress reduction exercise involving breath work and guided imagery, has been proposed to lower serum triglycerides, blood sugar, and C-reactive protein levels in patients with the metabolic syndrome, ultimately improving cardiovascular risk factors [141]. Moreover, disturbances in metabolism, like reductive stress caused by excess food and alcohol consumption, can be alleviated through exercise and nutritional interventions, highlighting the interconnectedness of stress, metabolism, and overall health [142].

HYDRATION

Hydration refers to the state of the body having sufficient fluids to support optimal cell metabolism, which is crucial for both adults and children [57]. Proper hydration is essential for maintaining health, reducing fatigue, and promoting overall well-being, as observed in studies on sedentary individuals and athletes [39].

Role of Water in Metabolic Processes

Water is important in various metabolic processes as a catalyst and regulator. It is essential for human metabolism, controlling blood volume, body temperature, and muscle function [143]. Metabolic water, produced during respiration, influences the isotopic signature within organisms, impacting oxygen-stable isotopes like phosphate [144]. Hydration levels have been linked to metabolic health, with studies showing that water intake can modulate metabolic disease risk by affecting glycemic control and vasopressin levels [75].

Hydration Strategies and Guidelines

Hydration strategies and guidelines are necessary for optimizing human performance in various operations. Special Operations Forces (SOF) emphasize the importance of intentional hydration planning to prevent performance degradation and exertional heat illness [145]. In the athletic realm, various methods exist for assessing hydration status, ranging from invasive to non-invasive techniques, with a combination approach recommended for increased accuracy [26]. It is noted that dehydration of $\geq$ 2% body mass can impair thermoregulatory function and exercise performance, highlighting the significance of maintaining adequate hydration levels, especially during longer or high-intensity activities [146]. However, challenges persist in establishing evidence-based guidelines for intrapartum hydration assessment and management, particularly during labor and birth, due to the lack of specific recommendations and potential harm to laboring women [147]. Therefore, a comprehensive approach integrating different assessment methods and tailored strategies is essential for effective hydration management across various domains.

INTEGRATIVE AND HOLISTIC APPROACHES

Understanding overall metabolic health involves considering various factors such as lifestyle interventions, metabolic responses to stress, and the impact of different oncogenes on metabolic pathways. These approaches mainly focus on treating the whole person through lifestyle modifications. These lifestyle modifications include vigorous physical activity, proper sleep, eating-based habits, *etc.* Additionally, these approaches emphasize the importance of addressing the root causes that impact a person's metabolic health [1 - 4].

Personalized Nutrition

Personalized nutrition, also known as precision nutrition, aids in developing specific nutrition or dietary recommendations to overcome metabolic disorders based on individual microbiota factors, genetic makeup, physical activity, food habits, and behaviour [22, 148, 149]. Under the spotlight of these factors, personalized nutrition aims to predict the type of diet a person requires according to their metabolic alterations, ultimately influencing the metabolic rate and improving overall health outcomes [148]. Integrating technologies like continuous glucose monitoring, genomic sequences, and microbiome analysis allows for the development of tailored dietary and lifestyle advice, showing promising results in managing chronic metabolic diseases like diabetes through personalized interventions [150].

Personalized nutrition involves tailoring dietary recommendations based on an individual's biology, lifestyle, behaviour, and environment to prevent and manage metabolic diseases [149]. This approach utilizes advancements in genomics, metabolomics, and gut microbiome technologies to provide personalized dietary and lifestyle advice, considering factors like genetic variants that influence nutrient metabolism [151]. The gut microbiota also plays a crucial role in modulating human physiology through diet-microbiota interactions, impacting digestion, absorption, and immune responses, thus influencing metabolism and overall health [40]. By integrating omics technologies and deep phenotyping, personalized nutrition strategies can be developed to improve metabolic balance, enhance nutrient metabolism, and address individual variations in responses to dietary inputs, ultimately promoting better health outcomes and metabolic well-being.

Genetic and Epigenetic Considerations in Dietary Planning

Genetic and epigenetic factors play crucial roles in influencing metabolism and can be considered in dietary planning to improve metabolic health. Nutrigenomics and nutrigenetics focus on gene-nutrient interactions, offering personalized nutritional recommendations based on genetic variations [27, 41]. Epigenetic modifications, such as histone acetylation and DNA methylation, can impact gene expression without altering DNA sequences, affecting metabolic processes and disease development [152]. Understanding how specific genetic variants influence responses to dietary components and how diet can modulate gene expression is essential in optimizing metabolic health through various interventions [27]. By considering genetic predispositions and epigenetic modifications, tailored dietary plans can be designed to promote optimal metabolism and prevent metabolic disturbances like Type 2 diabetes and cardiovascular diseases [27, 41, 152].

Role of Personalized Nutrition in Optimizing Metabolic Health

Personalized nutrition is important in regulating metabolic health through dietary recommendations to individual genetic and metabolic profiles. By integrating genetics and metabolomics data, personalized nutrition strategies can identify specific nutrient metabolism pathways affected by genetic variations, such as single nucleotide polymorphisms (SNPs) [153]. This approach enables the design of personalized diets that align with an individual's genetic makeup, influencing gene expression and health outcomes [154]. Studies have shown that personalized nutrition advice based on metabotypes can significantly improve dietary quality and metabolic health biomarkers, reducing cardiovascular disease risk and improving metabolic profiles [58]. Furthermore, interventions like medical nutrition therapy (MNT) within personalized nutrition frameworks have consistently enhanced outcomes related to glycaemic control, anthropometry, and blood lipids, emphasizing the efficacy of personalized approaches by enhancing metabolic health [69].

INTEGRATIVE MEDICINES

Combining Conventional and Alternative Therapies

Integrative medicine is vital in optimizing metabolism by combining conventional and alternative therapies to enhance patient outcomes and improve overall quality of life. Research highlights the potential of dietary supplements, herbal extracts, and phytochemicals in treating metabolic diseases, showcasing their insulin-sensitizing, antioxidant, and anti-inflammatory properties [70]. Additionally, an integrative metabolic therapeutic approach has been shown to improve functional capacity and quality of life in patients with heart failure and reduce ejection fraction [155]. Furthermore, conventional and complementary therapies in renal health have been emphasized, underlining the importance of patient-centered and holistic care models to enhance outcomes for individuals with renal conditions [4]. Integrative therapies, such as those used in addressing poor performance issues in horses, offer unique perspectives on disease patterns and therapeutic approaches, complementing conventional methods when needed [156]. Integrating conventional and alternative therapies in integrative medicine presents a promising avenue for optimizing metabolism and improving overall health outcomes.

CONCLUSION

Chronic kidney diseases (CKD) and cardiovascular diseases (CVD) are posing significant threats worldwide. As a result, active efforts have been made to reduce its progression. Metabolic diseases like obesity, uncontrolled diabetes, and hypertension tend to increase their progression. Nutritional and lifestyle modifications, including physical activity, weight loss, healthy dietary patterns, alcohol and smoking cessation, control diabetes and hypertension thereby delaying their progression. However, approved medications like pentoxifylline, RAAS blockers, finer enone, and SGLT2 inhibitors, while statins and β-blockers are used to reduce the progression of diseases related to the kidney and heart, respectively. The role of various other agents is under investigation through clinical trials.

Micronutrients, including vitamins, minerals, and antioxidants, are equally crucial. Their adequate intake, whether through diet or supplementation, is necessary for maintaining immune function, cognitive health, and bone strength. Deficiencies in these essential nutrients can lead to significant health issues. Beyond nutrition, lifestyle choices like regular physical activity, effective stress management, and sufficient sleep play vital roles in maintaining health. Regular exercise strengthens the body and enhances mental health by reducing stress and anxiety. Incorporating stress-reduction techniques such as yoga, meditation, and deep breathing fosters emotional resilience and mental clarity. Meanwhile, maintaining a consistent and restful sleep pattern is indispensable for sustaining energy levels and overall health.

In conclusion, making informed and proactive choices in diet and lifestyle can significantly lower the risk of chronic diseases, enhance metabolic processes, and promote long-term health and well-being. Evidence demonstrates that dietary patterns rich in fruits, vegetables, whole grains, and lean proteins, with less consumption of processed foods, refined sugars, and highly saturated fats, are pivotal in preventing chronic diseases such as obesity, diabetes, and cardiovascular disorders. These dietary habits contribute to physical health along with optimal metabolic function, ensuring the metabolic processes run efficiently.

REFERENCES

[1] K.N. Frayn, *Understanding human metabolism.* Cambridge University Press, 2022. [http://dx.doi.org/10.1017/9781009104463]

[2] L. Outland, "Redefining lifestyle to improve metabolic health", *Internet J Adv Nurs Pract,* vol. 17, 2018. [http://dx.doi.org/10.5580/IJANP.53209]

[3] C. Campbell, D. Grapov, O. Fiehn, C. J. Chandler, D. J. Burnett, E. C. Souza, G. A. Casazza, M. B. Gustafson, N. L. Keim, J. W. Newman, G. R. Hunter, J. R. Fernandez, W. T. Garvey, M.-E. Harper, C. L. Hoppel, J. K. Meissen, K. Take, and S. H. Adams, “Improved metabolic health alters host metabolism in parallel with changes in systemic xeno-metabolites of gut origin,” *PLoS ONE,* vol. 9, no. 1, 2014. [http://dx.doi.org/10.1371/journal.pone.0084260]

[4] Y. Kalariya, A. Kumar, A. Ullah, A. Umair, F. Neha, F. Madhurita, V. Varagantiwar, S. M. I. A. Jaffari, A. Ahmad, M. Aman, F. Sapna, G. Varrassi, S. Kumar, and M. Khatri, "Integrative medicine approaches: bridging the gap between conventional and renal complementary therapies", *Cureus*, vol. 15, no. 9, p. e46033, 2023. [http://dx.doi.org/10.7759/cureus.46033] [PMID: 37900457]

[5] T.C. Gheonea, C.N. Oancea, M. Mititelu, E.C. Lupu, C.B. IoniE>ŽŸ-MA(r)ndrican, and I. Rogoveanu, "Nutrition and Mental Well-Being: Exploring Connections and Holistic Approaches", *J. Clin. Med.,* vol. 12, no. 22, p. 7180, 2023.

[http://dx.doi.org/10.3390/jcm12227180] [PMID: 38002792]

[6] S. Espinosa-Salas, and M. Gonzalez-Arias, "Nutrition: micronutrient intake, imbalances, and interventions", In: *StatPearls* StatPearls Publishing, 2023.

[7] L. Bell, S. Edwards, and J. Grieger, "The relationship between dietary patterns and metabolic health in a representative sample of adult Australians", *Nutrients,* vol. 7, no. 8, pp. 6491-6505, 2015. [http://dx.doi.org/10.3390/nu7085295] [PMID: 26251918]

[8] L. Brown, H. Poudyal, and S.K. Panchal, "Functional foods as potential therapeutic options for metabolic syndrome", *Obes. Rev.,* vol. 16, no. 11, pp. 914-941, 2015. [http://dx.doi.org/10.1111/obr.12313] [PMID: 26345360]

[9] F. Magkos, M. Yannakoulia, J.L. Chan, and C.S. Mantzoros, "Management of the metabolic syndrome and type 2 diabetes through lifestyle modification", *Annu. Rev. Nutr.,* vol. 29, no. 1, pp. 223-256, 2009. [http://dx.doi.org/10.1146/annurev-nutr-080508-141200] [PMID: 19400751]

[10] J. M. Rippe and T. J. Angelopoulos, "Lifestyle strategies for risk factor reduction, prevention and treatment of cardiovascular disease," In: *Lifestyle Medicine*, 3rd ed., CRC Press, 2019, pp. 18. [http://dx.doi.org/10.1201/9781315201108-2]

[11] R. Dhiman, "Novel LC-MS/MS bioanalytical method validation for gabapentin in human plasma in accordance with current regulatory guidances," *Inter. J. Pharm. Chem. Sci.,* 2015.

[12] J.V. Woodside, D. McCall, C. McGartland, and I.S. Young, "Micronutrients: dietary intake v. supplement use", *Proc. Nutr. Soc.,* vol. 64, no. 4, pp. 543-553, 2005. [http://dx.doi.org/10.1079/PNS2005464] [PMID: 16313697]

[13] R. Dhiman, V. Durga, and K. Hamlet, "Determination of S-amlodipine in human plasma using tizanidine as internal standard by LC/MS/MS method", *J Bioanal Biomed,* vol. 7, 2015.

[14] R. Dhiman, V. Durga, and A. Rio, "Bioanalytical method for lornoxicam determination in human plasma by using piroxicam as internal standard by LC-MS/MS",", *J. Chem. Pharm. Res.,* vol. 7, pp. 54-60, 2015.

[15] R. Dhiman, D. Kumar, B. Kumar, and B. Pandey, "Quantitative determination of atorvastatin, ortho-hydroxy atorvastatin, para-hydroxy atorvastatin in human plasma using rosuvastatin as internal standard by LC-MS/MS", *Int J Pharm Chem Sci,* vol. 4, pp. 487-500, 2016.

[16] R. Dhiman, B. Kumar, D. Kumar, and A. Kumar, "Liquid-liquid extraction method for Ziprasidone (ZRS) bioanalysis by using ZRS-D8 (stable isotope) as internal standard", *J. Chem. Pharm. Res.,* vol. 8, no. 3, pp. 722-727, 2016.

[17] L.C. Tapsell, E.P. Neale, A. Satija, and F.B. Hu, "Foods, nutrients, and dietary patterns: interconnections and implications for dietary guidelines", *Adv. Nutr.,* vol. 7, no. 3, pp. 445-454, 2016. [http://dx.doi.org/10.3945/an.115.011718] [PMID: 27184272]

[18] J. Manivannan, T. Silambarasan, J. Shanthakumar, N. Suganya, and S. Kanchana, "Role of antioxidants in human health", In: *Omega-3 Fatty Acids: Keys Nutr Health*, 2016, pp. 501-512. [http://dx.doi.org/10.1007/978-3-319-40458-5_36]

[19] K. Gulati, R. Anand, and A. Ray, "Nutraceuticals as adaptogens: their role in health and disease", In: *Nutraceuticals.* Elsevier, 2016, pp. 193-205. [http://dx.doi.org/10.1016/B978-0-12-802147-7.00016-4]

[20] E. Parsi, N. Bitterlich, A. Winkelmann, D. RAsler, and C. Metzner, "Dietary intervention with a specific micronutrient combination for the treatment of patients with cardiac arrhythmias: the impact on insulin resistance and left ventricular function", *BMC Cardiovasc. Disord.,* vol. 18, no. 1, p. 220, 2018. [http://dx.doi.org/10.1186/s12872-018-0954-6] [PMID: 30509185]

[21] E.A. Vogel, and D.E. Ramo, "Smoking cessation, metabolic risk behaviors, and stress management

over time in a sample of young adult smokers", *Transl. Behav. Med.,* vol. 11, no. 1, pp. 189-197, 2021. [http://dx.doi.org/10.1093/tbm/ibz139] [PMID: 31595303]

[22] S. Asgary, M. Zeinalian, M. Eshaghi, M. Hadian, and H. Naji, "Personalized nutrition: An ancient concept in Iranian traditional medicine", *Int. J. Prev. Med.,* vol. 10, no. 1, p. 94, 2019. [http://dx.doi.org/10.4103/ijpvm.IJPVM_122_18] [PMID: 31360341]

[23] A.L. Tardy, E. Pouteau, D. Marquez, C. Yilmaz, and A. Scholey, "Vitamins and minerals for energy, fatigue and cognition: a narrative review of the biochemical and clinical evidence", *Nutrients,* vol. 12, no. 1, p. 228, 2020. [http://dx.doi.org/10.3390/nu12010228] [PMID: 31963141]

[24] I. Sequi-Dominguez, C. Alvarez-Bueno, V. Martinez-Vizcaino, R. Fernandez-Rodriguez, A. del Saz Lara, and I. Cavero-Redondo, "Effectiveness of mobile health interventions promoting physical activity and lifestyle interventions to reduce cardiovascular risk among individuals with metabolic syndrome: Systematic review and meta-analysis", *J. Med. Internet Res.,* vol. 22, no. 8, p. e17790, 2020. [http://dx.doi.org/10.2196/17790] [PMID: 32865503]

[25] Tremblay A., Bélanger M.-P., Dhaliwal R., Brauer P., Royall D., Mutch D.M., Rhéaume C., "Impact of a multidisciplinary intervention on physical fitness, physical activity habits and the association between aerobic fitness and components of metabolic syndrome in adults diagnosed with metabolic syndrome", *Arch Public Health,* vol. 78, no. 1, p. 22, 2020. [http://dx.doi.org/10.1186/s13690-020-0399-0]

[26] O.R. Barley, D.W. Chapman, and C.R. Abbiss, "Reviewing the current methods of assessing hydration in athletes", *J. Int. Soc. Sports Nutr.,* vol. 17, no. 1, p. 52, 2020. [http://dx.doi.org/10.1186/s12970-020-00381-6] [PMID: 33126891]

[27] M. Franzago, D. Santurbano, E. Vitacolonna, and L. Stuppia, "Genes and diet in the prevention of chronic diseases in future generations", In: *Int J Mol Sci,* vol. 21., 2020, no. 7, p. 2633. [http://dx.doi.org/10.3390/ijms21072633]

[28] S.N. Rai, N. Tiwari, P. Singh, D. Mishra, A.K. Singh, E. Hooshmandi, E. Vamanu, and M.P. Singh, ""Therapeutic potential of vital transcription factors in Alzheimer's and Parkinson's disease with particular emphasis on transcription factor EB mediated autophagy."", *Front. Neurosci.,* vol. 15, p. 777347, 2021. [http://dx.doi.org/10.3389/fnins.2021.777347] [PMID: 34970114]

[29] A. Kumar, M.K. Nandi, B. Kumar, A. Kumar, R. Kumar, A.K. Singh, and M. Singh, "Antiurolithiasis, antioxidant, anti-inflammatory, analgesic, and diuretic activity of ethanolic extract of seeds of Caesalpinia bonducella", *Int. J. Pharm. Investig.,* vol. 11, no. 3, pp. 306-311, 2021. [http://dx.doi.org/10.5530/ijpi.2021.3.54]

[30] T. Saito, P. Whatmore, J.F. Taylor, J.M.O. Fernandes, A.C. Adam, D.R. Tocher, M. Espe, and K.H. SkjAÝrven, "Micronutrient supplementation affects transcriptional and epigenetic regulation of lipid metabolism in a dose-dependent manner", *Epigenetics,* vol. 16, no. 11, pp. 1217-1234, 2021. [http://dx.doi.org/10.1080/15592294.2020.1859867] [PMID: 33315488]

[31] A. Ray, K. Gulati, S. Rehman, N. Rai, and R. Anand, "Role of nutraceuticals as adaptogens", In: *Nutraceuticals.* Elsevier, 2021, pp. 229-244. [http://dx.doi.org/10.1016/B978-0-12-821038-3.00016-1]

[32] A. Kumar, "Acute and subacute toxicity study of ethanolic extract of Calotropis procera (Aiton) Dryand flower in Swiss albino mice", *Phytomed Plus,* vol. 2, no. 2, p. 100224, 2022. [http://dx.doi.org/10.1016/j.phyplu.2022.100224]

[33] R. Garg, "COVID-19: A challenge towards the sustainability of health in platform era," *J. Pharm. Negat. Results*, pp. 1666–1678, 2022.

[34] S. Sairoz, K. Prabhu, R. G. Dastidar, A. R. Aroor, M. Rao, S. Shetty, V. G. Poojari, and V. B. S., "Micronutrients in adverse pregnancy outcomes," *F1000Research,* vol. 11, p. 1369, 2024.

[35] C.E. Guardiola-MArquez, M.T. Santos-RamA-rez, M.E. Segura-JimA(c)nez, M.L. Figueroa-Montes, and D.A. Jacobo-VelAzquez, "Fighting obesity-related micronutrient deficiencies through biofortification of agri-food crops with sustainable fertilization practices", *Plants,* vol. 11, no. 24, p. 3477, 2022.
[http://dx.doi.org/10.3390/plants11243477] [PMID: 36559589]

[36] M. Charkviani, C. Thongprayoon, S. Tangpanithandee, P. Krisanapan, J. Miao, M.A. Mao, and W. Cheungpasitporn, "Effects of Mediterranean diet, DASH diet, and Plant-based diet on outcomes among end stage kidney disease patients: a systematic review and meta-analysis", *Clin. Pract.,* vol. 13, no. 1, pp. 41-51, 2022.
[http://dx.doi.org/10.3390/clinpract13010004] [PMID: 36648844]

[37] A. Mukhametov, M. Yerbulekova, G. Aitkhozhayeva, G. Tuyakova, and D. Dautkanova, ""Effects of ω-3 fatty acids and ratio of ω-3/ω-6 for health promotion and disease prevention", ", *Food Sci. Technol. (Campinas),* vol. 42, p. e58321, 2022.
[http://dx.doi.org/10.1590/fst.58321]

[38] S. Rathore, B.K. Khuntia, M. Wadhawan, V. Sharma, and G. Sharma, "Herbal adaptogens: An integrative approach for enhancing performance and resilience in athletes", *Phytother. Res.,* vol. 36, no. 10, pp. 3678-3680, 2022.
[http://dx.doi.org/10.1002/ptr.7529] [PMID: 35751417]

[39] G. Marin-E~tefan, M. Horhocea, D. Horhocea, and M. Badea, "Hidratarea*f*?"o problemŽŸ de sŽŸnŽŸtate publicŽŸ|[Hydration-a public health issue]", *J. Med. Bras.,* vol. *f*?›*f*?›*f*?›, pp. 26-32, 2022.

[40] M. H. Sarfraz, A. Shahid, S. Asghar, B. Aslam, U. A. Ashfaq, H. Raza, M. A. Prieto, J. Simal-Gandara, F. J. Barba, M. S. R. Rajoka, M. Khurshid, and A. J. Nashwan, "Personalized nutrition, microbiota, and metabolism: A triad for eudaimonia," Front. Mol. Biosci., vol. 9, 2022.
[http://dx.doi.org/10.3389/fmolb.2022.1038830]

[41] A.K. Kiani, G. Bonetti, K. Donato, J. Kaftalli, K.L. Herbst, L. Stuppia, F. Fioretti, S. Nodari, M. Perrone, P. Chiurazzi, F. Bellinato, P. Gisondi, and M. Bertelli, "Polymorphisms, diet and nutrigenomics", *J. Prev. Med. Hyg.,* vol. 63, no. 2, suppl. 3, pp. E125-E141, 2022.
[PMID: 36479483]

[42] A.K. Alzubaidi, W.J. Al-Kaabi, A.A. Ali, S. Albukhaty, H. Al-Karagoly, G.M. Sulaiman, M. Asiri, and Y. Khane, "Green synthesis and characterization of silver nanoparticles using flaxseed extract and evaluation of their antibacterial and antioxidant activities", *Appl. Sci. (Basel),* vol. 13, no. 4, p. 2182, 2023.
[http://dx.doi.org/10.3390/app13042182]

[43] M. Singh, B. Kumar, A. K Singh, C. Shekhar Azad, M. K Yadav, A. Kumar, R. Kumar, A. Kumar, and P. Chaudhary, "Pharmacokinetic screening to Estimate the drug likeliness characteristics of selected Herbal Anticancer Drugs", *Res. J. Pharma. Technol.,* vol. 16, no. 7, pp. 3422-3426, 2023.
[http://dx.doi.org/10.52711/0974-360X.2023.00566]

[44] M.A. Alnawwar, M.I. Alraddadi, R.A. Algethmi, G.A. Salem, M.A. Salem, and A.A. Alharbi, "The effect of physical activity on sleep quality and sleep disorder: a systematic review", *Cureus,* vol. 15, no. 8, p. e43595, 2023.
[http://dx.doi.org/10.7759/cureus.43595] [PMID: 37719583]

[45] V. J. Clemente-Suárez, A. Martín-Rodríguez, L. Redondo-Flórez, C. López-Mora, R. Yañez-Sepúlveda, and J. F. Tornero-Aguilera, "New insights and potential therapeutic interventions in metabolic diseases", *Int. J. Mol. Sci.,* vol. 24, no. 13, p. 10672, 2023.
[http://dx.doi.org/10.3390/ijms241310672] [PMID: 37445852]

[46] L. S. Prado, K. P. Rodrigues, and M. L. Almeida, "Effect of a nutritional intervention associated with different models of physical training on health parameters of adult and elderly women: A randomized trial," *Preprint*, 2023.

[http://dx.doi.org/10.21203/rs.3.rs-2621949/v1]

[47] Sulfianti, L. Fitriani, Hasnidar, Mustar, and D. Yanti, "Efficacy of macro and micronutrient interventions in adolescent nutritional status: A systematic review", *Influence: Int. J. Sci. Rev.,* vol. 5, no. 2, pp. 176-183, 2023.
[http://dx.doi.org/10.54783/influencejournal.v5i2.147]

[48] V. Dixit, S.W. Joseph Kamal, P. Bajrang Chole, D. Dayal, K.K. Chaubey, A.K. Pal, J. Xavier, B.T. Manjunath, and R.K. Bachheti, "Functional foods: exploring the health benefits of bioactive compounds from plant and animal sources", *J. Food Qual.,* vol. 2023, no. 1, pp. 1-22, 2023.
[http://dx.doi.org/10.1155/2023/5546753]

[49] J. Iglesies-Grau, "Acute and subacute toxicity study of ethanolic extract of Calotropis procera (Aiton) Dryand flower in Swiss albino mice", *Phytomed Plus,* vol. 2, no. 2, p. 100224, 2022.
[http://dx.doi.org/10.1016/j.ajpc.2022.100397]

[50] W. Elks, A. Rooks, S. Schulte, K. Batra, J. Burke, and V. Jain, "Resistance training in patients after metabolic and bariatric surgery: Protocol for a systematic review", *JMIR Res. Protoc.,* vol. 12, p. e49513, 2023.
[http://dx.doi.org/10.2196/49513] [PMID: 37995123]

[51] R. Amici, M. Cerri, T. Hitrec, and G. Zoccoli, "Overview of physiological processes during sleep", In: *Encyclopedia of Sleep and Circadian Rhythms.,* C.A. Kushida, Ed., 2nd ed. Academic Press: Oxford, 2023, pp. 1-10.
[http://dx.doi.org/10.1016/B978-0-12-822963-7.00178-X]

[52] A. Sengupta, J.C. Tudor, D. Cusmano, J.A. Baur, T. Abel, and A.M. Weljie, "Sleep deprivation and aging are metabolically linked across tissues", *Sleep,* vol. 46, no. 11, 2023.
[http://dx.doi.org/10.1093/sleep/zsad246] [PMID: 37738102]

[53] D. Duan, L.J. Kim, J.C. Jun, and V.Y. Polotsky, "Connecting insufficient sleep and insomnia with metabolic dysfunction", *Ann. N. Y. Acad. Sci.,* vol. 1519, no. 1, pp. 94-117, 2023.
[http://dx.doi.org/10.1111/nyas.14926] [PMID: 36373239]

[54] P-P. Shital, and D.L. Bijayalakshmi, "Effectiveness of sleep hygiene strategies on sleep quality and level of stress among adolescents", *Imanagers J. Nurs.,* vol. 12, no. 4, pp. 4-12, 2023.
[http://dx.doi.org/10.26634/jnur.12.4.19337]

[55] K.Y. Kreitinger, and C.A. Kushida, "Quality improvement in sleep medicine", In: *Encyclopedia of Sleep and Circadian Rhythms.,* C.A. Kushida, Ed., 2nd ed. Academic Press: Oxford, 2023, pp. 618-624.
[http://dx.doi.org/10.1016/B978-0-12-822963-7.00189-4]

[56] W.C. Kuo, L.C. Bratzke, E.W. Hagen, L. Hale, R.L. Brown, J.H. Barnet, and P.E. Peppard, "Metabolic health disparities driven by financial stress: Behavioural adaptation or modification?", *Stress Health,* vol. 39, no. 3, pp. 614-626, 2023.
[http://dx.doi.org/10.1002/smi.3210] [PMID: 36413205]

[57] P. Carolina, and T. Arisandy, "Pengaruh pendidikan kesehatan terhadap pengetahuan anak usia sekolah dasar tentang pentingnya hidrasi: effectivenness health education on knowledge for children elementary school-age about the importance of hydration", *J Surya Medika,* vol. 9, no. 2, pp. 225-230, 2023.
[http://dx.doi.org/10.33084/jsm.v9i2.5702]

[58] E. Hillesheim, X. Yin, G.P. Sundaramoorthy, and L. Brennan, "Using a metabotype framework to deliver personalized nutrition improves dietary quality and metabolic health parameters: a 12-week randomized controlled trial", *Mol. Nutr. Food Res.,* vol. 67, no. 10, p. 2200620, 2023.
[http://dx.doi.org/10.1002/mnfr.202200620] [PMID: 37038841]

[59] A. Daneshvar, M. Farokhi, S. Bonakdar, and M. Vossoughi, "Synthesis and characterization of injectable thermosensitive hydrogel based on Pluronic-grafted silk fibroin copolymer containing hydroxyapatite nanoparticles as potential for bone tissue engineering", *Int. J. Biol. Macromol.,* vol.

277, p. 134412, 2024.
[http://dx.doi.org/10.1016/j.ijbiomac.2024.134412] [PMID: 39097043]

[60] A. Singh, R. Kumar, R. Patel, Trishna, R.N. Gautam, M.K. Bharty, and L.B. Prasad, "Dithiocarbamate-based novel anti-histaminic agents: synthesis, characterization, crystal structure and thermal study", *Dalton Trans.,* vol. 53, no. 33, pp. 14077-14088, 2024.
[http://dx.doi.org/10.1039/D4DT01777C] [PMID: 39113544]

[61] A. Hamid, R. Kumar, V. K. Chaturvedi, S. Dutt, G. Sulabh, V. Kumar, and D. P. Yadav, "Artificial intelligence (AI) based colonoscopy," In: *Nanobiotechnology and Artificial Intelligence in Gastrointestinal Diseases*, V. K. Chaturvedi, A. K. Singh, J. Singh, and D. P. Yadav, Eds. IOP Publishing Ltd, 2024, pp. 12-1, 12-23.
[http://dx.doi.org/10.1088/978-0-7503-6134-7ch12]

[62] M. Miranda-Tueros, J. Ramirez-PeAña, M. Cabanillas-Lazo, J. L. Paz-Ibarra, and I. Pinedo-Torres, "Effects of aerobic exercise on the components of metabolic syndrome in the older population with type 2 diabetes mellitus: a systematic review and meta-analysis", *Rev Peru Med Exp Salud Public,* vol. 41, no. 2, pp. 146-155, 2024.
[http://dx.doi.org/10.17843/rpmesp.2024.412.12751]

[63] A. M. Koman, K. Chamera-Cyrek, M. Pliszka, I. Janik, K. Gadżała, K. A. Palacz, K. B. Kułak, and I. Sztybór, "The beneficial effects of aerobic exercise on human systems and organs: a literature review", *J Educ Health Sport*, vol. 73, p. 51710, 2024.
[http://dx.doi.org/10.12775/JEHS.2024.73.51710]

[64] W. Huang, Y. He, W. Ruan, H. Zhai, Y. Lin, X. Dai, T. Wang, C. Huo, J. Zhuang, and J. Lu, "Association between metabolic flexibility and hepatic fat content in individuals with non-alcoholic fatty liver disease," *Preprint*, Research Square, 2024.
[http://dx.doi.org/10.21203/rs.3.rs-4336762/v1]

[65] E. Murru, C. Manca, G. Carta, M. Ruggiu, R. Solinas, R. Montisci, L. Hodson, D. Dearlove, M. P. Mollica, F. Tocco, and S. Banni, "Indirect calorimetry-based novel approach for evaluating metabolic flexibility and its association with circulating metabolic markers in middle-aged subjects", In: *Nutrients*, vol. 16, 2024, no. 4, p. 525.
[http://dx.doi.org/10.3390/nu16040525]

[66] A. Bhatnagar, A. Puppala, S. Rankawat, S. Ray, and S. Ray, "Role of circadian rhythms in metabolic syndrome", In: *Metabolic Syndrome.,* S. Mukhopadhyay, S. Mondal, Eds., Academic Press, 2024, pp. 199-218.
[http://dx.doi.org/10.1016/B978-0-323-85732-1.00006-2]

[67] D. Gubin, K. Danilenko, O. Stefani, S. Kolomeichuk, A. Markov, I. Petrov, K. Voronin, M. Mezhakova, M. Borisenkov, A. Shigabaeva, N. Yuzhakova, S. Lobkina, D. Weinert, and G. Cornelissen, "Blue light and temperature actigraphy measures predicting metabolic health are linked to melatonin receptor polymorphism," *Biology*, vol. 13, no. 1, p. 22, 2024.
[https://www.mdpi.com/2079-7737/13/1/22]

[68] J. G. Marques, M. Kuntic, R. Krishnankutty, G. R. Blanco, M. Malkov, K. Frenis, J. Wills, E. Shokry, F. L. Mow Chee, C. T. Taylor, T. Münzel, A. Daiber, and A. von Kriegsheim, "Noise stress induces cardiovascular metabolic shifts," *bioRxiv*, 2024.
[http://dx.doi.org/10.1101/2024.04.22.590539]

[69] S. Robertson, E. D. Clarke, M. Gómez-Martín, V. Cross, C. E. Collins, and J. Stanford, "Do precision and personalised nutrition interventions improve risk factors in adults with prediabetes or metabolic syndrome? a systematic review of randomised controlled trials", In: Nutrients vol. 16, 2024, no. 10, p. 1479.
[http://dx.doi.org/10.3390/nu16101479]

[70] P. Halagali, A. Inamdar, J. Singh, A. Anand, P. Sadhu, R. Pathak, H. Sharma, and D. Biswas, "Phytochemicals, herbal extracts, and dietary supplements for metabolic disease management", *Endocr. Metab. Immune Disord. Drug Targets,* vol. 24, 2024.

[http://dx.doi.org/10.2174/0118715303287911240409055710] [PMID: 38676520]

[71] M. Lihua, Z. Kaipeng, M. Xiyan, C. Yaowen, and Z. Tao, "Systematic review and meta-analysis of stress management intervention studies in patients with metabolic syndrome combined with psychological symptoms.," *Medicine (Baltimore).*, vol. 102, no. 42, p. e35558, Oct. 2023. [http://dx.doi.org/10.1097/MD.0000000000035558]

[72] S. Sharma, and M. Kavuru, "Sleep and metabolism: an overview", *Int. J. Endocrinol.,* vol. 2010, no. 1, p. 270832, 2010. [PMID: 20811596]

[73] H. Oster, and I. Chaves, "Effects of healthy lifestyles on chronic diseases: diet, sleep and exercise", *Nutrients,* vol. 15, p. 4627, 2023. [http://dx.doi.org/10.3390/nu15214627]

[74] K.I. Galaviz, K.M.V. Narayan, F. Lobelo, and M.B. Weber, "Lifestyle and the prevention of type 2 diabetes: a status report", *Am. J. Lifestyle Med.,* vol. 12, no. 1, pp. 4-20, 2018. [http://dx.doi.org/10.1177/1559827615619159] [PMID: 30202378]

[75] T. Vanhaecke, E. T. Perrier, and O. Melander, "A journey through the early evidence linking hydration to metabolic health", *Ann Nutr Metab,* vol. 76, no. Suppl. 1, pp. 4-9, 2021.

[76] H. Shi, Y. Ren, and Y. Jia, "Effects of nutritional interventions on the physical development of preschool children: a systematic review and meta-analysis", *Transl. Pediatr.,* vol. 12, no. 5, pp. 991-1003, 2023. [http://dx.doi.org/10.21037/tp-23-205] [PMID: 37305721]

[77] A. Kumar, P. Singh, R. Kumar, and S. Dutt, "Novel drug delivery systems for inflammatory bowel disease", In: *Nanobiotechnology and Artificial Intelligence in Gastrointestinal Diseases, V. K. Chaturvedi, A. K. Singh, J. Singh, and D. P. Yadav, Eds.,* IOP Publishing, 2024, pp. 4-1-4-24. [http://dx.doi.org/10.1088/978-0-7503-6134-7ch4]

[78] E. Nury, J. Stadelmaier, J. Morze, B. Nagavci, K. Grummich, G. Schwarzer, G. Hoffmann, C.M. Angele, J.M. Steinacker, J. Wendt, J. Conrad, D. Schmid, J.J. Meerpohl, and L. Schwingshackl, "Effects of nutritional intervention strategies in the primary prevention of overweight and obesity in school settings: systematic review and network meta-analysis", *BMJ Med.,* vol. 1, no. 1, p. e000346, 2022. [http://dx.doi.org/10.1136/bmjmed-2022-000346] [PMID: 36936562]

[79] L. Sobotka, "Metabolism of macronutrients", In: *Nutrition Neurol Disord: A Pract Guide*, 2017, pp. 1-17. [http://dx.doi.org/10.1007/978-3-319-53171-7_1]

[80] C. Du, J. Almotawa, C.E. Feldpausch, S.Y.L. Folk, H. Parag, and R.M. Tucker, "Effects of macronutrient intake on sleep duration and quality: A systematic review", *Nutr. Diet.,* vol. 79, no. 1, pp. 59-75, 2022. [http://dx.doi.org/10.1111/1747-0080.12671] [PMID: 33876534]

[81] J. Park, and O.Y. Kim, "Macronutrients modified dietary intervention in the management of overweight/obese children and adolescents: a systematic review", *Clin. Exp. Pediatr.,* vol. 67, no. 4, pp. 191-200, 2024. [http://dx.doi.org/10.3345/cep.2023.00262] [PMID: 37448128]

[82] D. Linsmayer, G.P. Eckert, J. Reiff, and D.F. Braus, "Nutrition, metabolism, brain and mental health", *Nervenarzt,* vol. 95, no. 7, pp. 667-680, 2024. [http://dx.doi.org/10.1007/s00115-024-01678-6] [PMID: 38884643]

[83] V. J. Clemente-Suárez, J. Mielgo-Ayuso, A. Martín-Rodríguez, D. J. Ramos-Campo, L. Redondo-Flórez, and J. F. Tornero-Aguilera, "The burden of carbohydrates in health and disease", *Nutrients,* vol. 14, no. 18, p. 3809, 2022. [http://dx.doi.org/10.3390/nu14183809] [PMID: 36145184]

[84] T. Fernandes, A. Faria, and H. Loureiro, "Sources of free and added sugars and their nutritional impact in diabetic patients", *Diabetology (Basel),* vol. 3, no. 4, pp. 634-638, 2022. [http://dx.doi.org/10.3390/diabetology3040049]

[85] B. K. Wölnerhanssen and A. C. Meyer-Gerspach, “Effekte von Zuckerkonsum auf die Gesundheit und mögliche Alternativen [Health effects of sugar consumption and possible alternatives],” *Ther. Umsch.,* vol. 76, no. 3, pp. 111–116, Sep. 2019, [PMID: 31498044]

[86] A. Laviano, "Living in interesting times – challenging protein metabolism in the era of the epidemiological shift", *Curr. Opin. Clin. Nutr. Metab. Care,* vol. 19, no. 1, pp. 37-38, 2016. [http://dx.doi.org/10.1097/MCO.0000000000000244] [PMID: 26560528]

[87] K. Katayama, "Zinc and protein metabolism in chronic liver diseases", *Nutr. Res.,* vol. 74, pp. 1-9, 2020. [http://dx.doi.org/10.1016/j.nutres.2019.11.009] [PMID: 31891865]

[88] E. Pasini, R. Aquilani, and F.S. Dioguardi, "The enemy within': How to identify chronic diseases induced-protein metabolism impairment and its possible pharmacological treatment", *Pharmacol. Res.,* vol. 76, pp. 28-33, 2013. [http://dx.doi.org/10.1016/j.phrs.2013.06.006] [PMID: 23827158]

[89] T. Jesmin, "Investigation of common disease regulatory network for metabolic disorders: A bioinformatics approach", *New Biol.,* vol. 6, no. 1, p. 28, 2016.

[90] H.B. Lovins, B.E. Bathon, S.R. Shaikh, and K.M. Gowdy, "Inhaled toxicants and pulmonary lipid metabolism: biological consequences and therapeutic interventions", *Toxicol. Sci.,* vol. 196, no. 2, pp. 141-151, 2023. [http://dx.doi.org/10.1093/toxsci/kfad100] [PMID: 37740395]

[91] J.F. Bogie, J. Guns, and S. Vanherle, "Lipid metabolism in neurodegenerative diseases", In: *Cellular Lipid in Health and Disease.* Elsevier, 2023, pp. 389-419. [http://dx.doi.org/10.1016/B978-0-323-95582-9.00008-5]

[92] L. Bordoni, M. Zec, N. Naumovski, and D. Sergi, “Editorial: The role of dietary fatty acids in metabolic health,” Front. Physiol., vol. 14, 2023. [http://dx.doi.org/10.3389/fphys.2023.1211151]

[93] A.A. Johnson, and A. Stolzing, "The role of lipid metabolism in aging, lifespan regulation, and age-related disease", *Aging Cell,* vol. 18, no. 6, p. e13048, 2019. [http://dx.doi.org/10.1111/acel.13048] [PMID: 31560163]

[94] M.A. DeBiasse, and D.W. Wilmore, "What is optimal nutritional support?", *New Horiz.,* vol. 2, no. 2, pp. 122-130, 1994. [PMID: 7922436]

[95] P.B. Soeters, "Macronutrient metabolism in starvation and stress", In: *The Importance of Nutrition as an Integral Part of Disease Management.* vol. 82. Karger Publishers, 2015, pp. 17-25. [http://dx.doi.org/10.1159/000381998]

[96] Y. Wan, F. Wang, J. Yuan, and D. Li, "Optimal dietary macronutrient distribution in China (ODMDC): a randomised controlled-feeding trial protocol", *Asia Pac. J. Clin. Nutr.,* vol. 26, no. 5, pp. 972-980, 2017. [PMID: 28802307]

[97] P. Turner, "Providing optimal nutritional support on the intensive care unit: key challenges and practical solutions", *Proc. Nutr. Soc.,* vol. 69, no. 4, pp. 574-581, 2010. [http://dx.doi.org/10.1017/S002966511000385X] [PMID: 20860859]

[98] V. P. F. Assumpção and O. Q. Assumpção, "Nutrological and metabolic approaches to the action of the some special micronutrients in heart failure and metabolic syndrome: a systematic review", *Int. J. Neurol.,* vol. 15, no. 3, 2022.

[http://dx.doi.org/10.54448/ijn22308]

[99] A. Tariq, F. Zeng, C. Graciano, and A. Ullah, “Regulation of metabolites by nutrients in plants,” In: *Plant Ionomics: Sensing, Signaling and Regulation,* Wiley, 2023, pp. 1–18.

[100] R. Verma, S. Chawla, and M. Dhankar, "Importance of micronutrient supplementation programme in childhood to reduce child mortality: the Haryana experience", *Int J Prev Med,* vol. 7, 2016. [http://dx.doi.org/10.4103/2008-7802.184501]

[101] N. Bomer, M.G. Pavez-Giani, N. Grote Beverborg, J.G.F. Cleland, D.J. van Veldhuisen, and P. van der Meer, "Micronutrient deficiencies in heart failure: Mitochondrial dysfunction as a common pathophysiological mechanism?", *J. Intern. Med.,* vol. 291, no. 6, pp. 713-731, 2022. [http://dx.doi.org/10.1111/joim.13456] [PMID: 35137472]

[102] Y. Li, S.S. Madduri, E.T. Rezeli, C. Santos, H. Freeman III, J. Peng, S.L. McRitchie, W. Pathmasiri, S.D. Hursting, S.J. Sumner, and D.A. Stewart, "Macronutrient-differential dietary pattern impacts on body weight, hepatic inflammation, and metabolism", *Front. Nutr.,* vol. 11, p. 1356038, 2024. [http://dx.doi.org/10.3389/fnut.2024.1356038] [PMID: 38868554]

[103] Z. Salehi, M. Askari, A. Jafari, B. Ghosn, P.J. Surkan, M.J. Hosseinzadeh-Attar, H. Pouraram, and L. Azadbakht, "Dietary patterns and micronutrients in respiratory infections including COVID-19: a narrative review", *BMC Public Health,* vol. 24, no. 1, p. 1661, 2024. [http://dx.doi.org/10.1186/s12889-024-18760-y] [PMID: 38907196]

[104] M.I. Sultan, S.A. Ibrahim, and R.F. Youssef, "Impact of a Mediterranean diet on prevention and management of urologic diseases", *BMC Urol.,* vol. 24, no. 1, p. 48, 2024. [http://dx.doi.org/10.1186/s12894-024-01432-9] [PMID: 38408996]

[105] S. Schlesinger, "Diet and diabetes prevention: Is a plant-based diet the solution?", *Diabetes Care,* vol. 46, no. 1, pp. 6-8, 2023. [http://dx.doi.org/10.2337/dci22-0041] [PMID: 36548706]

[106] T.V. Shelenga, "The potential of small grains crops in enhancing biofortification breeding strategies for human health benefit", *Agronomy (Basel),* vol. 11, no. 7, p. 1420, 2021. [http://dx.doi.org/10.3390/agronomy11071420]

[107] M. Jendyose, "Development of Functional Foods with Enhanced Health Benefits", *J. Food Sci.,* 2024.

[108] M. Sharma, K. Ojha, B. Singh, S. Gupta, and S.K. Pandey, "The role of functional foods and nutraceuticals in disease prevention and health promotion", *Eur. J. Nutr. Food Saf.,* vol. 16, no. 2, pp. 61-83, 2024. [http://dx.doi.org/10.9734/ejnfs/2024/v16i21388]

[109] N.V. Bolgova, S.O. Huba, V.V. Sokolenko, and A.K. Mazhara, "Study of the influence of vitamins on the fermentation process in the production of yogurt", *Sci. Mess. LNU Vet. Med. Biotechnol.,* vol. 25, no. 100, pp. 43-46, 2023. [http://dx.doi.org/10.32718/nvlvet-f10007]

[110] A. Mikulec, M. Zborowski, and A. Klimczak, "Functional food in the primary prevention of cardiovascular diseases", *J. Educ. Health Sport,* vol. 12, no. 7, pp. 848-863, 2022. [http://dx.doi.org/10.12775/JEHS.2022.12.07.085]

[111] J. B. Hoffman, "PCB disruption of gut and host health: Implications of prebiotic nutritional intervention," 2018.

[112] K. Tiihonen, T. Suomalainen, S. Tynkkynen, and N. Rautonen, "Effect of prebiotic supplementation on a probiotic bacteria mixture: comparison between a rat model and clinical trials", *Br. J. Nutr.,* vol. 99, no. 4, pp. 826-831, 2008. [http://dx.doi.org/10.1017/S0007114507825141] [PMID: 17868490]

[113] H. Norooziseyedhosseini, R. Imani, B. Sumra, M. Tariq, and M. Upadhyay, “The effects of omega-3 and omega-6 fatty acids on the mental and physical health of children, adolescents, and adults,” *Authorea*, 2023.

[http://dx.doi.org/10.22541/au.167655900.03369842/v1]

[114] S. Shackelford, "Lifestyle Intervention in Primary Care Patients with Metabolic Syndrome", *J. Clin. Lipidol.,* vol. 17, no. 4, p. e49, 2023.
[http://dx.doi.org/10.1016/j.jacl.2023.05.073]

[115] S. Park, J. Lee, J.W. Seok, C.G. Park, and J. Jun, "Comprehensive lifestyle modification interventions for metabolic syndrome: A systematic review and meta-analysis", *J. Nurs. Scholarsh.,* vol. 56, no. 2, pp. 249-259, 2024.
[http://dx.doi.org/10.1111/jnu.12946] [PMID: 38009414]

[116] O. Ramos-Lopez, "Epigenetic biomarkers of metabolic responses to lifestyle interventions", In: *Nutrients* vol. 15. , 2023, no. 19, p. 4251.
[http://dx.doi.org/10.3390/nu15194251]

[117] G. Wood, "Estimating the effect of aerobic exercise training on novel lipid biomarkers: a systematic review and multivariate meta-analysis of randomized controlled trials", *Sports Med,* vol. 53, no. 4, pp. 871-886, 2023.
[http://dx.doi.org/10.1007/s40279-023-01817-0]

[118] B. Olorunfemi Oyewole, O. Paul Olamide, A. Ebenezer, A. Precious Olayinka, and I. Opeyemi, "Lifestyle interventions to manage insulin resistance", In: *Hypoglycemia - New Insights* R. Dr. Alok and M. D. R. Shaginian Eds. Rijeka IntechOpen: Rijeka, 2024.

[119] A.D. Faigenbaum, and R.S. Lloyd, "Resistance training", In: *Oxford Textbook of Children's Sport and Exercise Medicine* Oxford University Press, 2023.
[http://dx.doi.org/10.1093/med/9780192843968.003.0035]

[120] T. John, and T. Dr, "Raising the bar for public health: resistance training and health benefits", *Int J Strength Cond,* vol. 3, no. 1, 2023.
[http://dx.doi.org/10.47206/ijsc.v3i1.195]

[121] V.V. Sverchkov, and E.V. Bykov, "Low-Intensity Resistance Training with Blood Flow Restriction Improves Metabolic Parameters in Men with Metabolic Syndrome", *Zh. Med. Biol. Issled.,* vol. 11, no. 3, pp. 310-320, 2023.
[http://dx.doi.org/10.37482/2687-1491-Z149]

[122] B.F. Palmer, and D.J. Clegg, "Metabolic Flexibility and Its Impact on Health Outcomes", *Mayo Clin. Proc.,* vol. 97, no. 4, pp. 761-776, 2022.
[http://dx.doi.org/10.1016/j.mayocp.2022.01.012] [PMID: 35287953]

[123] T. Sato, and S. Sato, "Circadian Regulation of Metabolism: Commitment to Health and Diseases", *Endocrinology,* vol. 164, no. 7, 2023.
[http://dx.doi.org/10.1210/endocr/bqad086] [PMID: 37253106]

[124] E. Scoditti, and S. Garbarino, *Nutrition at the Interface of Sleep and Circadian Rhythms: Implications for Health.* MDPI-Multidisciplinary Digital Publishing Institute, 2023.
[http://dx.doi.org/10.3390/books978-3-0365-6709-9]

[125] E. Grosjean, V. Simonneaux, and E. Challet, "Reciprocal interactions between circadian clocks, food intake, and energy metabolism", In: *Biology* vol. 12, 2023, no. 4, p. 539.
[http://dx.doi.org/10.3390/biology12040539]

[126] "Metabolic consequences of sleep disorders", *Clin Endocrinol Endocrine Surg,* no. 2, pp. 39-46, 2022.
[http://dx.doi.org/10.30978/CEES-2022-2-39]

[127] S. Zhang, "Metabolic flexibility during sleep", *Sci Rep,* vol. 11, no. 1, p. 17849, 2021.
[http://dx.doi.org/10.1038/s41598-021-97301-8]

[128] E. Humer, C. Pieh, and G. Brandmayr, "Metabolomics in sleep, insomnia and sleep apnea", In: *Int J Mol Sci* vol. 21, 2020, no. 19, p. 7244.
[http://dx.doi.org/10.3390/ijms21197244]

[129] A. Sengupta, and A.M. Weljie, "Metabolism of sleep and aging: Bridging the gap using metabolomics", *Nutr. Healthy Aging,* vol. 5, no. 3, pp. 167-184, 2019. [http://dx.doi.org/10.3233/NHA-180043] [PMID: 31984245]

[130] E.M. Rogers, N.F. Banks, and N.D.M. Jenkins, "The effects of sleep disruption on metabolism, hunger, and satiety, and the influence of psychosocial stress and exercise: A narrative review", *Diabetes Metab. Res. Rev.,* vol. 40, no. 2, p. e3667, 2024. [http://dx.doi.org/10.1002/dmrr.3667] [PMID: 37269143]

[131] F.M. Zuraikat, S. Jelic, and M.P. St-Onge, "Wake up! It's time to recognize the importance of sleep in metabolic health", *Obesity (Silver Spring),* vol. 31, no. 5, pp. 1188-1191, 2023. [http://dx.doi.org/10.1002/oby.23776] [PMID: 37140407]

[132] J. Lin, B. Wang, W. Zeng, S. Zhuang, M. Liu, and J. Yang, "Evidence-based interventions to improve sleep quality after thoracic surgery: a retrospective analysis of clinical studies", *Sleep Med,* vol. 121, pp. 85-93, 2024. [http://dx.doi.org/10.1016/j.sleep.2024.06.006]

[133] J.C. Ong, and C. Gamaldo, "Optimizing Behavioral Sleep Strategies", *Continuum (Minneap. Minn.),* vol. 26, no. 4, pp. 1075-1081, 2020. [http://dx.doi.org/10.1212/CON.0000000000000876] [PMID: 32756237]

[134] M. Lihua, Z. Kaipeng, M. Xiyan, C. Yaowen, and Z. Tao, "Systematic review and meta-analysis of stress management intervention studies in patients with metabolic syndrome combined with psychological symptoms", *Medicine (Baltimore),* vol. 102, no. 42, p. e35558, 2023. [http://dx.doi.org/10.1097/MD.0000000000035558] [PMID: 37861474]

[135] S.-H. Ro, J. Fay, C. I. Cyuzuzo, Y. Jang, N. Lee, H.-S. Song, and E. N. Harris, "SESTRINs: emerging dynamic stress-sensors in metabolic and environmental health", *Front. Cell Dev. Biol.,* vol. 8, Dec. 3, 2020. [http://dx.doi.org/10.3389/fcell.2020.603421]

[136] C. Macedo, J. Monnerat, B. Lucchetti, G. Teixeira, J. Mentzinger, H. Rocha, R. Medeiros, N. Rocha, D. de Souza, F. Sampaio, and B. Gregorio, "Impact of maternal stress on metabolism and penile morphology in young offspring rats", *Histol. Histopathol.,* vol. 39, no. 8, pp. 1009-1015, 2024. [PMID: 38221876]

[137] W.D. Lee, L. Liang, J. AbuSalim, C.S.R. Jankowski, L.Z. Samarah, M.D. Neinast, and J.D. Rabinowitz, "Impact of acute stress on murine metabolomics and metabolic flux", *Proc. Natl. Acad. Sci. USA,* vol. 120, no. 21, p. e2301215120, 2023. [http://dx.doi.org/10.1073/pnas.2301215120] [PMID: 37186827]

[138] S. Hewitt, E. HernAndez-Montes, A. Dhingra, and M. Keller, "Impact of heat stress, water stress, and their combined effects on the metabolism and transcriptome of grape berries", *Sci Rep,* vol. 13, no. 1, p. 9907, 2023. [http://dx.doi.org/10.1038/s41598-023-36160-x]

[139] O.G. Kurinna, "Effects of stressogenic factors on the course of metabolic-associated fatty liver disease. review", In: *Сучасна гастроентерологія*, 2023, pp. 69-73. [http://dx.doi.org/10.30978/MG-2023-2-69]

[140] M. Roberto Delgado, R. Oropeza Tena, F. J. Pedroza Cabrera, J. C. Verdugo Lucero, J. F. Enríquez Bielma, “Manejo del estrés para el control metabólico de personas con diabetes mellitus tipo 2,” *En-Claves Pensam.,* vol. 7, no. 13, pp. 67-87, 2013. [http://dx.doi.org/10.46530/ECDP.V0I13.190]

[141] J. M. Kennedy, and D. J. Miller, "Effect of stress reduction using the BREATHE technique on inflammatory markers and risk factors for atherosclerosis", *Clevel Clin J Med,* vol. 78, pp. S97-S97, 2011.

[142] C. D. Heer, and C. Brenner, "Letting off electrons to cope with metabolic stress", *Nat Metab,* vol. 2,

no. 6, pp. 485-486, 2020.
[http://dx.doi.org/10.1038/s42255-020-0207-8]

[143] O. Onur, "The physiological role of hydration in exercise performance", *Europ J Phys Educ Sport Sci,* 2019.
[http://dx.doi.org/10.46827/EJPE.V0I0.2291]

[144] A. Angert, T. Weiner, F. Tamburini, H. Zer, and N. Keren, "Does metabolic water control the isotopic composition of water in microbial cells?", In: *EGU Gen Assem Conf Abstr*, 2023, p. EGU-2218.
[http://dx.doi.org/10.5194/egusphere-egu23-2218]

[145] J. M. Scott, J. R. Linderman, and P. A. Deuster, "Hydration: Tactical and practical strategies," *J. Spec. Oper. Med.,* vol. 23, no. 1, pp. 88–91, 2023.
doi: 10.55460/QOBG-HTOX.

[146] R. W. Kenefick, "Drinking strategies: planned drinking versus drinking to thirst", *Sports Med,* vol. 48, pp. 31-37, 2018.
[http://dx.doi.org/10.1007/s40279-017-0844-6]

[147] L. Kearney, A. Craswell, N. Dick, D. Massey, and R. Nugent, "Evidence-based guidelines for intrapartum maternal hydration assessment and management: A scoping review", *Birth,* vol. 51, no. 2, pp. 253-263, 2024.
[http://dx.doi.org/10.1111/birt.12773] [PMID: 37803945]

[148] N. Torres, and A.R. Tovar, "The present and future of personalized nutrition", *Rev. Invest. Clin.,* vol. 73, no. 5, pp. 321-325, 2021.
[http://dx.doi.org/10.24875/RIC.21000346] [PMID: 34609363]

[149] I.N. Napolsky, and P.V. Popova, "Personalized nutrition for the prevention and treatment of metabolic diseases: opportunities and perspectives", *Russ. J. Pers. Med.,* vol. 2, no. 1, pp. 15-34, 2022.
[http://dx.doi.org/10.18705/2782-3806-2022-2-1-15-34]

[150] R. Chen, and G. Chen, "Personalized nutrition for people with diabetes and at risk of diabetes has begun", *Journal of Future Foods,* vol. 2, no. 3, pp. 193-202, 2022.
[http://dx.doi.org/10.1016/j.jfutfo.2022.06.001]

[151] F. Wang, J. Zheng, J. Cheng, H. Zou, M. Li, B. Deng, R. Luo, F. Wang, D. Huang, G. Li, R. Zhang, X. Ding, Y. Li, J. Du, Y. Yang, and J. Kan, "Personalized nutrition: a review of genotype-based nutritional supplementation", *Front Nutr*, vol. 9, no. Review, 2022.
[http://dx.doi.org/10.3389/fnut.2022.992986]

[152] M. Gómez de Cedrón, R. Moreno Palomares, and A. Ramírez de Molina, "Metabolo-epigenetic interplay provides targeted nutritional interventions in chronic diseases and ageing," Front. Oncol., vol. 13, 2023.
[http://dx.doi.org/10.3389/fonc.2023.1169168]

[153] Pary Hadi, "A Review Of The Methods And Approaches For Integrating Genetics And Metabolomics Into Personalized Nutrition", *J. Pharm. Negat. Results,* vol. *f?›f?›f?›*, pp. 3823-3830, 2023.
[http://dx.doi.org/10.47750/pnr.2023.14.02.447]

[154] D. Sonawdekar, D. Katore, M. Patil, and S. Pawar, "Nutrigenomics: Personalized nutrition for health and sustainability," In: *Futuristic Trends in Agriculture Engineering & Food Sciences*, vol. 3, no. 13, 2024.
[http://dx.doi.org/10.58532/V3BCAG13P1CH11]

[155] M. Jakovljevic, "Integrative metabolic therapeutic approach for symptomatic patients with left ventricular dilatation and reduced ejection fraction," *Clinical & Experimental Cardiology,* 2020.

[156] K. Henneman, "Optimizing health – integrative medicine & poor performance", *Vet. Clin. North Am. Equine Pract.,* vol. 38, no. 3, pp. 463-474, 2022.
[http://dx.doi.org/10.1016/j.cveq.2022.08.001] [PMID: 36244943]

CHAPTER 9

Future Directions and Research Frontiers

Sunil Dutt[1], **Shiwank Rana**[2], **Satish Patnaik**[4], **Rajesh Kumar**[1,*] and **Ravinesh Mishra**[3]

[1] *Maharaja Agrasen School of Pharmacy, Maharaja Agrasen University, Solan, Himachal Pradesh, India*

[2] *Dreamz College of Pharmacy, Sundernagar, Mandi, Himachal Pradesh, India*

[3] *School of Pharmacy and Emerging Sciences, Baddi University of Emerging Sciences & Technology, Baddi, Solan, Himachal Pradesh, India*

[4] *Stein Eye Institute, Geffen School of Medicine, University of California, Los Angeles, California, USA*

Abstract: Incomplete biochemical pathways lead to the build-up of metabolic intermediates, which is known as incomplete metabolism. This condition is significant for many physiological and pathological processes. This chapter highlights the importance of incomplete metabolism in understanding disease mechanisms, energy generation, and cellular adaptability as it examines future directions and research horizons in the field. Determining the underlying cause of incomplete metabolism requires advances in metabolomic profiling, enzyme functioning research, and the discovery of new biomarkers. Moreover, studying incomplete metabolic pathways in relation to illnesses like diabetes, cancer, and neurological problems might lead to the creation of focused treatment plans. Understanding how cells adapt to metabolic inefficiencies, especially in the mitochondria, might help us understand how cells survive under stressful circumstances.

Furthermore, new prospects for individualized therapy are presented by the interaction between environmental factors, the gut microbiota, and incomplete metabolism. Using creative methods, this investigation expands our understanding of basic biology and creates new avenues for therapeutic treatments targeted at enhancing the prognosis of metabolic and associated disorders. Incomplete metabolism is emerging as a major area with broad implications for fundamental science and therapeutic applications, as research into it continues.

Keywords: Biochemical pathways, Metabolomics, Personalized medicines, Pharmacogenomics, Transcriptomics.

* **Corresponding Author Rajesh Kumar:** Maharaja Agrasen School of Pharmacy, Maharaja Agrasen University, Solan, Himachal Pradesh, India; E-mail: rajdhiman60@gmail.com

INTRODUCTION

A person's health is directly impacted by incomplete metabolism. A compromised metabolism may increase the risk of cardiovascular disease. Both acquired and genetic factors may contribute to the pathophysiology of metabolic dysfunction. Indeed, type 2 diabetes and cardiovascular diseases are among the consequences closely associated with incomplete metabolism of blood glucose. However, poor iron and fat metabolism can lead to obesity [1, 2]. Early identification of clinical biomarkers and diagnosis can be crucial in managing negative health effects resulting from incomplete metabolism. Metabolome analysis is the technique to identify and analyze biological fluids and clinical biomarkers. In this way, we can monitor the metabolic function of a patient. We can find clinical indicators of metabolic dysfunction by combining specialized techniques like Gas and Liquid Chromatography (GC and LC) with Nuclear Magnetic Resonance (NMR) or Mass Spectroscopy (MS). For instance, upregulation of glucose, lactate, CH_3 lipids, and CH_2 lipids, and downregulation of alanine and choline in human serum, as revealed by NMR biomarker analysis, may contribute to the development of chronic illnesses like diabetes and obesity [3, 4]. Management of metabolic dysfunction requires a lifestyle change and an appropriate diet plan. Another novel strategy for treating metabolic dysfunction is gender-specific personalized nutrition and medicine, which involves tracking the composition of the microbiota, the rate of inflammation, oxidative stress, and the metabolism of fats and carbohydrates. Management of metabolic dysfunction may benefit from dietary pattern modifications that include or exclude particular foods. Omega-3, polyphenols, flavonoids, curcumin, chlorogenic acid, and pre- and probiotics may aid in the management of and protection against metabolic dysfunction when included in the diet [5 - 8].

Finally, we can say that an incomplete metabolism directly affects an individual's health because it can result in a number of chronic illnesses, including obesity, cardiovascular disease, diabetes, and more. Numerous studies on the treatment of metabolic disorders are still ongoing. Personalised medications, early biomarker detection and diagnosis, and appropriate dietary modifications are a few strategies that can be crucial in managing an incomplete metabolism.

Pharmacogenomics and Personalized Medicine

Genetic variables impact a patient's response to medication, influencing both the toxicity and effectiveness of the treatment. Using a person's genetic information, pharmacogenomics appears to be an effective approach that can predict a sub-therapeutic response to a drug and a higher risk of adverse effects, as shown in Fig. (**1**).

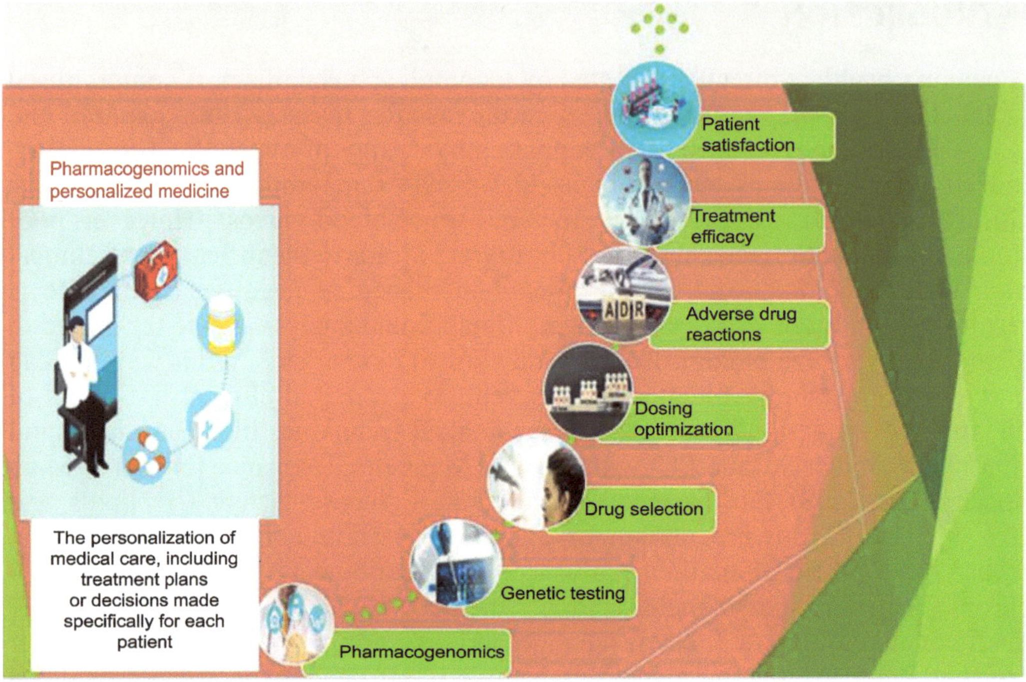

Fig. (1). Pharmacogenomics and personalized medicine.

Genetic technologies, such as pharmacogenomics and personalized medicine, are being applied to improve patient outcomes and guide medication selection. Pharmacogenomics and personalized medicine are extensively employed in the management of metabolic illnesses and disorders resulting from dysfunction in the metabolism [9, 10]. Personalized medicines for various metabolic conditions are discussed below.

Diabetes Mellitus

Diabetes mellitus (DM) is a condition caused by insufficient glucose metabolism in the blood. *AGTR1, IL6, NOS3,* and *TNFA* genes have been discovered to be frequently associated with various illnesses, according to a study. The *STAT3* polyautoimmunopathy, which encompasses intrauterine growth retardation, thyroid dysfunction, celiac disease, and persistent autoimmune neonatal diabetes, has been linked to mutations in the *STAT3* gene [11, 13]. An increased understanding of the need for personalized diagnosis in Type 1 diabetes has driven extensive research efforts toward the development of personalized therapies [14]. The most popular method of treating Type I diabetes is manual blood sugar testing, which is followed by several daily subcutaneous insulin injections. Conventional injections may be replaced with insulin pumps. Given the significant hereditary component in the development of Type 1 diabetes, gene therapy presents a viable treatment option for the disease in place of insulin administration [15]. The process of transferring or modifying genetic substances inside the cell as a therapeutic approach to treat the disease is known as gene therapy [16]. By altering the defective genes responsible for the advancement of the disease, it seeks to either prevent the onset of the disease or halt its progression.Another strategy is to use stem cells from various sources to explore their potential in β-cell regeneration. Different types of stem cells, including mesenchymal stem cells, haematopoietic stem cells, human pluripotent stem cells, and embryonic stem cells, are employed for this. Similar to this, the antidiabetic medication class thiazolidinediones affects the *PPARG* gene, which is positively linked to Type 2 diabetes [14, 17].

Cancer

Pharmacogenomics is a rapidly developing field that could have a huge impact on personalized medicine and cancer treatment. Health experts can choose a personalized dosage and treatment strategy by examining the genetic composition of the patient and the somatic tumours [21]. Genome profiling is used to identify cancer subgroups or biomarkers, allowing for the administration of targeted medications based on the target molecule. According to a study, the *EML4-ALK* gene is involved in Non-Small Cell Lung Cancer (NSCLC) and is linked to a number of growth-stimulatory signaling pathways, including the PI3K-AKT, (JAK3)-STAT3, and RAS-MEK-ERK pathways. Crisotinib, a targeted Tyrosine Kinase Inhibitor (TKI), was found to be particularly efficient at inhibiting ALK in patients with ALK-positive NSCLC [18]. Targeting the *HERS2* gene in breast cancer involves using *HERS2* and kinase inhibitors, such as lapatinib, pertuzumab, trastuzumab, and adotratuzumab emtansine. For better treatment outcomes, TAM, fulvestrant, and aromatase inhibitors are typically used to target Luminal A and Luminal B positive breast tumours in which expression of estrogen receptor is the main precipitating factor.

Comparable research has also demonstrated the need of using trastuzumab and pertuzumab in combination therapy to prevent the creation of HER2:HER3 dimers, the strongest heterodimer in the HER family [19, 20]. Patients with thyroid cancer have been shown to have the *BRAF V600E* gene involved. Vemurafenib inhibits the BRAF V600E mutation, which leads to notable outcomes. Thyroid cancer has been shown to express more growth factor receptors than normal tissue. Non-specific tyrosine kinase inhibitors include motesanib, sunitinib, and sorafenib, among other therapeutic agents, such as doxorubicin [21]. The development of hepatocellular carcinoma involves several genetic expressions and growth factors. An improved prognosis for hepatocellular carcinoma was achieved with the use of personalized medicine and pharmacogenomics in its therapy. Pathophysiological mechanisms relying on tyrosine kinases are important. Several inhibitors, including Sorafenib (SOR), Lenvatinib, Cabozantinib, and Regorafenib, have demonstrated encouraging outcomes in the treatment of hepatocellular carcinoma. Researchers have demonstrated that the vascular endothelial growth factor pathway gene is significantly expressed in hepatocellular carcinoma, suggesting that it may be an effective target for disease management. Moreover, research has found a link between hepatocellular cancer and overexpression of either c-MET or hepatocyte growth factor. An improved prognosis for hepatocellular carcinoma could result from gene therapy that targets hepatocyte growth factors [22].

Infectious Diseases

The nation with the highest number of tuberculosis sufferers worldwide is India. For an extended period, combination therapy for tuberculosis has been used. This therapy mostly consists of isoniazid, rifampicin, pyrazinamide, and ethambutol. Since first-line TB medicines are toxic, severe side effects may occur during treatment. N-acetyl transferase (NAT2) primarily metabolises isoniazid, a major medication used in therapy, by acetylation in the liver. Patients with lower levels of *NAT2* gene expression are more susceptible to harmful medication responses because their bodies eliminate drugs more slowly. In order to reduce side effects, patients can benefit from a personalized treatment plan that identifies gene expression.

Additionally, rifampicin, a significant TB medication, was found to be less effective in patients who expressed the *SLCO1B1* gene in an investigation [23]. According to a study, treating chronic Hepatitis C Virus (HCV) with boceprevir and telaprevir produced better outcomes than using standard hepatitis C virus treatment, which consists of ribavirin and pegylated interferon-alfa (PEG-IFN-α). The interleukin-28 B (*IL28B*) gene was found to be related to the response to standard therapy based on genome analysis of HCV patients. HCV is spontaneously cleared after personalized treatment with boceprevir, interferon-alfa, and ribavirin [24]. A rare and serious lung condition known as Chronic Cavitary Pulmonary Aspergillosis (CCPA), if left untreated, can lead to Aspergillus nodule, single aspergilloma, and chronic fibrosing pulmonary aspergillosis. Treatment options for persistent cavitary pulmonary aspergillosis include corticosteroid therapy, immunotherapy with interferon-γ, antifungal medications, and surgery. However, the treatment of CCPA involves the use of various personalised medicine techniques, such as drug sensitivity test-guided therapy [25].

Furthermore, every year, millions of people contract the influenza virus. Researchers have discovered that the influenza virus expresses two proteins, NS1 and NS2, which interact with host protein ADAR1, which has been shown to control viral replication. Personalised treatment aimed at this key protein may contribute to reduced viral contagiousness [26].

Cardiovascular Diseases

Since cardiovascular disease (CVD) is the leading cause of death, personalized medicine offers a novel approach to treating CVD because a patient's genetic makeup may impact how well a medication works [27]. Researchers have found that polymorphisms in patients' *CYP2C9* and *VKORC1* genes affect their response to warfarin. Thus, the expression of this specific gene in the patient determines both the therapeutic impact and any adverse effects. Conversely, polymorphisms in the patient's Arg389Gly and Ser49Gly genes impact β1 adrenergic receptor function, whilst β2 adrenergic receptors may be impacted by Gly16Arg and Gln27Glu. An additional investigation has demonstrated the significance of the Met235Thr and A1166C gene polymorphisms in patients, which impact the effectiveness and adverse events of Angiotensin Receptor Blockers (ARBs) and Angiotensin-Converting Enzyme (ACE) inhibitors. As we already know, statin use has grown significantly in popularity as a means of maintaining body lipid profiles. Pharmacogenomics is important when choosing a statin for a certain patient. Particularly, when treated with atorvastatin, the ABCG8 D19H polymorphism was found to be linked to a higher reduction in LDL in a study [28 - 32].

Inflammatory Diseases

There are numerous genetic markers linked to the aetiology of IBD. Certain medications react to particular markers, whereas others might not. Enhanced treatment outcomes could be significantly impacted by personalised treatment. For instance, Hsp10, Hsp60, and Hsp90 markers respond to 5-aminosalicylic acid (5-ASA therapy); RN6/2 GG (rs315951) and IL-1B-511 CC (rs16944) single nucleotide polymorphisms respond to corticosteroids; etrolizumab responds to Granzyme A and αE integrin gene and protein expression; and anti-tumor necrosis factor therapy responds to the genotypes of Caspase-9 TT and Fas ligand-843 CC/CT [33-41].

In summary, the fields of pharmacogenomics and personalised medicine hold great potential to transform the healthcare industry by facilitating the creation of customised treatment regimens that consider the unique genetic makeup of each patient [42]. By detecting genetic variables, such as variation in gene expression and medication target function, and applying genetic information into clinical practice, healthcare providers can enhance therapeutic efficacy, minimise adverse effects, and maximise treatment success [43]. With the advancement of science, personalised medicine is beginning to fulfil its promise of providing more accurate, efficient, and patient-centered care [44]. However, in order to fully benefit from this innovative approach, a number of problems need to be solved, including accessibility, ethical considerations, and the necessity of providing medical staff with extensive training [45].

Functions of Microbiome in Drug Metabolism

Thousands of bacterial species with a wide range of metabolic functions are found in the human gastrointestinal microbiome. Every person has a unique gut flora that can affect how they metabolize medications. These bacteria not only help metabolize drugs but also make them more absorbable.

They metabolize drugs that initially have no therapeutic impact to create metabolites that have therapeutic potential. The most common method of drug administration is oral, as we already know. The stomach and intestines absorb the majority of the medication, which is then sent to the liver for metabolism. The medication can be secreted back into the intestine by combining with bile after being metabolized, even in situations where alternative routes of administration are used [34, 35].

Drug metabolism and its Metabolites

Research has shown that different drug metabolites are produced by the interaction of drugs and microorganisms. These microorganisms include many metabolic enzymes that assist in breaking down medicines into more potent metabolites. A research was conducted to find out how gut bacteria bio-transformed 176 different medications. For instance, it was discovered that the gram-negative, rod-shaped, obligatory anaerobic bacterium *Bacteroides thetaiotaomicron* metabolizes the medication diltiazem. This demonstrates that the microbiome has an impact on the pharmacokinetics and pharmacodynamics of drugs. A person with these microbes may react differently than a person without them [36]. Previous research has demonstrated the connection between a given drug's metabolism and a particular microorganism. They not only metabolise the medicine, but they can also make it more or less effective. Levodopa absorption may be impacted by *Helicobacter pylori*, but *E. coli*, *Enterobacter*, *Proteus species*, *Klebsiella species*, and *Bacteroides species* may enhance the effects of sulindac by breaking it down into more active components. *Escherichia* and metformin have already been found to have a link in another study, which increases the hypoglycemic impact of the medication [37, 39].

Impact of Microbiome's Alteration

Both drug metabolism and effectiveness can be significantly impacted by changes in the microbiome.gut bacteria largely hydrolyse drugs, prodrugs, and xenobiotic conjugates. The glucuronidase enzymes produced by these microorganisms enable the conversion of conjugated substances into glucuronic acids. Drug metabolism may vary if certain microorganisms are absent, leading to different responses in individuals with altered microbiomes. Pharmacogenomics plays a crucial role in explaining how a drug's adverse effects can differ between individuals. However, variations in gut microbiota may also significantly influence a person's response to medication [40,41].

Finally, it may be concluded that there is a connection between drug metabolism processes and the gut microbiota. These bacteria have enzymes that can transform a drug to make it more active or alter the extent of its absorption. Alterations in these microorganisms could result in inadequate drug metabolism, which would cause variations in the pharmacokinetic and pharmacodynamic characteristics of drugs.

Chronic Diseases and the Effect of Incomplete Metabolism

Chronic Diseases

In the current period, chronic illnesses constitute a major cause of death. Some chronic diseases include metabolic syndrome, obesity, insulin resistance, prediabetes, Type 2 diabetes, non-alcoholic fatty liver disease, and coronary heart disease. The development of chronic diseases is linked to environmental factors, such as air pollution, poor food, genetic susceptibility, or a combination of these [42, 43].

Effects of Incomplete Metabolism

Some scientists have proposed that a key contributing factor to chronic diseases is physical inactivity. A study found a direct correlation between physical activity like cycling, running, *etc*., and Cardiovascular and Respiratory Fitness (CRF). In a study, it was discovered that among healthy individuals experiencing extreme physical inactivity or bed rest, there was a deviation in both metabolic response and CRF. Nevertheless, inactivity can also trigger metabolic syndrome. Genetics, age, gender and diet can also be the reason for metabolic dysfunction. Chronic conditions like diabetes, dyslipidaemia, visceral obesity, and hypertension may result from this [44]. Incomplete metabolism may also aggravate chronic kidney diseases.

The incomplete metabolism of kidney-regulated chemicals may lead to increased stress on the kidneys. For example, the normal Albumin-to-Creatinine Ratio (ACR) is a marker of kidney function. Elevated ACR levels due to metabolic dysfunction may lead to chronic kidney disease or even kidney failure [45]. Lipid metabolism in the human body is essential for hormone synthesis, energy generation, and fat storage. According to several studies, the ability to metabolize fats declines with age. Malfunction of lipid metabolism can result in obesity, nonalcoholic fatty liver disease, cardiovascular disease, type 2 diabetes, and other conditions [46 - 48]. Obesity associated with chronic illness has been linked to dysregulation of iron metabolism. Fat molecules deposited in adipose tissue during obesity cause increased regulation of inflammatory mediators such as tumor necrosis factor (TNF-α), interleukins (IL-6), and leptin, which in turn leads to overexpression of hepcidin. This overexpression of hepcidin may cause iron deficiency [49,50].

Ultimately, it can be concluded that chronic illness and insufficient metabolism are related. As we have already covered, incomplete metabolism can result in metabolite deposition and buildup, which can trigger a number of chronic disorders. However, a chronic illness can also result in metabolic dysfunction, which can then trigger the onset of more chronic illnesses.

Production and Development of Safer Drugs

New Drug Development

The process of developing new drugs and bringing them to market is incredibly time-consuming and expensive. The majority of medications that fail during development have issues with their efficacy and safety. More than 30% of drugs fail during the preclinical trial phase because of safety concerns, and 20% fail during the clinical phase [51 - 53].

Drug Development and Production of Safer Drugs

Safety needs to come first when developing novel drugs because toxicity and adverse drug reactions are the main reasons why new medications fail, even after they are brought to the market. Ideally, toxicity risks should be identified early in the lead selection process using tried-and-true techniques and research designs. To achieve this, a variety of methods are employed, including in silico tools to filter out undesirable lead molecules, *in vitro* testing to screen out weak drug candidates, and *in vivo* testing to evaluate the safety, efficacy, and metabolic properties of drug molecules [54 - 58].

Drug Metabolism and Safer Drugs

The complex process of drug pharmacokinetics in the human body involves the absorption of the drug from the site of administration, its distribution to the intended tissues, its metabolism through biotransformation, and its excretion. Screening these metabolites for toxicity could lead to the development of safer medications. A toxic metabolite that is created during drug metabolism might have harmful effects or render a drug unfit for use. Every metabolite has a soft spot, which manifests as an enzyme or functional group that might have a harmful effect. Enzymes inherited by the human body must be able to identify and catalyze these soft spots. The development of a safer drug may benefit from metabolic switching and the masking of soft spots. Since pro-drugs and active metabolites have functional groups that are easily cleaved by the body's enzymes, they are also being used to address issues with metabolic toxicity and absorption [59 - 61].

To sum up, safety should be prioritized when developing a novel drug. A drug's metabolism is a key factor in determining its safety. For example, incomplete metabolism of a medication can produce harmful metabolites that have detrimental effects. These metabolites may not only raise safety concerns but also interfere with the drug's effectiveness. In an effort to create a safer medication, several novel techniques are being used, and regulatory bodies are attempting to come to an agreement regarding benefit-risk assessment.

Cutting-Edge Biomarkers to Monitor Metabolism

Information on particular metabolic pathways can be obtained by metabolic biomarkers. Monitoring of metabolic biomarkers is necessary for a number of metabolic diseases, including diabetes, obesity, and others. These metabolic biomarkers could aid in illness prediction and early diagnosis [62]. A few of these markers, which are important for tracking metabolism and associated illnesses, are discussed in the section below:

Creatine and Glycogen

When exercising vigorously, the body produces certain metabolites that can be interpreted as biomarkers to track metabolism. Indications of overexertion include low levels of creatine and glycogen stored in muscle and an accumulation of ions, such as potassium, chloride, and lactic acid [63, 64].

Lactate and Pyruvate

The oxidative phosphorylation system, a component of the mitochondrial respiratory chain in the inner membrane, is responsible for ATP production in the organelles known as mitochondria. Multiple markers can be used to track abnormalities in mitochondrial energy metabolism. Two byproducts released by the mitochondrial energy metabolism pathway are lactate and pyruvate. The accumulation of these metabolites indicates defects in the metabolic system. Similarly, elevated levels of serum biomarkers such as 3-methylglutaconic acid, alanine, methylmalonic acid, and ethylenemalonic acid suggest a malfunction in the mitochondrial energy metabolism system [65 - 67].

Leptin and Adiponectin

The human body releases the hormone leptin, which aids in the long-term maintenance of ideal body weight. Conversely, the hormone adiponectin plays a significant role in inflammation and in regulating insulin sensitivity. Elevated levels of leptin are associated with a number of metabolic illnesses, such as obesity, insulin resistance, myocardial infarction, and congestive heart failure. Conversely, a low level of adiponectin indicates a higher chance of developing Type 2 diabetes and other metabolic illnesses in the future [68, 69].

Ghrelin

Ghrelin is a hormone that is released by the pancreas, stomach, and small intestine that aids in sustaining hunger. It has the potential to be a biomarker for the identification of disorders of metabolism. According to certain research, decreased ghrelin levels have been linked to metabolic syndromes, such as insulin resistance, hypertension, and obesity [70].

Uric acid

It has been demonstrated that uric acid, the byproduct of purine synthesis, has pro-inflammatory and pro-oxidant properties inside cells. Researchers have discovered that uric acid is a biomarker for oxidative damage in diseases, such as diabetes, atherosclerosis, ischaemic liver, and chronic heart failure [71].

Cytokines (IL-6, IL-10 and TNF-α)

Interleukin-6 (IL-6)

The majority of IL-6, a pro-inflammatory cytokine, is secreted by M1 macrophages. Increased IL-6 levels have been linked to several metabolic disorders, including hypertension, hypertriglyceridemia, and high fasting glucose [72, 73].

Interleukin-10 (IL-10)

Monocytes release interleukin-10 (IL-10), a cytokine that is mostly anti-inflammatory. A study established a link between low levels of IL-10 and Type 2 diabetes and insulin resistance. It was discovered that elevated levels of blood glucose, triglycerides, cholesterol, and low levels of IL-10 were associated [74].

Tumor Necrosis Factor-Alpha (TNF-α)

The pro-inflammatory cytokine Tumor Necrosis Factor-alpha (TNF-α) is released by visceral adipose tissue and is associated with various metabolic disorders. Numerous metabolic diseases, such as insulin resistance, hypertriglyceridemia, and central obesity, have been associated with elevated TNF-α levels [75].

ASPECTS OF THE ENVIRONMENT AND LIFESTYLE

Diet and Nutrition: Nutrient Deficiency

Enzyme activity can be hampered by a lack of vital nutrients (such as vitamins and minerals), which can result in insufficient macronutrient metabolism. For instance, hereditary metabolic abnormalities of glycogen metabolism, known as glycogenoses, or Glycostor Diseases (GSDs), are caused by deficiencies in enzymes or transporters involved in the production or breakdown of glycogen, resulting in abnormal glycogen use and/or storage. One instance of GSD is thought to occur for every 2000–43000 live births. GSD occurs in more than 20 varieties, including subtypes. This diverse collection of uncommon illnesses, which are categorized according to the impaired enzyme and tissues impacted, is an example of inborn mistakes in the metabolism of carbohydrates because glycogen is particularly rich in the liver and muscle, and GSDs typically affect these tissues or both. However, in addition to the liver and skeletal muscle, multiorgan involvement, including the heart, kidney, and/or brain, may be observed, contingent on the impacted enzyme and its expression in different tissues [76]. A major cause of chronic liver disease globally, Metabolic Dysfunction-associated Steatohepatitis (MASH), is becoming more common in the context of the obesity pandemic. Due to its correlation with an increased risk of developing end-stage liver disease, MASH is also a prime candidate for liver transplantation. One of the primary challenges in managing MASH is the lack of approved medications. In the absence of pharmacological treatments, lifestyle interventions, particularly regular physical activity and a healthy diet, are currently recommended [77].

Nutrients provide energy for metabolism, which is a strictly controlled process. Thus, the energy level of the organism is directly impacted by the nutrient status. Cellular energy state, characterized by low ATP levels and an elevated ADP:AMP ratio, is a prerequisite for AMPK activation [100]. It is triggered by energy stress, which is defined by low ATP levels. By turning on catabolic pathways to produce ATP and turning off anabolic pathways that would otherwise drain ATP, AMPK activation reorganizes cellular metabolism [78].

Overindulgence in Calories

Consuming too many calories, especially in the form of refined carbohydrates and fats, can overload metabolic pathways and cause a buildup of unfinished metabolic products [14]. For example, dietary fructose intake significantly increases the risk of hepatic insulin resistance through the intricate interaction of several metabolic pathways, some of which are unrelated to increased caloric intake and weight gain. Aside from having a significant impact on insulin action, circulating macronutrients like amino acids also operate as metabolic signals in the hypothalamus, assisting in the control of blood glucose levels *via* altering hepatic glucose metabolism. When the levels of macronutrients are high enough to trigger a negative glucoregulatory mechanism in the MBH, resulting in an inhibition, signals originating in the central nervous system are sent to the liver through the vagus nerve [79].

PHYSICAL ACTIVITY

Sedentary Lifestyle

Exercise improves glucose metabolism and lowers metabolic rate, which leads to partial oxidation of lipids and carbs. Research on the effects of increasing physical activity, whether conducted as part of a structured exercise intervention or in a cross-sectional cohort, has demonstrated a significant influence on the risk of cardiometabolic diseases. Regular exercise helps manage lipid disorders by reducing triglyceride levels, increasing HDL cholesterol, promoting weight loss, and lowering blood pressure.

With high rates of morbidity and mortality in patients with underlying chronic diseases like hypertension, diabetes, Cardiovascular Disease (CVD), and Chronic Obstructive Pulmonary Disease (COPD), physical inactivity increases the risk of all-cause mortality, overall poor health, and shortened life expectancy. This is especially true when the patient's BMI is above 30, indicating obesity. Moreover, engaging in intense or severe physical exercise improves general health conditions in addition to stimulating bodily systems. For example, engaging in intense aerobic or endurance exercises (such as swimming or brisk running/walking for 150 minutes a week) can lower the risk of cardiovascular illnesses and enhance the lipid profile in addition to lowering the risk of high blood pressure and Type 2 diabetes [79, 80].

Excessive or Imbalanced Activity

Without sufficient recuperation, high-intensity exercise, even though beneficial, might result in incomplete metabolic processes due to oxidative stress. Elevated levels of physical activity often promote a stronger and more efficient metabolism, which has several positive health effects. To minimize any potential side effects, exercise must be balanced with a healthy diet and adequate recovery time. Increased physical activity lowers the risk of insulin resistance and associated metabolic diseases by improving insulin sensitivity [79].

Toxin Exposure

Exposure to environmental pollutants, including pesticides and heavy metals, can disrupt metabolic enzymes, leading to incomplete detoxification. Smoking and alcohol have the potential to affect liver function, which may result in insufficient metabolism of toxins and medications. Children's neurotoxicity can stem from various factors, such as prescription pharmaceuticals, illegal substances, external toxins and traumas, hereditary or acquired diseases, and metabolic anomalies that cause the accumulation of toxic metabolites. In agricultural and indoor environments, pesticides, such as Organophosphate (OP), Organochlorine (OC), and Carbamate (CB) chemicals are frequently used. While OC is neurotoxic and implicated in ion channel modification, OP and CB act as Acetylcholinesterase (AChE) inhibitors, impacting several organs, including the brain, muscles, liver, pancreas, and both peripheral and central nervous systems. Numerous studies have linked acute and long-term pesticide exposure to oxidative stress, metabolic abnormalities, hyperglycemia, and other metabolic issues related to diabetes and other conditions [81].

Chronic Stress

Extended periods of stress can cause hormonal imbalances (*e.g.*, cortisol) that impact metabolism and result in the partial breakdown of carbohydrates and lipids. The limbic forebrain, the hypothalamus, and the brainstem, for example, are among the brain circuits that are altered during the physiological stress response. These circuits influence attentiveness to sensory cues and elicit emotional reactions, such as fear and avoidance [81].

Sleep Deprivation

Sleep deprivation disrupts circadian rhythms, which impacts glucose metabolism and other metabolic functions. Sleep is physiologically characterized by a lack of response to external stimuli. In vertebrates, sleep is marked by a specific pattern of synchronized electrical activity in the brain's cortical region. Polysomnography (PSG) is commonly used to assess sleep; it simultaneously records cortical activity using Electroencephalogram (EEG), eye movements using Electrooculogram (EOG), and skeletal muscle activity using Electromyogram (EMG). Rapid Eye Movement (REM) sleep and non-REM (NREM) sleep are the two primary stages of sleep in humans and other mammals. In both human and animal models of fear conditioning, Sleep Deprivation (SD) negatively affects the extinction of fear after traumatic exposure, contributing to the development of Post-traumatic Stress Disorder (PTSD). Furthermore, there is a strong correlation between sleep deprivation and psychological disorders, such as anxiety and depression [82].

MICROBIOME HEALTH

Gut Dysbiosis

An imbalance in gut flora might cause insufficient nutrition absorption and digestion, which can impact metabolism as a whole. Many variables affect the composition and predominance of different bacterial species, including the method of feeding (breastfeeding *vs.* milk replacers), the style of delivery (normal birth *vs.* caesarean section), and the mother's diet during pregnancy. Nonetheless, as people age, the composition of their microbiome constantly shifts. Generally, microbial profiles adjust to reflect differences in microbiome function between newborns and adults. For instance, as a person ages, microbiota involved in fat metabolism gradually replace those involved in the metabolism of amino acids and vitamins, which predominate in the guts of newborns [83].

Exposure to Environmental Toxins

Due to the harmful effects on metabolic processes, populations residing in locations with high levels of air, water, or soil pollution may have a higher incidence of metabolic illnesses, such as obesity, Type 2 diabetes, and cardiovascular diseases or other illnesses. A high blood urate level, or hyperuricaemia, is a crucial component that leads to the deposition of Monosodium Urate (MSU) crystals and accelerates the evolution of gout, a chronic illness characterized by inflammatory arthritis. Gout often develops as a consequence of a combination of variables, including genetics, environment, and food. Recent research has indicated that air pollutants, including Carbon

Monoxide (CO), Sulfur Dioxide (SO_2), and particulate matter, may raise the incidence of gout hospitalizations, and variations in humidity and temperature may have an impact on Uric Acid (UA) levels. Gout also exhibits a seasonal tendency. Research has shown that environmental stimuli can either trigger or hasten the synthesis and release of pro-inflammatory mediators, which can lead to imbalanced oxidative stress and systemic inflammation. These variables can then contribute to the overall process or a specific link in the gout process [84].

Exposure to industrial chemicals and pesticides can cause endocrine disruptions that lead to metabolic imbalances in populations. These imbalances include increased fat accumulation and altered glucose metabolism. Pesticides significantly affect not only the metabolism of tryptophan but also that of other amino acids. Changes in amino acid levels can impact vital metabolic processes in the brain, as they serve as building blocks for various pathways, including the synthesis of neurotransmitters, purines, and pyrimidines [85].

Nutritional Exposure

People who have poor access to nutrient-dense food may be malnourished, which can result in impaired metabolism and greater rates of vitamin deficiencies. The most effective non-pharmacological strategy to promote metabolic health is Calorie Restriction (CR). Calorie restriction has benefits beyond weight loss. Numerous studies have consistently shown that CR causes an increase in energy expenditure greater than what can be accounted for by the decrease in metabolic mass (fat-free mass and fat mass). This disproportionate decrease in metabolic rate, referred to as metabolic adaptation, reduces oxidative damage and delays age-related losses in physiological performance, according to accepted theories of mammalian aging [86].

Eating too few calories, however, can cause a major decrease in metabolism. Although a calorie deficit is necessary for weight loss, consuming too few calories can be counterproductive. When calorie intake is drastically lowered, the body senses food scarcity and slows the rate at which it burns calories. Controlled studies in both lean and overweight individuals confirm that consuming fewer than 1,000 calories per day can significantly reduce metabolic rate [87].

High-calorie, Low-Nutrient Diets

Communities exposed to diets rich in processed foods and sugars are at greater risk of obesity and related metabolic diseases due to the consumption of empty calories and a lack of essential nutrients. Daily calorie burn may significantly decline if you lead a sedentary lifestyle. Remarkably, many individuals spend most of their day sitting, which can be detrimental to overall health and metabolic rate. Even simple physical activities like standing up, cleaning, and using the stairs can help burn calories, although engaging in exercise like working out or playing sports has a more significant impact. Non-exercise Activity Thermogenesis (NEAT) refers to this kind of physical activity [88].

Socioeconomic Factors

Based on an extensive body of reliable research, a disadvantaged socioeconomic position (SEP) is linked to an elevated risk of mortality and morbidity, including diabetes, cancer, heart disease, and frailty. However, intermediate risk factors such as smoking and obesity only partiallyexplain the gradient between SEP and health outcomes. Assessments of several physiological systems have revealed that SEP appears to have a unique effect on biomarkers related to inflammatory and metabolic systems [89]. Due to income disparities, environments with low Socioeconomic Status (SES) can be stressful, depleting available reserves and hindering the development of reserve capacity or a "bank" of resources. As a result, individuals with lower SES may experience more negative emotions, which can severely impact their health. Moreover, the reserve capacity concept has been tested by the authors in relation to metabolic syndrome [90].

Education and Awareness

The incidence of metabolic disorders, such as disturbed glucose metabolism, obesity (both general and abdominal), elevated blood pressure, dyslipidemias, insulin resistance, high blood sugar levels, and hyperuricemia, has significantly increased over the past few decades. These conditions are major risk factors for serious diseases, including type 2 Diabetes Mellitus (DM2), Cardiovascular Disease (CVD), and stroke. Given its objectivity, ease of measurement, and ability to reflect various social determinants of health, such as individual behavior, living and working conditions, and access to resources and economic and social opportunities, educational level may be a particularly suitable indicator. There is strong evidence that improved health outcomes are typically connected with increased educational attainment [91, 92]. However, the majority of research shows that the related groups' educational range is rather constrained, with either few individuals at one end or none at all.

Furthermore, most research examining the relationship between education and health status has focused on only one or a limited number of health metrics or indicators. These include conditions like type 2 diabetes (DM2), cardiovascular disease (CVD), and metabolic syndrome (MS), as well as biomarkers such as White Blood Cell Count (WBC), waist circumference, and Body Mass Index (BMI). Other terms related to insulin resistance include high-density lipoprotein (HDL), low-density lipoprotein (LDL), glycosylated hemoglobin, homeostatic model assessment of insulin resistance, systolic and diastolic blood pressure, and systemic pulse pressure (PP). For instance, research has explored the relationship between BMI and educational level, which is pertinent to the current study [93, 94]. Moreover, as examined, BMI is generally found to be inversely related to educational level.

GLOBALIZATION AND CULTURAL INFLUENCE

Trends of Westernization of Diets

People accustomed to traditional, healthier diets may see an increase in the prevalence of metabolic illnesses as Western diets high in fats, sugars, and processed foods, continue to expand globally. Numerous chronic non-infectious illnesses, such as coronary heart disease, cerebrovascular disease, various cancers, diabetes mellitus, dental caries, and multiple bone and joint disorders, have been consistently linked to poor dietary habits, according to studies. However, the likelihood of chronic illnesses is not determined solely by an adult's lifestyle. Health outcomes are influenced by diet and nutrition throughout the life cycle [95]. Moreover, general physical activity levels among people are declining as sedentary activities like desk work and screen time become more prevalent worldwide, which negatively impacts metabolic health [96].

Effects of Enzyme and Epigenetic Exposure

Exposure to environmental and lifestyle factors may have an amplified effect in populations with a high frequency of genetic abnormalities influencing metabolism, potentially resulting in a greater incidence of metabolic illnesses. Epigenetic alterations in populations can also result from exposure to environmental factors such as pollutants or malnutrition, which may induce hereditary metabolic abnormalities in subsequent generations. The increased incidence of non-communicable diseases, including diabetes, obesity, and cardiovascular disorders, can result from epidemiological shifts in communities due to widespread exposure to adverse environmental and lifestyle factors. Populations with significant exposure to variables that negatively influence metabolism may place a greater strain on healthcare systems due to the higher demand for treatment and management of chronic metabolic conditions [97].

ETHICS AND REGULATORY CONCERNS

Environmental Justice

Marginalized communities often face higher exposure to environmental toxins. Variations in meteorological patterns are linked to an increase in allergens and disease vectors, which, in turn, is associated with a rise in the occurrence of respiratory ailments and disorders like asthma. Climate change will lead to more weather variations and natural calamities. Additionally, it is linked to an increase in occurrences like winter inversion, which happens during the winter and significantly raises air pollution levels by trapping a layer of cold air beneath a layer of warm air. It is a major ethical issue to ensure that all people have fair access to clean air, water, nutritious food, and safe environments. Reducing inequities in exposure to hazardous substances is a key objective of the current policy [98].

Identifying Genetic Predispositions

Through the study of an organism's whole genome, a process known as genomics, scientists are able to pinpoint genetic variations linked to metabolic diseases. For instance, certain individuals may be predisposed to diabetes, obesity, or other metabolic disorders due to specific gene abnormalities. Recent advances in genetics and genomics are thoroughly and methodically characterizing the genetic foundations of major metabolic disorders and traits, demonstrating the intrinsic complexity of homeostatic regulatory systems and the various ways they can malfunction [99]. In addition, genomic science is the study of an organism's complete genome (whole genome sequencing). About 20,000 genes are encoded by the 3 billion DNA base pairs that make up the haploid genome of *Homo sapiens*. These genes constitute the coding regions, which make up about 1–2% of the entire genome. The remaining 98–99% consists of non-coding regions, which are significant both structurally and functionally (Fig. **2**) [100].

Personalized Medicine

Treatments and interventions may be customized to a patient's unique needs by understanding their genetic composition, which can improve the management of metabolic illnesses. Pharmacogenomics is the study of individual drug responses using genetic information. Clinical decisions can be guided by genetics when a patient's gene variation is linked to a specific pharmacological response. This may involve adjusting the medication or its dosage, for example. The same approach used to evaluate gene variations associated with diseases is used to assess gene variants that influence a person's drug response: scientists identify genetic loci linked to known drug responses and test those with unknown responses. Today, whole-genome Single Nucleotide Polymorphism (SNP) profiling and multigene analysis are among the modern techniques used in clinical drug discovery and development [101].

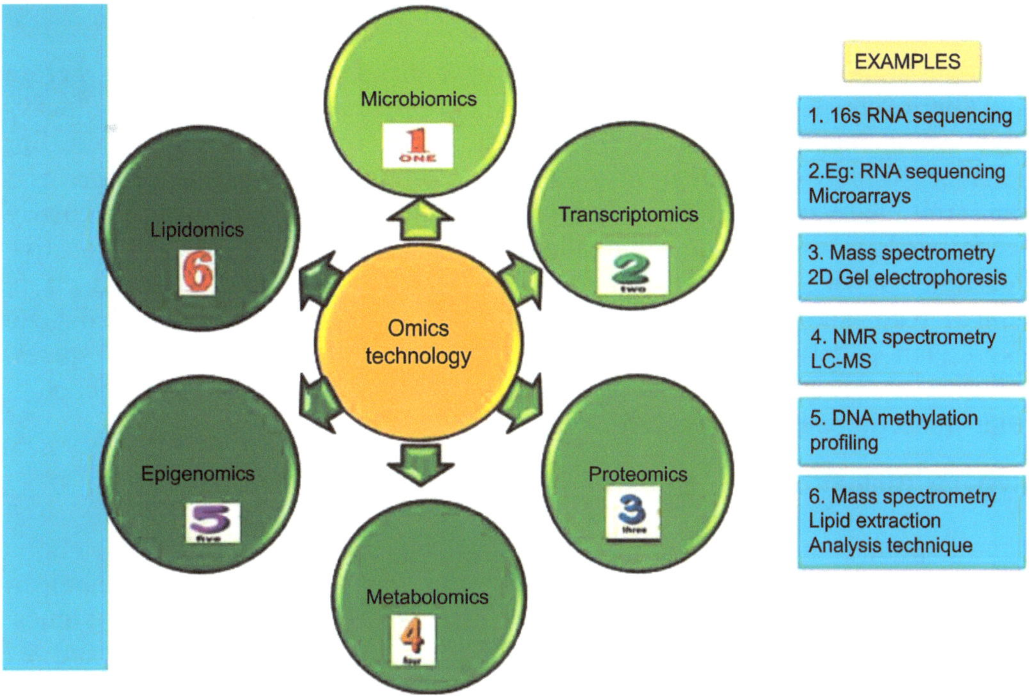

Fig. (2). Omics technology with examples.

Transcriptomics

The study of all RNA transcripts (transcriptomes) generated by an organism's or cell's genome, with a particular emphasis on those associated with metabolic activities, is known as transcriptomics in metabolism. Researchers in this field can investigate how different diets, diseases, or environmental changes affect gene expression and, in turn, how those changes influence metabolic pathways [102].

Transcriptomics data have been integrated with Genome-Wide Association Studies (GWASs) to help understand the molecular mechanisms underlying diseases and traits. However, the utility of metabolomics, when integrated with transcriptomics and disease GWASs, to understand molecular mechanisms for metabolite levels or diseases has not been thoroughly evaluated [103].

Gene Expression Profiling

Transcriptomics analyzes the complete set of RNA transcripts produced by the genome under specific conditions. This technique helps identify which genes are actively being expressed in response to environmental exposures or lifestyle changes [104].

Understanding Pathways

By comparing gene expression profiles under different conditions (*e.g.*, before and after exercise or in the presence of toxins), researchers can identify pathways involved in metabolic processes and understand how they are regulated. For instance, in breeding strategies for *Jatropha curcas*, it is crucial to comprehend transcriptional variation during different stages of seed maturation, particularly regarding low levels of anti-nutritional factors, high-quality oil, and bioactive constituents. These traits may qualify the crop for use in biodiesel production, animal feed, and pharmaceutical applications. In this study, the global gene expression patterns of *J. curcas* seeds at six distinct developmental stages, collected at the same time point on a single plant, were determined using genome-wide transcriptome analysis [105].

Systems Biology

Transcriptomics has the potential to provide comprehensive models of metabolic networks and predict the effects of changes in gene expression on metabolism when combined with other 'omics' data, such as proteomics or metabolomics. Metabolomics can help uncover additional factors driving diabetes and its progression by providing insights into underlying metabolic pathways. Novel biomarkers for Type 2 diabetes (T2D) and associated complications, including branched-chain amino acids, phenylalanine metabolites, and metabolites involved in energy and lipid metabolism, have been identified through metabolomics in epidemiological studies. Additionally, metabolomics has been used to investigate potential drug-modulated pathways. By integrating metabolomics with other omics approaches, a complete metabolic network can be constructed, facilitating causal inference and enabling a systems biology perspective on diabetes research [106].

Disease Research

Transcriptomic investigations can show how metabolic gene expression is changed in diseases like diabetes, obesity, or cancer, which can help identify biomarkers or potential treatment targets. Transcriptomics has been a standard research area since before the Human Genome Project, particularly in the study of human disorders. The transcriptome is the complete set of all RNA transcripts produced by the genome in a specific tissue or cell type, at a particular developmental stage, and under certain physiological or pathological conditions. Therefore, transcriptome analysis enables us to comprehend not only the structure and function of genes, but also the control of gene expression, genome plasticity, and the human genome at the transcription level [107].

Role of Transcriptomics and Pharmacogenomics in Personalized Medicine

Transcriptomic data may help customize medicine based on the unique gene expression profile of each patient, especially in metabolic disease cases.The impact of precision medicine on illnesses and its therapeutic applications is widely documented nowadays. An example of pharmacogenetics in action is the numerous trials conducted to genotype *VKORC1* and *CYP2C9* to modify the dosage of warfarin. Ultimately, these studies produced successful outcomes, including techniques for automated dosage estimation in warfarin-treated patients.

In contrast to pharmacogenetics, which commonly refers to genes directly involved in drug metabolism, pharmacogenomics encompasses all genomic interactions related to drug metabolism (Fig. **2**). Therefore, the goal of both pharmacogenetics and pharmacogenomics is to ascertain a patient's genotype [108, 109].

Proteomics

There are countless functions that proteins carry out for the cell. The entire collection of proteins in a cell is referred to as its proteome, and proteomics is the study of protein structure, function, and the roles that each protein plays inside the cell. The proteome is extremely dynamic and occasionally changes in response to various environmental factors. Understanding how proteins interact with one another, perform their functions, and contribute to biological processes is the aim of the field of proteomics.

Protein "expression profiling" is an application of proteomics in which proteins are identified in an organism at a certain point in time based on their expression in response to a stimulus. Additionally, proteomics may be used to create a protein-network map that shows how different proteins interact in a given biological system [102].

In order to distinguish between a genetically modified organism and a natural type, proteomics may also be used to map protein modification. The study of protein-protein interactions involved in plant defense responses is another use for it. An application of proteomics is known as protein "expression profiling" where proteins are identified at a certain time in an organism as a result of the expression to a stimulus. Protcomics can also be used to develop a protein-network map where interaction among proteins can be determined for a particular living system.

In order to distinguish between a genetically modified organism and a natural type, proteomics may also be used to map protein modification. Another application is studying protein-protein interactions involved in plant defense responses [110].

Metabolomics

One of the newest "omics" sciences is metabolomics. The whole collection of low molecular weight substances in a sample is referred to as the metabolome. These substances directly impact the phenotype of the cell and are the substrates and byproducts of enzymatic processes. Therefore, the goal of metabolomics is to ascertain the composition of these chemicals in a sample at a given time and under specific environmental conditions [106].

While proteomics and genomics have revealed a great deal about genotype, they have revealed very little about phenotype. Chemicals with low molecular weight have the closest connection to phenotype. Metabolomics can detect changes in thousands of molecule levels between a healthy and sick plant. Additionally, this technique may be used to compare the nutritional value of conventional and genetically modified foods.

Metabolic profiling is the in-depth study of small metabolites created during metabolic processes and is known as metabolomics. This method can show how exposures or lifestyle choices impact metabolic pathways and provides a clear readout of metabolic activity. Alternatively, pathway mapping in metabolomics provides targets for therapeutic intervention by identifying bottlenecks or dysregulated processes that may contribute to metabolic diseases [111].

Epigenomics

One important regulatory mechanism of gene transcription is explained by epigenomics, which describes changes in the control of gene activity that occur without altering genetic sequences. It involves characterizing DNA/RNA modifications, including DNA/RNA methylation, as well as higher-order chromatin structure, which is also a component of the DNA-DNA interactome. Epigenomics is the study of changes in gene expression caused by modifications like DNA methylation and histone modification rather than sequence variations in DNA. Alterations in metabolism can be brought about by lifestyle choices and environmental influences. Through epigenetic inheritance, epigenomic research can investigate the long-term effects of exposures on not only the current generation but also subsequent generations. This provides valuable insights into lifestyle and environmental influences [112, 113].

Applications in Metabolic Health

With its origins in early metabolite analysis, the rapidly growing area of metabolomics is today predominantly utilized to find disease biomarkers, including those for diabetes. Diabetes causes abnormalities in the metabolism of sugar, protein, fat, water, and electrolytes, which adversely affect the liver, skeletal muscle, adipose tissue, and other organs. By tracking hundreds of metabolites throughout cells, organs, or whole organisms, metabolomics provides a thorough understanding, or "snapshot," of the metabolic landscape [114]. This approach can identify the composition of metabolites as well as how those patterns change over time or in response to specific disturbances. The discovery of disease biomarkers has been greatly facilitated by recent advances in genomic technologies and molecular procedures, which have also enhanced our understanding of the pathophysiology of many common diseases and signaled the advent of precision medicine. Combining omics techniques, personalized nutrition can lead to tailored dietary recommendations based on an individual's genetic, epigenetic, and metabolic profiles, optimizing nutrition for metabolic health. Additionally, omics techniques can guide public health initiatives by evaluating how lifestyle interventions (such as exercise and stress management) and environmental factors, such as pollution and diet, affect metabolic health at the molecular level [115].

CONCLUSION

In conclusion, research on the potential future paths and frontiers of incomplete metabolism might greatly expand our understanding of a wide range of biological processes and disease mechanisms. By examining the underlying molecular mechanisms and pinpointing important metabolic intermediaries, scientists might find new biomarkers and potential treatment targets. Incomplete metabolism is linked to the pathophysiology of metabolic, neurological, and cancer-related disorders, opening new opportunities for targeted treatments and personalized therapies.

Furthermore, studying how cells adapt to metabolic defects and how incomplete metabolism interacts with the microbiota offers promisingavenues for novel treatment approaches. Integrating modern technologies such as AI-driven modeling and single-cell metabolomics will enhance the precision and breadth of these investigations. Furthermore, considering the evolutionary and environmental factorsinfluencing metabolism can offer more comprehensive insights into the ways in which organisms respond to and overcome metabolic barriers.

Taken together, further investigation into incomplete metabolism has enormous potential for advancing our understanding of basic biology and for developing cutting-edge treatment approaches that will eventually enhance human health and disease outcomes.

REFERENCES

[1] X. Li, Y. Zhai, J. Zhao, H. He, Y. Li, Y. Liu, A. Feng, L. Li, T. Huang, A. Xu, and J. Lyu, "Impact of Metabolic Syndrome and It's Components on Prognosis in Patients With Cardiovascular Diseases: A Meta-Analysis", *Front. Cardiovasc. Med.,* vol. 8, p. 704145, 2021. [http://dx.doi.org/10.3389/fcvm.2021.704145] [PMID: 34336959]

[2] A.P. Sobolev, A. Ciampa, C. Ingallina, L. Mannina, D. Capitani, I. Ernesti, E. Maggi, R. Businaro, M. Del Ben, P. Engel, A.M. Giusti, L.M. Donini, and A. Pinto, "Blueberry-based meals for obese patients with metabolic syndrome: A multidisciplinary metabolomic pilot study", *Metabolites,* vol. 9, no. 7, p. 138, 2019. [http://dx.doi.org/10.3390/metabo9070138] [PMID: 31295937]

[3] D. Ambroselli, F. Masciulli, E. Romano, G. Catanzaro, Z.M. Besharat, M.C. Massari, E. Ferretti, S. Migliaccio, L. Izzo, A. Ritieni, M. Grosso, C. Formichi, F. Dotta, F. Frigerio, E. Barbiera, A.M. Giusti, C. Ingallina, and L. Mannina, "New advances in metabolic syndrome, from prevention to treatment: The role of diet and food", *Nutrients,* vol. 15, no. 3, p. 640, 2023. [http://dx.doi.org/10.3390/nu15030640] [PMID: 36771347]

[4] C. Bruzzone, R. Gil-Redondo, M. Seco, R. Barragán, L. de la Cruz, C. Cannet, H. Schäfer, F. Fang, T. Diercks, M. Bizkarguenaga, B. González-Valle, A. Laín, A. Sanz-Parra, O. Coltell, A.L. de Letona, M. Spraul, S.C. Lu, E. Buguianesi, N. Embade, Q.M. Anstee, D. Corella, J.M. Mato, and O. Millet, "A molecular signature for the metabolic syndrome by urine metabolomics", *Cardiovasc. Diabetol.,* vol. 20, no. 1, p. 155, 2021. [http://dx.doi.org/10.1186/s12933-021-01349-9] [PMID: 34320987]

[5] S.M. Grundy, J.I. Cleeman, C.N.B. Merz, H.B. Brewer Jr, L.T. Clark, D.B. Hunninghake, R.C. Pasternak, S.C. Smith Jr, and N.J. Stone, "Implications of Recent Clinical Trials for the National Cholesterol Education Program Adult Treatment Panel III Guidelines", *Circulation,* vol. 110, no. 2, pp. 227-239, 2004. [http://dx.doi.org/10.1161/01.CIR.0000133317.49796.0E] [PMID: 15249516]

[6] P. Christensen, T. Meinert Larsen, M. Westerterp-Plantenga, I. Macdonald, J.A. Martinez, S. Handjiev, S. Poppitt, S. Hansen, C. Ritz, A. Astrup, L. Pastor-Sanz, F. Sandø-Pedersen, K.H. Pietiläinen, J. Sundvall, M. Drummen, M.A. Taylor, S. Navas-Carretero, T. Handjieva-Darlenska, S. Brodie, M.P. Silvestre, M. Huttunen-Lenz, J. Brand-Miller, M. Fogelholm, and A. Raben, "Men and women respond differently to rapid weight loss: Metabolic outcomes of a multi-centre intervention study after a low-energy diet in 2500 overweight, individuals with pre-diabetes (PREVIEW)", *Diabetes Obes. Metab.,* vol. 20, no. 12, pp. 2840-2851, 2018. [http://dx.doi.org/10.1111/dom.13466] [PMID: 30088336]

[7] K. Pigsborg, and F. Magkos, "Metabotyping for precision nutrition and weight management: Hype or hope?", *Curr. Nutr. Rep.,* vol. 11, no. 2, pp. 117-123, 2022. [http://dx.doi.org/10.1007/s13668-021-00392-y] [PMID: 35025088]

[8] A. Agodi, A. Maugeri, S. Kunzova, O. Sochor, H. Bauerova, N. Kiacova, M. Barchitta, and M. Vinciguerra, "Association of dietary patterns with metabolic syndrome: Results from the kardiovize brno 2030 study", *Nutrients,* vol. 10, no. 7, p. 898, 2018. [http://dx.doi.org/10.3390/nu10070898] [PMID: 30011827]

[9] D.T. Wake, N. Ilbawi, H.M. Dunnenberger, and P.J. Hulick, "Pharmacogenomics", *Med. Clin. North Am.,* vol. 103, no. 6, pp. 977-990, 2019. [http://dx.doi.org/10.1016/j.mcna.2019.07.002] [PMID: 31582008]

[10] R.M. Weinshilboum, and L. Wang, "Pharmacogenomics: Precision medicine and drug response", *Mayo Clin. Proc.,* vol. 92, no. 11, pp. 1711-1722, 2017.

[http://dx.doi.org/10.1016/j.mayocp.2017.09.001] [PMID: 29101939]

[11] M.J. Haller, M.A. Atkinson, and D. Schatz, "Type 1 diabetes mellitus: etiology, presentation, and management", *Pediatr. Clin. North Am.,* vol. 52, no. 6, pp. 1553-1578, 2005. [http://dx.doi.org/10.1016/j.pcl.2005.07.006] [PMID: 16301083]

[12] S.S. Shoily, T. Ahsan, K. Fatema, and A.A. Sajib, "Common genetic variants and pathways in diabetes and associated complications and vulnerability of populations with different ethnic origins", *Sci. Rep.,* vol. 11, no. 1, p. 7504, 2021. [http://dx.doi.org/10.1038/s41598-021-86801-2] [PMID: 33820928]

[13] M.J. Redondo, A.K. Steck, and A. Pugliese, "Genetics of type 1 diabetes", *Pediatr. Diabetes,* vol. 19, no. 3, pp. 346-353, 2018. [http://dx.doi.org/10.1111/pedi.12597] [PMID: 29094512]

[14] R. Dhiman, B. Kumar, D. Kumar, and A. Kumar, "Liquid-liquid extraction method for Ziprasidone (ZRS) bioanalysis by using ZRS-D8 (stable isotope) as internal standard", *J Chem Pharm Res,* vol. 8, no. 3, pp. 722-727, 2016.

[15] A. Singh, L.B. Prasad, K. Shiv, R. Kumar, and S. Garai, *J. Mol. Struct.,* vol. 1288, p. 135835, 2023. Synthesis, characterization, and *in vitro* antibacterial and cytotoxic study of Co(II), Ni(II), Cu(II), and Zn(II) complexes of N-(4-methoxybenzyl) N-(phenylethyl) dithiocarbamate ligand. [http://dx.doi.org/10.1016/j.molstruc.2023.135835]

[16] A. Singh, R. Kumar, K. Shiv, S.K. Pandey, M.K. Bharty, R.J. Butcher, and L.B. Prasad, "Synthesis, crystal structure and screening for anticonvulsant and antianxiety activities of three new Ni(II), Cu(II), and Zn(II) dithiocarbamate complexes", *J. Mol. Struct.,* vol. 1298, p. 137052, 2024. [http://dx.doi.org/10.1016/j.molstruc.2023.137052]

[17] A.A.S. Akil, E. Yassin, A. Al-Maraghi, E. Aliyev, K. Al-Malki, and K.A. Fakhro, "Diagnosis and treatment of type 1 diabetes at the dawn of the personalized medicine era", *J. Transl. Med.,* vol. 19, no. 1, p. 137, 2021. [http://dx.doi.org/10.1186/s12967-021-02778-6] [PMID: 33794915]

[18] B. McAdams, and A. Rizvi, "An overview of insulin pumps and glucose sensors for the generalist", *J. Clin. Med.,* vol. 5, no. 1, p. 5, 2016. [http://dx.doi.org/10.3390/jcm5010005] [PMID: 26742082]

[19] M.A. Atkinson, and E.H. Leiter, "The NOD mouse model of type 1 diabetes: As good as it gets?", *Nat. Med.,* vol. 5, no. 6, pp. 601-604, 1999. [http://dx.doi.org/10.1038/9442] [PMID: 10371488]

[20] P. Venkatachalapathy, S. Padhilahouse, M. Sellappan, T. Subramanian, S.J. Kurian, S.S. Miraj, M. Rao, A.A. Raut, R.K. Kanwar, J. Singh, S. Khadanga, S. Mondithoka, and M. Munisamy, "Pharmacogenomics and personalized medicine in type 2 diabetes mellitus: Potential implications for clinical practice", *Pharm. Genomics Pers. Med.,* vol. 14, pp. 1441-1455, 2021. [http://dx.doi.org/10.2147/PGPM.S329787] [PMID: 34803393]

[21] A. Singh, R. Kumar, R. Patel, Trishna, R. N. Gautam, M. K. Bharty, and L. B. Prasad, "Dithiocarbamate-based novel anti-histaminic agents: synthesis, characterization, crystal structure and thermal study", In: *Dalt. Trans*, no. 33, pp. 14077□14088, 2024. [http://dx.doi.org/10.1039/D4DT01777C]

[22] J. Patel, "Cancer pharmacogenomics, challenges in implementation, and patient-focused perspectives", *Pharmacogenomics Pers Med,* vol. 9, pp. 65-77, 2016. [http://dx.doi.org/10.2147/PGPM.S62918]

[23] S. Jeibouei, "Personalized medicine in breast cancer: pharmacogenomics approaches", *Pharmacogenomics Pers Med,* vol. 12, pp. 59-73, 2019. [http://dx.doi.org/10.2147/PGPM.S167886]

[24] D. Dizon, K. Reynolds, S. Sarangi, and A. Bardia, "Precision medicine and personalized breast cancer:

combination pertuzumab therapy", *Pharm. Genomics Pers. Med.,* vol. 7, pp. 95-105, 2014. [http://dx.doi.org/10.2147/PGPM.S37100] [PMID: 24715764]

[25] F. Khatami, B. Larijani, S. Nikfar, M. Hasanzad, K. Fendereski, and S.M. Tavangar, "Personalized treatment options for thyroid cancer: current perspectives", *Pharm. Genomics Pers. Med.,* vol. 12, pp. 235-245, 2019. [http://dx.doi.org/10.2147/PGPM.S181520] [PMID: 31571972]

[26] "COVID-19: a challenge towards the sustainability of health in platform era", In: *NeuroQuantology* vol. 20. , 2022, no. 11, pp. 2503-2519.

[27] M. Younis, and H. Harashima, "Understanding gene involvement in hepatocellular carcinoma: Implications for gene therapy and personalized medicine", *Pharm. Genomics Pers. Med.,* vol. 17, pp. 193-213, 2024. [http://dx.doi.org/10.2147/PGPM.S431346] [PMID: 38737776]

[28] R. Kumar, S. Dutt, A. Goyal, A. Kumar, and B. Kumar, "Nanotechnology in gastrointestinal endoscopy," In; *Nanobiotechnology and Artificial Intelligence in Gastrointestinal Diseases*, V. K. Chaturvedi, A. K. Singh, J. Singh, and D. P. Yadav, Eds., IOP Publishing Ltd, 2024, pp. 5-1, 5-32. [http://dx.doi.org/10.1088/978-0-7503-6134-7ch5]

[29] S. Swaminathan, and G. Ramachandran, "Role of pharmacogenomics in the treatment of tuberculosis: a review", *Pharm. Genomics Pers. Med.,* vol. 5, pp. 89-98, 2012. [http://dx.doi.org/10.2147/PGPM.S15454] [PMID: 23226065]

[30] F. Habersetzer, C. Leboeuf, M. Doffoël, and T. Baumert, "Boceprevir and personalized medicine in hepatitis C virus infection", *Pharm. Genomics Pers. Med.,* vol. 5, pp. 125-137, 2012. [http://dx.doi.org/10.2147/PGPM.S24259] [PMID: 23226068]

[31] A. Hamid, R. Kumar, V. K. Chaturvedi, S. Dutt, G. Sulabh, V. Kumar, and D. P. Yadav, "Artificial intelligence (AI) based colonoscopy," In: *Nanobiotechnology and Artificial Intelligence in Gastrointestinal Diseases*, V. K. Chaturvedi, A. K. Singh, J. Singh, and D. P. Yadav, Eds., IOP Publishing Ltd, 2024, pp. 12-1, 12-23. [http://dx.doi.org/10.1088/978-0-7503-6134-7ch12]

[32] D.W. Denning, J. Cadranel, C. Beigelman-Aubry, F. Ader, A. Chakrabarti, S. Blot, A.J. Ullmann, G. Dimopoulos, and C. Lange, "Chronic pulmonary aspergillosis: rationale and clinical guidelines for diagnosis and management", *Eur. Respir. J.,* vol. 47, no. 1, pp. 45-68, 2016. [http://dx.doi.org/10.1183/13993003.00583-2015] [PMID: 26699723]

[33] B. de Chassey, A. Aublin-Gex, A. Ruggieri, L. Meyniel-Schicklin, F. Pradezynski, N. Davoust, T. Chantier, L. Tafforeau, P.E. Mangeot, C. Ciancia, L. Perrin-Cocon, R. Bartenschlager, P. André, and V. Lotteau, "The interactomes of influenza virus NS1 and NS2 proteins identify new host factors and provide insights for ADAR1 playing a supportive role in virus replication", *PLoS Pathog.,* vol. 9, no. 7, p. e1003440, 2013. [http://dx.doi.org/10.1371/journal.ppat.1003440] [PMID: 23853584]

[34] W. Zhang, Dolan, and Huang, "Integrating epigenomics into pharmacogenomic studies", *Pharm. Genomics Pers. Med.,* vol. 2008, no. 1, pp. 7-14, 2008. [http://dx.doi.org/10.2147/PGPM.S4341] [PMID: 20622972]

[35] A. Kumar, P. Singh, R. Kumar, and S. Dutt, "Novel drug delivery systems for inflammatory bowel disease," In: *Nanobiotechnology and Artificial Intelligence in Gastrointestinal Diseases*, V. K. Chaturvedi, A. K. Singh, J. Singh, and D. P. Yadav, Eds., IOP Publishing Ltd, 2024, pp. 4-1, 4-24. [http://dx.doi.org/10.1088/978-0-7503-6134-7ch4]

[36] M.J. Rieder, A.P. Reiner, B.F. Gage, D.A. Nickerson, C.S. Eby, H.L. McLeod, D.K. Blough, K.E. Thummel, D.L. Veenstra, and A.E. Rettie, "Effect of VKORC1 haplotypes on transcriptional regulation and warfarin dose", *N. Engl. J. Med.,* vol. 352, no. 22, pp. 2285-2293, 2005. [http://dx.doi.org/10.1056/NEJMoa044503] [PMID: 15930419]

[37] D. Rosskopf, and M.C. Michel, "Pharmacogenomics of G protein-coupled receptor ligands in

cardiovascular medicine", *Pharmacol. Rev.,* vol. 60, no. 4, pp. 513-535, 2008. [http://dx.doi.org/10.1124/pr.108.000612] [PMID: 19074621]

[38] M. Scharplatz, M.A. Puhan, J. Steurer, and L.M. Bachmann, *BMC Med. Genet.,* vol. 5, no. 1, p. 23, 2004. What is the impact of the *ACE* gene insertion/deletion (I/D) polymorphism on the clinical effectiveness and adverse events of *ACE* inhibitors? – Protocol of a systematic review. [http://dx.doi.org/10.1186/1471-2350-5-23] [PMID: 15361261]

[39] K. Kajinami, M.E. Brousseau, C. Nartsupha, J.M. Ordovas, and E.J. Schaefer, "ATP binding cassette transporter G5 and G8 genotypes and plasma lipoprotein levels before and after treatment with atorvastatin", *J. Lipid Res.,* vol. 45, no. 4, pp. 653-656, 2004. [http://dx.doi.org/10.1194/jlr.M300278-JLR200] [PMID: 14703505]

[40] C. Barone, S.S. Mousa, and S. Mousa, "Pharmacogenomics in cardiovascular disorders: Steps in approaching personalized medicine in cardiovascular medicine", *Pharm. Genomics Pers. Med.,* vol. 2, pp. 59-67, 2009. [http://dx.doi.org/10.2147/PGPM.S5805] [PMID: 23226035]

[41] J. Yamamoto-Furusho, "Pharmacogenetics in inflammatory bowel disease: understanding treatment response and personalizing therapeutic strategies", *Pharm. Genomics Pers. Med.,* vol. 10, pp. 197-204, 2017. [http://dx.doi.org/10.2147/PGPM.S109648] [PMID: 28603427]

[42] A. Kumar, B. Kumar, R. Kumar, A. Kumar, M. Singh, V. Tiwari, A. Trigunayat, P. Paul, and P. Singh, "Acute and subacute toxicity study of ethanolic extract of Calotropis procera (Aiton) Dryand flower in Swiss albino mice", *Phytomed. Plus,* vol. 2, no. 2, p. 100224, 2022. [http://dx.doi.org/10.1016/j.phyplu.2022.100224]

[43] R. Dhiman, "Novel LC-MS/MS bioanalytical method validation for gabapentin in human plasma in accordance with current regulatory guidances", *Int J Pharm Chem Sci,* vol. 4, 2015no. 4, pp. 462-468. Available from: http://www.ijpcsonline.com/files/08-10-15/07-1035.pdf

[44] H. Pandey, S. Srivastava, R. Kumar, and Y. B. Tripathi, "Preclinical acute and repeated dose toxicity of *Pueraria tuberosa* (PTWE) on Charles Foster rats," Int. J. Pharm. Sci. Res., pp. 4572–4581, 2018. Available from: https://ijpsr.com/bft-article/preclinical-acute-and-repeated-dose-toxicity-of--ueraria-tuberosa-ptwe-on-charles-foster-rats/

[45] R. Dhiman, D. Kumar, B. Kumar, and B.L. Pandey, "Quantitative determination of atorvastatin, ortho-hydroxy atorvastatin, para-hydroxy atorvastatin in human plasma using rosuvastatin as internal standard by LC-MS / MS", *Int. J. Pharm. Chem. Sci.,* vol. 4, no. 4, pp. 487-500, 2016.

[46] J. Qin, R. Li, J. Raes, M. Arumugam, K.S. Burgdorf, C. Manichanh, T. Nielsen, N. Pons, F. Levenez, T. Yamada, D.R. Mende, J. Li, J. Xu, S. Li, D. Li, J. Cao, B. Wang, H. Liang, H. Zheng, Y. Xie, J. Tap, P. Lepage, M. Bertalan, J.M. Batto, T. Hansen, D. Le Paslier, A. Linneberg, H.B. Nielsen, E. Pelletier, P. Renault, T. Sicheritz-Ponten, K. Turner, H. Zhu, C. Yu, S. Li, M. Jian, Y. Zhou, Y. Li, X. Zhang, S. Li, N. Qin, H. Yang, J. Wang, S. Brunak, J. Doré, F. Guarner, K. Kristiansen, O. Pedersen, J. Parkhill, J. Weissenbach, P. Bork, S.D. Ehrlich, and J. Wang, "A human gut microbial gene catalogue established by metagenomic sequencing", *Nature,* vol. 464, no. 7285, pp. 59-65, 2010. [http://dx.doi.org/10.1038/nature08821] [PMID: 20203603]

[47] T. Kimura, K. Sudo, Y. Kanzaki, K. Miki, Y. Takeichi, Y. Kurosaki, and T. Nakayama, "Drug absorption from large intestine: physicochemical factors governing drug absorption", *Biol. Pharm. Bull.,* vol. 17, no. 2, pp. 327-333, 1994. [http://dx.doi.org/10.1248/bpb.17.327] [PMID: 8205133]

[48] M. Zimmermann, M. Zimmermann-Kogadeeva, R. Wegmann, and A.L. Goodman, "Mapping human microbiome drug metabolism by gut bacteria and their genes", *Nature,* vol. 570, no. 7762, pp. 462-467, 2019. [http://dx.doi.org/10.1038/s41586-019-1291-3] [PMID: 31158845]

[49] H. Hashim, S. Azmin, H. Razlan, N.W. Yahya, H.J. Tan, M.R.A. Manaf, and N.M. Ibrahim, *PLoS*

One, vol. 9, no. 11, p. e112330, 2014. Eradication of *Helicobacter pylori* infection improves levodopa action, clinical symptoms and quality of life in patients with Parkinson's disease. [http://dx.doi.org/10.1371/journal.pone.0112330] [PMID: 25411976]

[50] H.A. Strong, A.G. Renwick, C.F. George, Y.F. Liu, and M.J. Hill, "The reduction of sulphinpyrazone and sulindac by intestinal bacteria", *Xenobiotica,* vol. 17, no. 6, pp. 685-696, 1987. [http://dx.doi.org/10.3109/00498258709043976] [PMID: 3630204]

[51] Y. Lee, A.H. Kim, E. Kim, S. Lee, K.S. Yu, I.J. Jang, J.Y. Chung, and J.Y. Cho, "Changes in the gut microbiome influence the hypoglycemic effect of metformin through the altered metabolism of branched-chain and nonessential amino acids", *Diabetes Res. Clin. Pract.,* vol. 178, p. 108985, 2021. [http://dx.doi.org/10.1016/j.diabres.2021.108985] [PMID: 34329692]

[52] S.J. Pellock, and M.R. Redinbo, "Glucuronides in the gut: Sugar-driven symbioses between microbe and host", *J. Biol. Chem.,* vol. 292, no. 21, pp. 8569-8576, 2017. [http://dx.doi.org/10.1074/jbc.R116.767434] [PMID: 28389557]

[53] A. Pant, T.K. Maiti, D. Mahajan, and B. Das, "Human Gut Microbiota and Drug Metabolism", *Microb. Ecol.,* vol. 86, no. 1, pp. 97-111, 2023. [http://dx.doi.org/10.1007/s00248-022-02081-x] [PMID: 35869999]

[54] F.W. Booth, C.K. Roberts, and M.J. Laye, "Lack of exercise is a major cause of chronic diseases", *Compr. Physiol.,* vol. 2, no. 2, pp. 1143-1211, 2012. [http://dx.doi.org/10.1002/j.2040-4603.2012.tb00425.x] [PMID: 23798298]

[55] A. Villanueva-Olivo, H. Rodríguez-Rocha, A. García-García, O. Saucedo-Cárdenas, R. Montes De Oca-Luna, and M. J. Loera-Arias, "Chronic diseases: Origin and cell mechanisms involved", *Cell. Mol. Biol.,* vol. 69, no. 15, pp. 26-37, 2023. [http://dx.doi.org/10.14715/cmb/2023.69.15.5] [PMID: 38279502]

[56] R.H. Olsen, R. Krogh-Madsen, C. Thomsen, F.W. Booth, and B.K. Pedersen, "Metabolic responses to reduced daily steps in healthy nonexercising men", *JAMA,* vol. 299, no. 11, pp. 1261-1263, 2008. [http://dx.doi.org/10.1001/jama.299.11.1259] [PMID: 18349087]

[57] "Changes in albumin excretion in the diabetes prevention program", *Diabetes Care,* vol. 32, no. 4, pp. 720-725, 2009. [http://dx.doi.org/10.2337/dc08-1400] [PMID: 19131464]

[58] B.N. Ames, "Prolonging healthy aging: Longevity vitamins and proteins", *Proc. Natl. Acad. Sci. USA,* vol. 115, no. 43, pp. 10836-10844, 2018. [http://dx.doi.org/10.1073/pnas.1809045115] [PMID: 30322941]

[59] H. Moreau, R. Laugier, Y. Gargouri, F. Ferrato, and R. Verger, "Human preduodenal lipase is entirely of gastric fundic origin", *Gastroenterology,* vol. 95, no. 5, pp. 1221-1226, 1988. [http://dx.doi.org/10.1016/0016-5085(88)90354-X] [PMID: 3169491]

[60] P. Bergman, and S. Brighenti, "Targeted Nutrition in Chronic Disease", *Nutrients,* vol. 12, no. 6, p. 1682, 2020. [http://dx.doi.org/10.3390/nu12061682] [PMID: 32516882]

[61] L. Zhao, X. Zhang, Y. Shen, X. Fang, Y. Wang, and F. Wang, "Obesity and iron deficiency: a quantitative meta-analysis", *Obes. Rev.,* vol. 16, no. 12, pp. 1081-1093, 2015. [http://dx.doi.org/10.1111/obr.12323] [PMID: 26395622]

[62] F. Qiu, L. Wu, G. Yang, C. Zhang, X. Liu, X. Sun, X. Chen, and N. Wang, "The role of iron metabolism in chronic diseases related to obesity", *Mol. Med.,* vol. 28, no. 1, p. 130, 2022. [http://dx.doi.org/10.1186/s10020-022-00558-6] [PMID: 36335331]

[63] K.I. Kaitin, and J.A. DiMasi, "Pharmaceutical innovation in the 21st century: new drug approvals in the first decade, 2000-2009", *Clin. Pharmacol. Ther.,* vol. 89, no. 2, pp. 183-188, 2011. [http://dx.doi.org/10.1038/clpt.2010.286] [PMID: 21191382]

[64] I. Kola, and J. Landis, "Can the pharmaceutical industry reduce attrition rates?", *Nat. Rev. Drug*

Discov., vol. 3, no. 8, pp. 711-716, 2004.
[http://dx.doi.org/10.1038/nrd1470] [PMID: 15286737]

[65] J. Arrowsmith, "Phase III and submission failures: 2007–2010", *Nat. Rev. Drug Discov.,* vol. 10, no. 2, p. 87, 2011.
[http://dx.doi.org/10.1038/nrd3375] [PMID: 21283095]

[66] M. Singh, B. Kumar, A. K Singh, C. Shekhar Azad, M. K Yadav, A. Kumar, R. Kumar, A. Kumar, and P. Chaudhary, "Pharmacokinetic screening to Estimate the drug likeliness characteristics of selected Herbal Anticancer Drugs", *Res. J. Pharm. Technol.,* vol. 16, no. 7, pp. 3422-3426, 2023.
[http://dx.doi.org/10.52711/0974-360X.2023.00566]

[67] K.M. Giacomini, R.M. Krauss, D.M. Roden, M. Eichelbaum, M.R. Hayden, and Y. Nakamura, "When good drugs go bad", *Nature,* vol. 446, no. 7139, pp. 975-977, 2007.
[http://dx.doi.org/10.1038/446975a] [PMID: 17460642]

[68] R. Ulrich, and S.H. Friend, "Toxicogenomics and drug discovery: will new technologies help us produce better drugs?", *Nat. Rev. Drug Discov.,* vol. 1, no. 1, pp. 84-88, 2002.
[http://dx.doi.org/10.1038/nrd710] [PMID: 12119613]

[69] G.D. Veith, "On the nature, evolution and future of quantitative structure-activity relationships (QSAR) in toxicology", *SAR QSAR Environ. Res.,* vol. 15, no. 5-6, pp. 323-330, 2004.
[http://dx.doi.org/10.1080/10629360412331297380] [PMID: 15669692]

[70] E.L. Roggen, *Front. Pharmacol.,* vol. 2, p. 3, 2011. *In vitro* toxicity testing in the twenty-first century.
[http://dx.doi.org/10.3389/fphar.2011.00003] [PMID: 21713102]

[71] J. Bolker, "There's more to life than rats and flies", *Nature,* vol. 491, no. 7422, pp. 31-33, 2012.
[http://dx.doi.org/10.1038/491031a] [PMID: 23128209]

[72] B. Kumar, R. Kumar, H. Patel, and A. Kumar, "Development and Validation of an RP-HPLC Method for Simultaneous Estimation of Amlodipine besylate and Indapamide in Bulk and Tablet Dosage Form", *Indo Am. J. Pharm. Sci.,* vol. 4, no. 11, pp. 4278-4287, 2017.

[73] SB. Kumar, R. Kumar, A. Kumar, and R. Patel, "Simultaneous RP-HPLC method development and validation of fenofibrate and rosuvastatin calcium in bulk and tablet dosage form," *Indo Am. J. Pharm. Sci.,* vol. 4, no. 11, pp. 4288-4297, 2017.

[74] Z. Zhang, M. Zhu, and W. Tang, "Metabolite identification and profiling in drug design: current practice and future directions", *Curr. Pharm. Des.,* vol. 15, no. 19, pp. 2220-2235, 2009.
[http://dx.doi.org/10.2174/138161209788682460] [PMID: 19601824]

[75] M. Theodosiou, A. Amir-Aslani, and B. Mégarbane, "Emerging technologies and challenges for better and safer drugs", *Biotechnol. Lett.,* vol. 36, no. 4, pp. 677-684, 2014.
[http://dx.doi.org/10.1007/s10529-013-1408-y] [PMID: 24243233]

[76] J. van Gelder, S. Deferme, L. Naesens, E. De Clercq, G. van den Mooter, R. Kinget, and P. Augustijns, "Intestinal absorption enhancement of the ester prodrug tenofovir disoproxil fumarate through modulation of the biochemical barrier by defined ester mixtures", *Drug Metab. Dispos.,* vol. 30, no. 8, pp. 924-930, 2002.
[http://dx.doi.org/10.1124/dmd.30.8.924] [PMID: 12124311]

[77] K. Ajay, N. Manmath Kumar, K. Brijesh, K. Ashutosh, K. Rajesh, K. Vikas, and S. Ambrish Kumar, *J. Appl. Pharm. Sci.,* vol. 12, no. 2, pp. 187-197, 2022. Toxicity (acute and subacute) assessment and *in-vivo* antiurolithiatic activity of ethanolic extract of *Caesalpinia bonducella* seed in albino Wistar rat.
[http://dx.doi.org/10.7324/JAPS.2022.120220]

[78] R. Kumar, B. Kumar, and A. Kumar, "Stability indicating method development and validation of abiraterone in human plasma by using LC-MS/MS", In: vol. 3, 2017, no. 5, pp. 12-26.

[79] R. Kumar, B. Kumar, and A. Kumar, "Pharmacokinetic profile of antitubercular drugs and their clinical applications in therapeutic drug monitoring", *Asian J Pharm,* vol. 3, no. 5, pp. 145-154, 2017. Available from: http://ajpp.in/uploaded/p71.pdf

[80] M. Stuckey, R. Fulkerson, E. Read, E. Russell-Minda, C. Munoz, P. Kleinstiver, and R. Petrella, "Remote monitoring technologies for the prevention of metabolic syndrome: the diabetes and technology for increased activity (data) study", *J. Diabetes Sci. Technol.,* vol. 5, no. 4, pp. 936-944, 2011.
[http://dx.doi.org/10.1177/193229681100500417] [PMID: 21880237]

[81] A. Kumar, B. Kumar, R. Kumar, A. Kumar, M. Singh, P. Singh, V. Tiwari, and A. Trigunayat, "Assessment of pain-ameliorative potential of ethanolic extract of *Calotropis procera* and its active components characterization by HPTLC and GC-MS analysis," *Sys. Rev. Pharm.,* vol. 13, no. 9, pp. 581–586, 2022.

[82] H. Ishii, and Y. Nishida, "Effect of lactate accumulation during exercise-induced muscle fatigue on the sensorimotor cortex", *J. Phys. Ther. Sci.,* vol. 25, no. 12, pp. 1637-1642, 2013.
[http://dx.doi.org/10.1589/jpts.25.1637] [PMID: 24409038]

[83] H. Roschel, B. Gualano, M. Marquezi, A. Costa, and A.H. Lancha Jr, "Creatine supplementation spares muscle glycogen during high intensity intermittent exercise in rats", *J. Int. Soc. Sports Nutr.,* vol. 7, no. 1, p. 6, 2010.
[http://dx.doi.org/10.1186/1550-2783-7-6] [PMID: 20205834]

[84] M. Adeva-Andany, M. López-Ojén, R. Funcasta-Calderón, E. Ameneiros-Rodríguez, C. Donapetry-García, M. Vila-Altesor, and J. Rodríguez-Seijas, "Comprehensive review on lactate metabolism in human health", *Mitochondrion,* vol. 17, pp. 76-100, 2014.
[http://dx.doi.org/10.1016/j.mito.2014.05.007] [PMID: 24929216]

[85] B. Scheiner, G. Lindner, T. Reiberger, B. Schneeweiss, M. Trauner, C. Zauner, and G.C. Funk, "Acid-base disorders in liver disease", *J. Hepatol.,* vol. 67, no. 5, pp. 1062-1073, 2017.
[http://dx.doi.org/10.1016/j.jhep.2017.06.023] [PMID: 28684104]

[86] S. Boenzi, and D. Diodato, "Biomarkers for mitochondrial energy metabolism diseases", *Essays Biochem.,* vol. 62, no. 3, pp. 443-454, 2018.
[http://dx.doi.org/10.1042/EBC20170111] [PMID: 29980631]

[87] R.D. Vijaya Durga, "Determination of S-amlodipine in human plasma using tizanidine as internal standard by LC/MS/MS method", *J. Bioanal. Biomed.,* vol. 7, no. 3, pp. 91-96, 2015.
[http://dx.doi.org/10.4172/1948-593X.1000130]

[88] C.M. Ghantous, Z. Azrak, S. Hanache, W. Abou-Kheir, and A. Zeidan, "Differential Role of Leptin and Adiponectin in Cardiovascular System", *Int. J. Endocrinol.,* vol. 2015, pp. 1-13, 2015.
[http://dx.doi.org/10.1155/2015/534320] [PMID: 26064110]

[89] J. Spranger, A. Kroke, M. Möhlig, M.M. Bergmann, M. Ristow, H. Boeing, and A.F.H. Pfeiffer, "Adiponectin and protection against type 2 diabetes mellitus", *Lancet,* vol. 361, no. 9353, pp. 226-228, 2003.
[http://dx.doi.org/10.1016/S0140-6736(03)12255-6] [PMID: 12547549]

[90] B. Fagerberg, L.M. Hultén, and J. Hulthe, "Plasma ghrelin, body fat, insulin resistance, and smoking in clinically healthy men: the atherosclerosis and insulin resistance study", *Metabolism,* vol. 52, no. 11, pp. 1460-1463, 2003.
[http://dx.doi.org/10.1016/S0026-0495(03)00274-9] [PMID: 14624407]

[91] L. Billiet, S. Doaty, J.D. Katz, and M.T. Velasquez, "Review of hyperuricemia as new marker for metabolic syndrome", *ISRN Rheumatol.,* vol. 2014, pp. 1-7, 2014.
[http://dx.doi.org/10.1155/2014/852954] [PMID: 24693449]

[92] T.W. Weiss, H. Arnesen, and I. Seljeflot, "Components of the Interleukin-6 transsignalling system are associated with the metabolic syndrome, endothelial dysfunction and arterial stiffness", *Metabolism,* vol. 62, no. 7, pp. 1008-1013, 2013.
[http://dx.doi.org/10.1016/j.metabol.2013.01.019] [PMID: 23428306]

[93] K. Indulekha, J. Surendar, and V. Mohan, "High sensitivity C-reactive protein, tumor necrosis factor-

α, interleukin-6, and vascular cell adhesion molecule-1 levels in Asian Indians with metabolic syndrome and insulin resistance (CURES-105)", *J. Diabetes Sci. Technol.,* vol. 5, no. 4, pp. 982-988, 2011.
[http://dx.doi.org/10.1177/193229681100500421] [PMID: 21880241]

[94] E. van Exel, J. Gussekloo, A.J.M. de Craen, M. Frölich, A. Bootsma-van der Wiel, and R.G.J. Westendorp, "Low production capacity of interleukin-10 associates with the metabolic syndrome and type 2 diabetes : the Leiden 85-Plus Study", *Diabetes,* vol. 51, no. 4, pp. 1088-1092, 2002.
[http://dx.doi.org/10.2337/diabetes.51.4.1088] [PMID: 11916930]

[95] A.R. Aroor, S. McKarns, V.G. DeMarco, G. Jia, and J.R. Sowers, "Maladaptive immune and inflammatory pathways lead to cardiovascular insulin resistance", *Metabolism,* vol. 62, no. 11, pp. 1543-1552, 2013.
[http://dx.doi.org/10.1016/j.metabol.2013.07.001] [PMID: 23932846]

[96] A. Kumar, M.K. Nandi, B. Kumar, A. Kumar, R. Kumar, A.K. Singh, and M. Singh, *Int. J. Pharm. Investig.,* vol. 11, no. 3, pp. 306-311, 2021. Antiurolithiasis, antioxidant, anti-inflammatory, analgesic, and diuretic activity of ethanolic extract of seeds of *Caesalpinia bonducella.*
[http://dx.doi.org/10.5530/ijpi.2021.3.54]

[97] R. Dhiman, V. Durga, K. Jayasankar, and S. Raj, "Solid phase extraction: Potential carbocisteine bioanalysis using carbocisteine 13c3 stable isotope: LC-MS/MS technique with pharmacokinetics on healthy human subjects", *Pharm. Lett.,* vol. 7, no. 8, pp. 126-134, 2015.

[98] E. Gümüş, and H. Özen, "Glycogen storage diseases: An update", *World J. Gastroenterol.,* vol. 29, no. 25, pp. 3932-3963, 2023.
[http://dx.doi.org/10.3748/wjg.v29.i25.3932] [PMID: 37476587]

[99] H. Ghazanfar, N. Javed, A. Qasim, G.S. Zacharia, A. Ghazanfar, A. Jyala, E. Shehi, and H. Patel, "Metabolic dysfunction-associated steatohepatitis and progression to hepatocellular carcinoma: A literature review", *Cancers (Basel),* vol. 16, no. 6, p. 1214, 2024.
[http://dx.doi.org/10.3390/cancers16061214] [PMID: 38539547]

[100] R. Dhiman, V. Durga, and A. J. Rio, "Bioanalytical method for lornoxicam determination in human plasma by using piroxicam as internal standard by LC-MS/MS", *J Chem Pharm Res,* vol. 7, no. 8, pp. 54-60, 2015.

[101] C.L. Lyons, and H.M. Roche, "Nutritional Modulation of AMPK-Impact upon Metabolic-Inflammation", *Int. J. Mol. Sci.,* vol. 19, no. 10, p. 3092, 2018.
[http://dx.doi.org/10.3390/ijms19103092] [PMID: 30304866]

[102] R. Kumar, B. Kumar, A. Kumar, A. Kumar, and M. Singh, *Open Access J Neurol Disord Journal of Neurological Disorders Commentary HPTLC and GC-MS Analyses of Biologically Active Extracts and Fractions from Premna Latifolia Roxb Leaves,* 2021.

[103] R. Kumar, B. Kumar, A. Kumar, A. Kumar, and M. Singh, "GC-MS analysis of Phytocomponents in the Methanol Extract of Premna latifolia Roxb", *Pharmacognosy Res.,* vol. 14, no. 1, pp. 19-23, 2021.
[http://dx.doi.org/10.5530/pres.14.1.4]

[104] J. Rossen, A. Yngve, M. Hagströmer, K. Brismar, B.E. Ainsworth, C. Iskull, P. Möller, and U.B. Johansson, "Physical activity promotion in the primary care setting in pre- and type 2 diabetes - the Sophia step study, an RCT", *BMC Public Health,* vol. 15, no. 1, p. 647, 2015.
[http://dx.doi.org/10.1186/s12889-015-1941-9] [PMID: 26164092]

[105] H. Pandey, S. Srivastava, A. Kumar, R. Kumar, and Y. Bhusan Tripathi, "Pharmacokinetics Study Of Puerarin Absorption In Blood After Consumption Of Pueraria Tuberosa Water Extract (PTWE) By Rats", *Explor. Anim. Med. Res.,* vol. 9, no. 2, pp. 188-196, 2019.

[106] R. Kumar, A. Kumar, A. Kumar, and M. Singh, "HPTLC and GC-MS analyses of biologically active extracts and fractions from Premna latifolia Roxb leaves reveal anti-inflammatory and analgesic effects," *Sys. Rev. Pharm.,* vol. 13, no. 8, pp. 811-818, 2021.

[107] R.R. Wolfe, "The underappreciated role of muscle in health and disease", *Am. J. Clin. Nutr.,* vol. 84, no. 3, pp. 475-482, 2006.
[http://dx.doi.org/10.1093/ajcn/84.3.475] [PMID: 16960159]

[108] I. Bezruk, A. Materiienko, S. Gubar, K. Proskurina, L. Budanova, L. Ivanauskas, and V. Georgiyants, *Nat. Prod. Res.,* vol. 36, no. 4, pp. 1014-1019, 2022. Estimation of the influence of the environmental factors on the accumulation of phytochemicals and antioxidant capacity in the ivy leaves (*Hedera helix* L.).
[http://dx.doi.org/10.1080/14786419.2020.1843029] [PMID: 33146030]

[109] F. Atrooz, and S. Salim, "Sleep deprivation, oxidative stress and inflammation", *Adv. Protein Chem. Struct. Biol.,* vol. 119, pp. 309-336, 2020.
[http://dx.doi.org/10.1016/bs.apcsb.2019.03.001] [PMID: 31997771]

[110] A. El-Sayed, L. Aleya, and M. Kamel, "Microbiota's role in health and diseases", *Environ. Sci. Pollut. Res. Int.,* vol. 28, no. 28, pp. 36967-36983, 2021.
[http://dx.doi.org/10.1007/s11356-021-14593-z] [PMID: 34043164]

[111] Z.-D. Wu, X.-K. Yang, Y.-S. He, J. Ni, J. Wang, K.-J. Yin, J.-X. Huang, Y. Chen, Y.-T. Feng, P. Wang, and H.-F. Pan, "Environmental factors and risk of gout", *Environ Res,* vol. 212, p. 113377, 2022.
[http://dx.doi.org/10.1016/j.envres.2022.113377]

[112] J.A. Rodrigues, R.K. Narasimhamurthy, M.B. Joshi, H.S. Dsouza, and K.D. Mumbrekar, "Pesticides exposure-induced changes in brain metabolome: implications in the pathogenesis of neurodegenerative disorders", *Neurotox. Res.,* vol. 40, no. 5, pp. 1539-1552, 2022.
[http://dx.doi.org/10.1007/s12640-022-00534-2] [PMID: 35781222]

[113] J. Most, and L.M. Redman, "Impact of calorie restriction on energy metabolism in humans", *Exp. Gerontol.,* vol. 133, p. 110875, 2020.
[http://dx.doi.org/10.1016/j.exger.2020.110875] [PMID: 32057825]

[114] J. Gornall, and R.G. Villani, "Short-term changes in body composition and metabolism with severe dieting and resistance exercise", *Int. J. Sport Nutr.,* vol. 6, no. 3, pp. 285-294, 1996.
[http://dx.doi.org/10.1123/ijsn.6.3.285] [PMID: 8876348]

[115] P.A. Villablanca, J.R. Alegria, F. Mookadam, D.R. Holmes Jr, R.S. Wright, and J.A. Levine, "Nonexercise activity thermogenesis in obesity management", *Mayo Clin. Proc.,* vol. 90, no. 4, pp. 509-519, 2015.
[http://dx.doi.org/10.1016/j.mayocp.2015.02.001] [PMID: 25841254]

CHAPTER 10

Supporting Caregivers and Patients in Managing Incomplete Metabolism

Ashima Thakur[2], **Ashok Kumar**[2], **Rahul Sharma**[3], **Pooja Thakur**[4], **Pankaj Kalia**[4], **Rajesh Kumar**[4] and **Shilpa Chandel**[1, 2,*]

[1] *Department of Pharmacy, Banasthali Vidyapith, Banasthali, Rajasthan, India*

[2] *Faculty of Pharmaceutical Sciences, The ICFAI University, Baddi, Himachal Pradesh, India*

[3] *Amar Shaheed Baba Ajit Singh Jujhar Singh Memorial College of Pharmacy Bela, Punjab, India*

[4] *Maharaja Agrasen School of Pharmacy, Maharaja Agrasen University, Solan, Himachal Pradesh, India*

Abstract: An effective solution that is revolutionizing the current healthcare environment focuses on empowering caregivers and patients through enhanced care delivery. The proposed plan emphasizes the active involvement of healthcare professionals (*e.g.*, nurses, doctors, pharmacists) and patients, promoting the development of specific skills, awareness, and access to health-related resources and tools to support effective self-management. A more collaborative and holistic approach to care has replaced the traditional defensive model, giving equal importance to the role of the caregiver and the involvement of the patient in the healing process. Adult education related to health and well-being is essential for both caregivers and patients, as it transfers knowledge about diseases, their treatments, and necessary lifestyle adjustments. Various channels, including personal consultations, health and disease-related pamphlets and magazines, as well as digital health tools, such as telemedicine and mobile health applications, enable caregivers to interact with patients more effectively.

Another notable approach gaining attention today is the creation of support networks, groups of caregivers who assist patients suffering from depression and anxiety by offering emotional and physical support. These networks play a vital role in reducing suicide cases and improving the overall well-being of affected individuals. They help foster a sense of responsibility, build a resilient community, and provide a platform for caregivers to share their experiences in managing such patients. Caregivers can be empowered through rigorous healthcare-related training and should be equipped with all the essential tools needed for effective caregiving. Ultimately, by providing caregivers with the necessary skills and resources, the quality of life and well-being of

* **Corresponding author Shilpa Chandel:** Department of Pharmacy, Banasthali Vidyapith, Banasthali, Rajasthan, India, Faculty of Pharmaceutical Sciences, The ICFAI University, Baddi, Himachal Pradesh, India; E-mail: shilpachandel65@gmail.com

Rajesh Kumar (Ed.)

both patients and caregivers can be significantly improved. Therefore, greater focus should be placed on building a patient-centric healthcare structure.

Keywords: Anxiety, Community, Caregivers, Depression, Healthcare, Patients, Support network.

INTRODUCTION

In today's era, it is crucial to empower both patients and caregivers, as prioritizing health is of utmost importance. Empowering caregivers and patients involves developing the necessary skill sets, knowledge, and confidence to manage health and various medical conditions effectively [1 - 6]. Several critical aspects must be considered, such as the type of illness, and the expectations of physicians, nurses, pharmacists, and other medical professionals in delivering effective treatment.

Empowering medical professionals leads to improved health outcomes, increased patient satisfaction, and more effective use of healthcare resources [7]. Similarly, empowering patients fosters stronger alliances with caregivers, encourages treatment adherence, promotes preventive health behaviors, and enhances the overall quality of the healthcare experience. These efforts help cultivate a collaborative network among caregivers, patients, and healthcare providers, ultimately advancing a more patient-centric care model and contributing to a healthier, more resilient community.

In recent years, the healthcare system has undergone a significant transformation, especially in the relationship between patients and medical professionals [8]. In the past, an overprotective model of care excluded patients from participating in their treatment and recovery, placing full responsibility on healthcare providers. According to this paternalistic model, physicians were viewed as the sole decision-makers, assumed to know what was best for the patient. As a result, patients were often left feeling passive, disempowered, and excluded from the decision-making process. However, over time, advancements in healthcare have highlighted the limitations and shortcomings of such an approach, as illustrated in Fig. (**1**) [9, 10].

Paternalistic care was extremely prevalent in ancient times. Hippocrates, considered the father of modern medicine, was a native of Greece and was highly admired in both Greece and Rome. In addition to treating illnesses, he also guided people of that era on healthy eating habits [11, 12]. During the medieval period, doctors held a prestigious position closely tied to religious beliefs. As a result, they were often regarded as god-like figures or representatives of divine authority, granting them unquestioned power and sovereign knowledge [13].

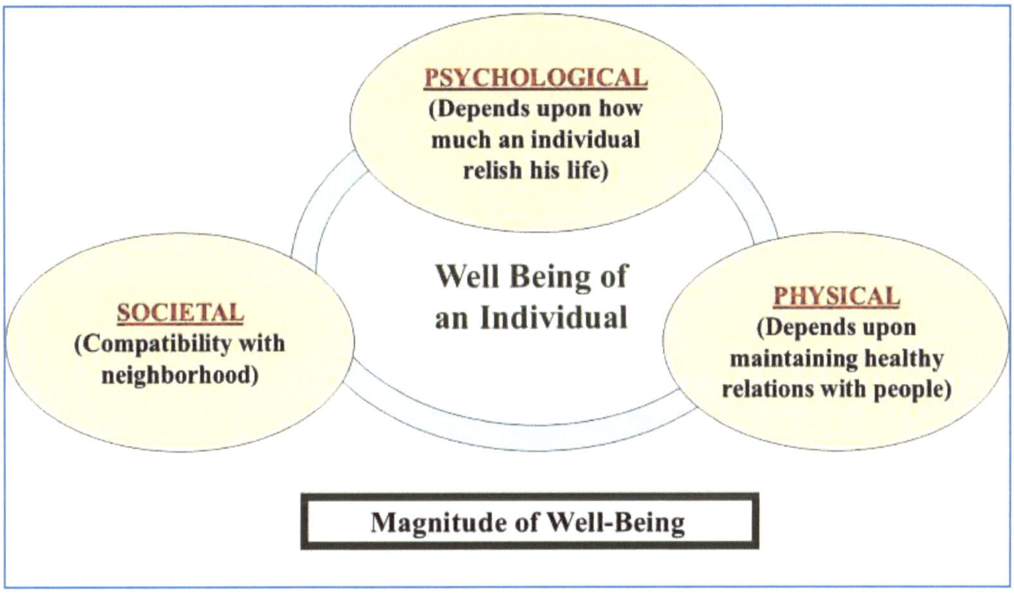

Fig. (1). This figure illustrates that an individual's well-being depends on physical, psychological, and societal factors.

By the 19th century, the paternalistic model of care had become more orthodox and rigid. The growing number of medical institutions and professional organizations, along with advancements in the medical field, reinforced the authority of physicians. This further solidified their role as the sole decision-makers in patient care [14].

CONDEMNATION OF PATERNALISTIC CARE MODELS

The condemnation of paternalistic care models began in the 20th century as people increasingly advocated for their societal rights [15, 16]. Numerous social movements, including feminist, civil rights, and disability rights movements, highlighted the crucial importance of personal autonomy and the right to independence. These movements not only challenged the exclusion of patients from their own care and treatment decisions but also promoted meaningful patient involvement in health-related choices [17 - 20]. Various bioethicists researching social, ethical, and legal issues in biomedicine and biomedical research argued that patients have the right to be fully informed about their treatments and medications. Furthermore, if patients disagree with their treatment plans, they should have the right to oppose a physician's decision. This shift empowered patients to express their opinions and make choices openly and confidently in collaboration with their caregivers [21 - 23].

Over time, patients gained confidence in the healthcare system and became more informed and assertive. Many lawsuits filed in various countries helped strengthen patients' rights and sovereignty. The legal obligation to inform patients about their treatment, including providing comprehensive information about available options, fostered a sense of confidence and security in patients [16, 24 - 26].

These developments created an environment conducive to adopting patient-centric strategies, leading to the evolution of a collaborative care model. In this model, patients and physicians work together to reach shared decisions. It preserves patient autonomy, fosters a healthier environment, and ensures patients are actively involved and informed about their treatment [2, 27 - 29].

ESSENTIAL FEATURES OF THE SYNERGETIC CARE MODEL

The essential features of the synergetic care model include collaborative decision-making, where patients are encouraged to actively participate in their treatment by expressing their preferences, values, and concerns. Decisions are made through a combination of patient and caregiver input, resulting in more informed and effective treatment plans [28, 30]. This model also involves a multidisciplinary team, comprising doctors, nurses, pharmacists, and other healthcare professionals, who collaborate to provide comprehensive care and support to the patient.

Through this approach, patients receive comprehensive care that promotes faster healing and better overall treatment [31 - 33]. The synergetic care model enables patients to understand their treatment options and possible outcomes by providing them with the necessary information and support to make informed decisions. This includes knowledge about their illness, potential treatments, and possible consequences [34, 35]. The shift from a paternalistic care model to a synergetic care model has led to a significant transformation in the healthcare system [2 - 6, 10, 16, 21, 28, 29, 32, 34 - 73]. This transformation has brought many benefits, such as improving patient satisfaction. When patients understand their condition and treatment, they feel more content. Moreover, being able to express their views and opinions makes patients feel heard and more actively engaged in the decision-making process.

The synergetic care model also builds a sense of trust and satisfaction towards healthcare professionals [37, 47]. Moreover, various studies have demonstrated that patients who are actively involved in their treatment plans tend to experience better health outcomes and faster recovery. This involvement also helps them adhere to their medication and diet plans, which accelerates recovery. Additionally, it aids physicians in the treatment process by ensuring full cooperation from the patient [74, 75].

Constructing a Support System

Constructing a support system for patients is pivotal in creating a positive and comforting environment for them. Such support helps patients navigate health-related challenges by providing a sense of community, understanding, and shared experience. Various support systems and social networks play a crucial role in this process. These systems come in many forms, including individual meetings, online platforms, printed materials, and more [66, 76].

Various Challenges Related to the Synergetic Care Model

Although the synergetic care model has multifaceted advantages, it requires consideration of several factors, such as effective communication skills, emotional competence, and the ability to manage differing opinions between caregivers and patients. Furthermore, it demands the application of multiple resources and requires ample time [40, 41]. It is extremely important to educate everyone about well-being and health. Such education involves strategies to provide knowledge and cultivate the necessary skills related to health and healthy eating habits.

This helps patients manage and live with long-term chronic diseases [77, 78]. Educating healthcare providers and patients enables them to better cope with serious illnesses as a whole. Such education involves equipping patients and caregivers with the necessary skills, knowledge, and trust in the healthcare system [66, 76]. There are multiple ways to empower patients, including enhancing their understanding of their medical conditions and ailments. This awareness facilitates early diagnosis and timely treatment. Patients become more vigilant about their medications and take precautionary measures proactively. They begin adopting healthier lifestyles and naturally shift to better diets. Moreover, this knowledge encourages patients to engage actively with their treatment plans and enables them to ask relevant questions about their conditions. Additionally, it empowers them to participate in decision-making processes regarding their treatment [42, 43, 51, 79, 80].

Another important aspect is developing essential skills in patients, which is increasingly vital today. Beyond knowledge, patients need practical tips and techniques to manage their conditions. For example, diabetes is one of the most prevalent non-communicable diseases worldwide, and patients suffering from it should be skilled in using a glucometer to regularly monitor their blood glucose levels. It is also important for patients to recognize hyperglycemic and hypoglycemic conditions, enabling them to seek timely treatment [4, 42, 43]. Additionally, building patients' confidence helps them better understand and cope with their health conditions. This confidence encourages adherence to treatment plans and self-management [46, 55, 81, 82]. Furthermore, empowering caregivers is equally important to achieve desired health outcomes. Providing caregivers with adequate support and training is crucial, as they play a pivotal role in the healthcare system, especially in managing disabilities, terminal illnesses, and chronic diseases.

When caregivers are equipped with the necessary skills and knowledge, they can provide effective care, administer proper treatment, and recognize any decline in the patient's condition [83, 84]. A skilled and knowledgeable caregiver is always prepared to manage challenges encountered during caregiving, which helps to reduce suffering and exhaustion.

By understanding the patient's condition, caregivers can anticipate their needs and respond more efficiently [39]. Moreover, it is crucial for caregivers to be well-informed and skilled, as this enables them to contribute valuable insights about the patient's condition and participate effectively in decision-making. Additionally, a knowledgeable and skilled caregiver can provide proper care, a healthy diet, and appropriate medication to the patient [85, 86].

VARIOUS TECHNIQUES TO IMPART HEALTH EDUCATION

There are numerous techniques through which health education can be imparted to patients and caregivers, including personal consultations, sharing evidence and real-time examples, and fostering strong trust in the healthcare system to help manage serious ailments [80]. Caregivers can educate patients through direct interactions, making it easier for them to understand important information. Additionally, patients can ask questions directly and clarify any health-related doubts. This approach also helps build strong trust and confidence in the healthcare system among patients [80, 87, 88].

The technique of providing practical examples and real-time experiences involves giving patients hands-on training for emergency situations. For instance, diabetic patients must be taught how to administer injections so they can do it themselves when needed. Additionally, patients should know how to use relevant medical devices [45, 89 - 91]. Furthermore, one-to-one communication between a physician and a patient helps build strong trust in the healthcare system. When patients feel heard regarding their ailments and concerns, they are more likely to follow the prescribed routine and instructions given by their physician.

Effective communication regarding health education between caregivers and patients is pivotal [90, 91]. Additionally, caregivers can enhance patients' knowledge through booklets, pamphlets, brochures, hoardings, and similar materials, which are among the easiest and most effective ways to disseminate health education [3, 45]. These materials can be widely distributed and provide appropriate, comprehensive information about health topics [92]. Moreover, printed materials help reinforce people's knowledge. If patients forget details about their medication or condition, they can refer back to these resources anytime to refresh their understanding. This approach is particularly helpful for older adults, who are more prone to forgetfulness and missed doses, which could otherwise lead to serious health complications [93, 94].

Furthermore, printed materials are extremely cost-effective, as they can be produced in the required quantities on a limited budget to impart health education to society [94] effectively. Nowadays, social media is one of the most effective and convenient mediums for disseminating health education. There are numerous digital platforms, including websites, mobile apps, social networks, and radio, that have great potential to disseminate health-related knowledge. Caregivers, by remaining in one place, can guide a large portion of the population on topics such as healthy diets, healthy lifestyles, and critical diseases [87, 88].

Digital platforms also facilitate interaction between caregivers and patients through various modes, including interactive video sessions, Zoom meetings, quiz competitions, and more. Apps designed for interaction, like Zoom, also set reminders to attend sessions. Moreover, some apps are built to track users' health metrics as well [95]. In today's era, advancements in technology have reached a new level, making it easier for almost everyone to afford televisions, radios, mobile phones, and other devices. This accessibility enables healthcare providers to disseminate health knowledge more effectively to society (Fig. **2**) [95 - 97].

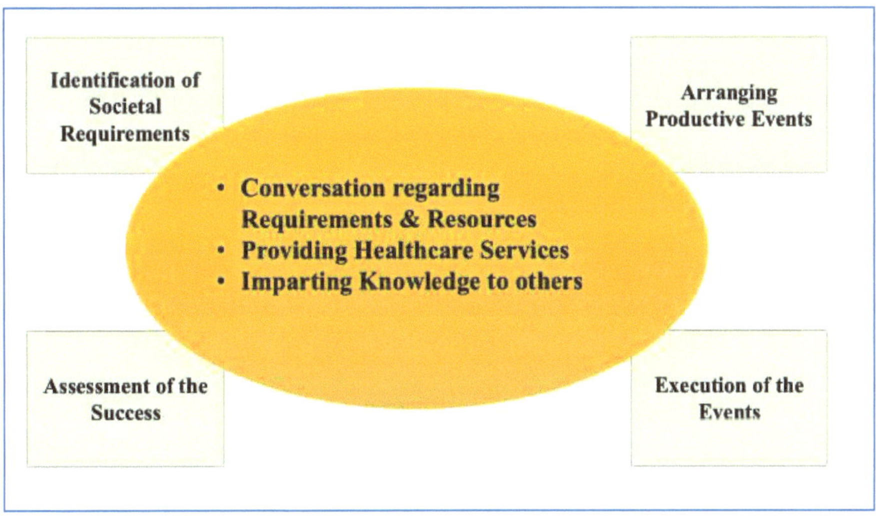

Fig. (2). This figure depicts the role of healthcare educators.

Encouragement of Patient Sovereignty and Collaborative Decision-Making

In today's healthcare system, it is essential to respect patients' autonomy so that they can actively participate in the decision-making process regarding their treatment. This involvement enhances their confidence and provides the satisfaction of being heard [48, 60]. There are various effective methods to involve patients in their treatment and care plans, one of which is collaborative decision-making. In this process, the caregiver informs the patient about the potential risks and benefits of the treatment.

In this process, the caregiver also takes into account the usefulness and preferences of the patient [48, 98]. This technique involves preparing the patient for collaborative decision-making by briefing them on all aspects and the importance of the process. Caregivers are encouraged to actively engage patients by prompting them to express their choices and preferences. This approach also raises the patient's awareness of various treatment alternatives, equipping them with knowledge about each option along with its potential benefits and drawbacks.

This ensures that the patient gains thorough knowledge and insight into all the potential aspects of their treatment [98]. Additionally, detailed discussions occur between the caregiver and patient, during which the caregiver may utilize pamphlets, informative videos, and social media platforms to enhance the patient's awareness of their condition and potential treatment options [48, 95]. After this comprehensive dialogue, the caregiver, considering both the patient's needs and their own professional expertise, formulates an appropriate treatment plan [48, 98]. The plan is then implemented and regularly evaluated, allowing for any necessary modifications based on the patient's condition and preferences [99]. For patients who are initially indecisive about their treatment or required lifestyle changes, caregivers provide ongoing motivation and counseling to help them remain committed to their care plan [100].

It is the moral duty of healthcare providers to listen attentively to patients' concerns, address their issues, and create a healthy and supportive environment that promotes a quick recovery from illness [45]. Additionally, caregivers should inform patients about their current health conditions and motivate them to follow a proper treatment regimen, maintain a healthy lifestyle, and adopt a nutritious diet to support early recovery [45, 101]. Caregivers also have a responsibility to encourage patients to make positive changes in their lives and to reassure them that such changes will lead to favorable outcomes. They should offer consistent motivation, especially when patients feel discouraged, to help them remain hopeful and committed to achieving healthier habits [94].

Such meetings are held between physicians, patients, and the patients' family members [5, 102]. Various factors must be considered when dealing with patients. First, it is important to establish clear, realistic, and attainable objectives that are jointly set by the patient and caregiver, ensuring they can be successfully achieved over time without significant obstacles [89]. Second, a detailed and well-structured plan should be developed to achieve these objectives, involving the entire healthcare team and considering timeframes and desired outcomes [89]. Lastly, the patient's condition should be systematically assessed on a regular basis, and any necessary adjustments to the care plan should be made accordingly [56, 58].

CASE STUDIES

The empowerment of patients and caregivers is pivotal in today's healthcare system, as it leads to improved health outcomes. A study in the literature focusing on patients with chronic kidney disease revealed that a substantial number of healthcare providers had taken significant initiatives to empower these patients. The strategy included educating them about the disease, available treatment options, and the importance of adopting a healthy lifestyle [61]. Various methods were employed to support this empowerment, including seminars and workshops where patients were informed about chronic kidney disease, necessary dietary modifications, treatment regimens, and the importance of managing blood pressure. Additionally, patients were given access to social media platforms, digital tools, and printed materials to enhance their understanding of the condition [62].

These efforts yielded several positive outcomes; for instance, patients participating in such programs demonstrated notable health improvements, including reduced blood pressure, better-controlled blood sugar levels, and improved kidney function [62]. Furthermore, those who participated in the empowerment strategies exhibited increased confidence and significantly higher morale. Patients who underwent this training were also more likely to adhere to their prescribed treatment plans and maintain a healthy diet [61, 62].

This strategy reduced the frequency of hospital visits by patients, as they were able to manage their care more effectively on their own. Patients became capable of identifying the signs and symptoms of kidney disease [63]. In another study focused on empowering caregivers in the management of Alzheimer's disease, conducted by a private organization, an initiative was launched to support both patients and families affected by the condition. The primary aim of the training program was to equip caregivers with the necessary skills, knowledge, and emotional support [6, 63].

The program included comprehensive caregiver training, covering key themes such as a detailed explanation of the progression of Alzheimer's disease, strategies for managing challenging patient behaviors, and effective knowledge transmission techniques [103, 104].

This study improved caregivers' skills in delivering care to patients, as evidenced by a significant increase in their ability to manage challenging behaviors, such as aggression and wandering. Caregivers became more confident and capable in handling patients suffering from Alzheimer's disease [105]. Numerous training sessions were organized to enhance caregiver skills, which proved effective in reducing caregiver stress by teaching strategies to maintain a healthy work-life balance. Caregivers also learned how to prioritize their own needs and health while fulfilling their responsibilities [103, 105].

As a result of having experienced and empowered caregivers, patients are more likely to feel secure, and the frequency of hospital visits is reduced. Overall, the well-being of patients improves when they are under the care of caregivers who are well-equipped [38].

In another study, it was revealed that for patients with mental illnesses, a personalized care team typically includes a doctor, a care manager, and a psychiatrist. In addition to these core members, nursing officers, pharmacists, psychologists, social workers, and other health specialists may also be included as needed. The study concluded that utilizing such multidisciplinary strategies benefits both caregivers and patients. The collaboration of all these professionals leads to skillful and comprehensive decision-making, contributing to the development of an extensive patient care strategy [57, 106].

Multiple studies conducted by Carney R.M., Grote N.K., Rollman B.L., *et al.* (including Rollman B.L., 2017) demonstrated that when caregivers collaborate with patients, a shared decision-making process emerges. In this process, the patient's opinion is valued and presented to the rest of the team, including doctors, psychologists, and other healthcare professionals. Such collaborative decisions are highly acceptable and tend to produce better outcomes. This approach offers several advantages, including improved treatment adherence, increased patient satisfaction, and effective responses to the unique challenges faced by patients with psychological conditions [44, 52, 59, 107].

In a research study, a patient with type 2 diabetes was struggling to manage his blood sugar levels despite undergoing regular medical checkups. A caregiver, who was a diabetes nurse educator, worked closely with the patient to enhance his understanding of the disease. The caregiver provided personalized education on maintaining a healthy diet, the importance of regular exercise, and proper glucose monitoring. Additionally, she utilized digital tools to track the patient's blood sugar levels and helped him set realistic, achievable health goals.

Consequently, the patient gained confidence in managing his blood sugar levels independently, which led to improved control and a reduction in emergency room visits. All the discussed case studies demonstrate that caregivers have the potential to empower their patients through knowledge, guidance on a healthy diet, support, and essential skills. They can assist patients with chronic ailments throughout their lives, helping to increase lifespan and improve overall health. Moreover, a collaborative healthcare system produces better outcomes and enhances patient satisfaction, providing a sense of fulfillment for both patients and caregivers.

CHALLENGES

Empowering patients and caregivers is an intricate process that requires addressing multiple complex challenges before successful implementation. These challenges can be overcome by providing health education to society, promoting cultural competency, supporting emotional well-being, tackling comprehensive barriers, and creating a healthy and positive environment for both patients and caregivers, where they can express themselves freely without feeling burdened. Such efforts are vital for the well-being and improvement of both patients and caregivers [108].

CONCLUSION

Empowerment of patients and caregivers is a crucial element of today's modern healthcare system, as it significantly enhances patient satisfaction and improves health outcomes. Equipping patients and caregivers with the necessary skills and knowledge boosts their confidence and helps overcome challenges, such as gaps in health education, cultural barriers, and other complex obstacles. Moreover, effective empowerment involves building trust, fostering strong relationships, and recognizing the unique needs of both patients and caregivers.

REFERENCES

[1] C. Bergerum, J. Thor, K. Josefsson, and M. Wolmesjö, "How might patient involvement in healthcare quality improvement efforts work—A realist literature review," *Health Expectations*, vol. 22, no. 3, pp. 283–296, 2019.
[http://dx.doi.org/10.1111/hex.12900]

[2] S. Reilly, C. Planner, L. Gask, M. Hann, S. Knowles, B. Druss, and H. Lester, "Collaborative care approaches for people with severe mental illness", *Cochrane Libr.,* no. 11, p. CD009531, 2013.
[http://dx.doi.org/10.1002/14651858.CD009531.pub2] [PMID: 24190251]

[3] R.M. Parker, D.W. Baker, M.V. Williams, and J.R. Nurss, "The test of functional health literacy in adults", *J. Gen. Intern. Med.,* vol. 10, no. 10, pp. 537-541, 1995.

[http://dx.doi.org/10.1007/BF02640361] [PMID: 8576769]

[4] E.H. Wagner, B.T. Austin, C. Davis, M. Hindmarsh, J. Schaefer, and A. Bonomi, "Improving chronic illness care: translating evidence into action", *Health Aff. (Millwood),* vol. 20, no. 6, pp. 64-78, 2001. [http://dx.doi.org/10.1377/hlthaff.20.6.64] [PMID: 11816692]

[5] I.S. Kickbusch, "Health literacy: addressing the health and education divide", *Health Promot. Int.,* vol. 16, no. 3, pp. 289-297, 2001. [http://dx.doi.org/10.1093/heapro/16.3.289] [PMID: 11509466]

[6] S. Argüelles, D.A. Loewenstein, C. Eisdorfer, and T. Argüelles, "Caregivers' judgments of the functional abilities of the Alzheimer's disease patient: impact of caregivers' depression and perceived burden", *J. Geriatr. Psychiatry Neurol.,* vol. 14, no. 2, pp. 91-98, 2001. [http://dx.doi.org/10.1177/089198870101400209] [PMID: 11419574]

[7] T. Scott Duncan, J. Engström, S. Riggare, M. Hägglund, and S. Koch, "Meeting the burden of self-management: Qualitative study investigating the empowering behaviors of patients and informal caregivers", *J. Particip. Med.,* vol. 14, no. 1, p. e39174, 2022. [http://dx.doi.org/10.2196/39174] [PMID: 36383418]

[8] N. Eriksson, E. Söderström, and R.-M. Åhlfeldt, "Patient empowerment and its connection to trust," In: *Proc. 18th Int. Symp. Health Inf. Manag. Res.,* 2020. [http://dx.doi.org/10.15626/ishimr.2020.06]

[9] J. Boivin, L. Bunting, E. Koert, C. ieng U, and C. Verhaak, "Perceived challenges of working in a fertility clinic: a qualitative analysis of work stressors and difficulties working with patients", *Hum. Reprod.,* vol. 32, no. 2, pp. 403-408, 2017. [http://dx.doi.org/10.1093/humrep/dew326] [PMID: 28057875]

[10] M. McMullan, "Patients using the Internet to obtain health information: How this affects the patient–health professional relationship", *Patient Educ. Couns.,* vol. 63, no. 1-2, pp. 24-28, 2006. [http://dx.doi.org/10.1016/j.pec.2005.10.006] [PMID: 16406474]

[11] L. Santacroce, L. Bottalico, and I. Charitos, "Greek medicine practice at ancient rome: the physician molecularist asclepiades", *Medicines (Basel),* vol. 4, no. 4, p. 92, 2017. [http://dx.doi.org/10.3390/medicines4040092] [PMID: 29231878]

[12] M. D. Grmek, Ed., *Storia del pensiero medico occidentale, vol. II: Dal Rinascimento all'inizio dell'Ottocento,* Versione Cartacea, 1996.

[13] J.A. Alford, "Medicine in the Middle Ages: the theory of a profession", *Centen. Rev.,* vol. 23, no. 4, pp. 377-396, 1979. [PMID: 11614331]

[14] H. Bloch, "Medicine and science in the 19th and 20th centuries", *J. Natl. Med. Assoc.,* vol. 80, no. 2, pp. 229-232, 1988. [PMID: 3071610]

[15] J. Boyle, *The Public Domain: Enclosing the Commons of the Mind.,* 2008.

[16] M. Young, and M.A. Smith, "Standards and evaluation of healthcare quality, safety, and personcentered care", In: *StatPearls,* StatPearls Publishing, 2024.

[17] R. M. Caron, K. Noel, R. N. Reed, J. Sibel, and H. J. Smith, "Health promotion, health protection, and disease prevention: challenges and opportunities in a dynamic landscape", *AJPM Focus,* vol. 3, no. 1, p. 100167, 2024. [http://dx.doi.org/10.1016/j.focus.2023.100167]

[18] A.S. Latif, "The importance of understanding social and cultural norms in delivering quality health care—A personal experience commentary", *Trop. Med. Infect. Dis.,* vol. 5, no. 1, p. 22, 2020. [http://dx.doi.org/10.3390/tropicalmed5010022] [PMID: 32033381]

[19] S. E. Lee, M. H. Lee, A. B. Peters, and S. H. Gwon, "Assessment of patient safety and cultural

competencies among senior baccalaureate nursing students," *Int. J. Environ. Res. Public Health,* vol. 17, no. 12, p. 4225, 2020
[http://dx.doi.org/10.3390/ijerph17124225]

[20] M.A. Conejo, "At the intersection of feminist and disability rights movements. From equality in difference to human diversity claims", *Disabil. Intersect. Statuses,,* vol. 7, pp. 23-45, 2013.
[http://dx.doi.org/10.1108/S1479-3547(2013)0000007004]

[21] Sugarman J, Sulmasy DP, Eds., *Methods in Medical Ethics.* Georgetown University Press; 2010.

[22] M.G. Singer, "The Many Methods of Sidgwick's Ethics", *Monist,* vol. 58, no. 3, pp. 420-448, 1974.
[http://dx.doi.org/10.5840/monist197458326]

[23] A.A. Kon, "The role of empirical research in bioethics", *Am. J. Bioeth.,* vol. 9, no. 6-7, pp. 59-65, 2009.
[http://dx.doi.org/10.1080/15265160902874320] [PMID: 19998120]

[24] L.O. Gostin, E.A. Friedman, S. Hossain, J. Mukherjee, S. Zia-Zarifi, C. Clinton, U. Rugege, P. Buss, M. Were, and A. Dhai, "Human rights and the COVID-19 pandemic: a retrospective and prospective analysis", *Lancet,* vol. 401, no. 10371, pp. 154-168, 2023.
[http://dx.doi.org/10.1016/S0140-6736(22)01278-8] [PMID: 36403583]

[25] D.J. Mazur, "Informed consent: court viewpoints and medical decision making", *Med. Decis. Making,* vol. 6, no. 4, pp. 224-230, 1986.
[http://dx.doi.org/10.1177/0272989X8600600407] [PMID: 3773652]

[26] "INTERNATIONAL code of medical ethics (World Medical Associaion, London, October 1949)", *Madjalah Kedokt. Indones.,* vol. 9, p. 462, 1959.
[PMID: 13857590]

[27] J. Whitfield, S. Owens, A. Bhat, and B. Felker, "Successful ingredients of effective Collaborative Care programs in low- and middle-income countries: A rapid review," *Cambridge University Press,* vol. 10, p. e11, 2023.
[http://dx.doi.org/10.1017/gmh.2022.60]

[28] D. Coates, D. Coppleson, and V. Schmied, "Integrated physical and mental healthcare: an overview of models and their evaluation findings", *Int. J. Evid.-Based Healthc.,* vol. 18, no. 1, pp. 38-57, 2020.
[http://dx.doi.org/10.1097/XEB.0000000000000215] [PMID: 31904722]

[29] J. Unützer, W. Katon, C.M. Callahan, J.W. Williams Jr, E. Hunkeler, L. Harpole, M. Hoffing, R.D. Della Penna, P.H. Noël, E.H.B. Lin, P.A. Areán, M.T. Hegel, L. Tang, T.R. Belin, S. Oishi, and C. Langston, "Collaborative care management of late-life depression in the primary care setting: a randomized controlled trial", *JAMA,* vol. 288, no. 22, pp. 2836-2845, 2002.
[http://dx.doi.org/10.1001/jama.288.22.2836] [PMID: 12472325]

[30] W. Katon, "Collaborative depression care models: from development to dissemination", *Am. J. Prev. Med.,* vol. 42, no. 5, pp. 550-552, 2012.
[http://dx.doi.org/10.1016/j.amepre.2012.01.017] [PMID: 22516497]

[31] A.O. Adewuya, T. Adewumi, O. Momodu, O. Olibamoyo, O. Adesoji, A. Adegbokun, S. Adeyemo, O. Manuwa, and D. Adegbaju, "Development and feasibility assessment of a collaborative stepped care intervention for management of depression in the mental health in primary care (MeHPriC) project, Lagos, Nigeria", *Psychol. Med.,* vol. 49, no. 13, pp. 2149-2157, 2019.
[http://dx.doi.org/10.1017/S0033291718002970] [PMID: 30326980]

[32] A. O. Adewuya, O. Momodu, O. Olibamoyo, A. Adegbaju, O. Adesoji, and A. Adegbokun, "The effectiveness and acceptability of mobile telephone adherence support for management of depression in the mental health in primary care (MeHPriC) project, Lagos, Nigeria: a pilot cluster randomised controlled trial", *J Affect Disord,* vol. 253, pp. 118-125, 2019.
[http://dx.doi.org/10.1016/j.jad.2019.04.025]

[33] A.O. Adewuya, B.A. Ola, O. Coker, O. Atilola, A. Fasawe, and T. Ajomale, "A stepped care

intervention for non-specialist health workers' management of depression in the Mental Health in Primary Care (MeHPriC) project, Lagos, Nigeria: A cluster randomised controlled trial", *Gen. Hosp. Psychiatry,* vol. 60, pp. 76-82, 2019.
[http://dx.doi.org/10.1016/j.genhosppsych.2019.07.012] [PMID: 31351240]

[34] C.G. Kemp, L.C.M. Johnson, R. Sagar, S. Poongothai, N. Tandon, R.M. Anjana, S. Aravind, G.R. Sridhar, S.A. Patel, K. Emmert-Fees, D. Rao, K.M.V. Narayan, V. Mohan, M.K. Ali, and L.A. Chwastiak, "Effect of a collaborative care model on anxiety symptoms among patients with depression and diabetes in India: The INDEPENDENT randomized clinical trial", *Gen. Hosp. Psychiatry,* vol. 74, pp. 39-45, 2022.
[http://dx.doi.org/10.1016/j.genhosppsych.2021.11.003] [PMID: 34883269]

[35] P. Rimal, N. Choudhury, P. Agrawal, M. Basnet, B. Bohara, D. Citrin, S. K. Dhungana, B. Gauchan, P. Gupta, T. K. Gupta, S. Halliday, B. Kadayat, R. Mahar, D. Maru, V. Nguyen, S. Poudel, A. Raut, J. Rawal, S. Sapkota, D. Schwarz, R. Schwarz, S. Shrestha, S. Swar, A. Thapa, P. Thapa, R. White, and B. Acharya, "Collaborative care model for depression in rural Nepal: a mixed-methods implementation research study," *Global Health*, vol. 11, no. 8, p. e048481, 2021.
[http://dx.doi.org/10.1136/bmjopen-2020-048481]

[36] E. Bagalman, "HHS roadmap for behavioral health integration", In: *ASPE* vol. 1, 2022, pp. 1-9.

[37] R. Dhiman, "Novel LC-MS/MS Bioanalytical Method Validation for Gabapentin In Human Plasma In Accordance with Current Regulatory Guidances", *Int. J. Pharm. Chem. Sci.,* vol. 4, no. 4, pp. 462-468, 2015.

[38] K. Sullivan, and F. O'Conor, "Providing education about Alzheimer's disease", *Aging Ment. Health,* vol. 5, no. 1, pp. 5-13, 2001.
[http://dx.doi.org/10.1080/13607860020020582] [PMID: 11513014]

[39] K. Donelan, C.A. Hill, C. Hoffman, K. Scoles, P.H. Feldman, C. Levine, and D. Gould, "Challenged to care: informal caregivers in a changing health system", *Health Aff. (Millwood),* vol. 21, no. 4, pp. 222-231, 2002.
[http://dx.doi.org/10.1377/hlthaff.21.4.222] [PMID: 12117133]

[40] P. Bower, S. Gilbody, D. Richards, J. Fletcher, and A. Sutton, "Collaborative care for depression in primary care", *Br. J. Psychiatry,* vol. 189, no. 6, pp. 484-493, 2006.
[http://dx.doi.org/10.1192/bjp.bp.106.023655] [PMID: 17139031]

[41] J. Unützer, M. Schoenbaum, B.G. Druss, and W.J. Katon, "Transforming mental health care at the interface with general medicine: report for the presidents commission", *Psychiatr. Serv.,* vol. 57, no. 1, pp. 37-47, 2006.
[http://dx.doi.org/10.1176/appi.ps.57.1.37] [PMID: 16399961]

[42] T.J. Paterick, G.V. Carson, M.C. Allen, and T.E. Paterick, "Medical informed consent: general considerations for physicians", *Mayo Clin. Proc.,* vol. 83, no. 3, pp. 313-319, 2008.
[http://dx.doi.org/10.4065/83.3.313] [PMID: 18315998]

[43] E.S. Ford, M.M. Bergmann, J. Kröger, A. Schienkiewitz, C. Weikert, and H. Boeing, "Healthy living is the best revenge: findings from the European Prospective Investigation Into Cancer and Nutrition-Potsdam study", *Arch. Intern. Med.,* vol. 169, no. 15, pp. 1355-1362, 2009.
[http://dx.doi.org/10.1001/archinternmed.2009.237] [PMID: 19667296]

[44] B.L. Rollman, B.H. Belnap, M.S. LeMenager, S. Mazumdar, P.R. Houck, P.J. Counihan, W.N. Kapoor, H.C. Schulberg, and C.F. Reynolds III, "Telephone-delivered collaborative care for treating post-CABG depression: a randomized controlled trial", *JAMA,* vol. 302, no. 19, pp. 2095-2103, 2009.
[http://dx.doi.org/10.1001/jama.2009.1670] [PMID: 19918088]

[45] R. R. Cottrell and J. F. McKenzie, *Health Promotion and Education Research Methods: Using the Five-Chapter Thesis,* vol. 12, no. 2013, pp. 8-9, Sudbury, Mass: Jones and Bartlett Publishers, 2011.

[46] B.A. Weidmer, C. Brach, and R.D. Hays, "Development and evaluation of CAHPS survey items assessing how well healthcare providers address health literacy", *Med. Care,* vol. 50, no. 9, suppl.

Suppl. 2, pp. S3-S11, 2012.
[http://dx.doi.org/10.1097/MLR.0b013e3182652482] [PMID: 22895227]

[47] J.C. Huffman, S.K. Niazi, J.R. Rundell, M. Sharpe, and W.J. Katon, "Essential articles on collaborative care models for the treatment of psychiatric disorders in medical settings: a publication by the academy of psychosomatic medicine research and evidence-based practice committee", *Psychosomatics,* vol. 55, no. 2, pp. 109-122, 2014.
[http://dx.doi.org/10.1016/j.psym.2013.09.002] [PMID: 24370112]

[48] J.J. Hopla, "Self-efficacy as an evaluation measure for programs in support of online learning literacies for undergraduates," *JOUR,* vol. 8, pp. 307–322, 2005.

[49] R. Dhiman, V. Durga, and K. Hamlet, "Determination of S-amlodipine in human plasma using tizanidine as internal standard by LC/MS/MS method", *J Bioanal Biomed,* vol. 7, 2015.

[50] R. Dhiman, V. Durga, and A. Rio, "Bioanalytical method for lornoxicam determination in human plasma by using piroxicam as internal standard by LC-MS/MS", *J. Chem. Pharm. Res.,* vol. 7, pp. 54-60, 2015.

[51] W. A. Psek, R. A. Stametz, L. D. Bailey-Davis, D. Davis, J. Darer, W. A. Faucett, D. L. Henninger, D. C. Sellers, and G. Gerrity, "Operationalizing the learning health care system in an integrated delivery system", eGEMs (Generating Evidence & Methods to improve patient outcomes), vol. 3, no. 1, p. 6, 2015.
[http://dx.doi.org/10.13063/2327-9214.1122]

[52] N.K. Grote, W.J. Katon, J.E. Russo, M.J. Lohr, M. Curran, E. Galvin, and K. Carson, ""Collaborative Care for Perinatal Depression in Socioeconomically Disadvantaged Women: A Randomized Trial"", *Depress. Anxiety,* vol. 32, no. 11, pp. 821-834, 2015.
[http://dx.doi.org/10.1002/da.22405] [PMID: 26345179]

[53] R. Dhiman, D. Kumar, B. Kumar, and B. Pandey, "Quantitative determination of atorvastatin, ortho-hydroxy atorvastatin, para-hydroxy atorvastatin in human plasma using rosuvastatin as internal standard by LC-MS/MS", *Int J Pharm Chem Sci,* vol. 4, pp. 487-500, 2016.

[54] R. Dhiman, B. Kumar, D. Kumar, and A. Kumar, "Liquid-liquid extraction method for Ziprasidone (ZRS) bioanalysis by using ZRS-D8 (stable isotope) as internal standard", *J. Chem. Pharm. Res.,* vol. 8, no. 3, pp. 722-727, 2016.

[55] D.M. Williams, and R.E. Rhodes, "The confounded self-efficacy construct: conceptual analysis and recommendations for future research", *Health Psychol. Rev.,* vol. 10, no. 2, pp. 113-128, 2016.
[http://dx.doi.org/10.1080/17437199.2014.941998] [PMID: 25117692]

[56] J.L. Wolff, B.C. Spillman, V.A. Freedman, and J.D. Kasper, "A national profile of family and unpaid caregivers who assist older adults with health care activities", *JAMA Intern. Med.,* vol. 176, no. 3, pp. 372-379, 2016.
[http://dx.doi.org/10.1001/jamainternmed.2015.7664] [PMID: 26882031]

[57] J.J.J. van Dongen, M.A. van Bokhoven, R. Daniëls, T. van der Weijden, W.W.G.P. Emonts, and A. Beurskens, "Developing interprofessional care plans in chronic care: a scoping review", *BMC Fam. Pract.,* vol. 17, no. 1, p. 137, 2016.
[http://dx.doi.org/10.1186/s12875-016-0535-7] [PMID: 27655185]

[58] C. Riffin, P.H. Van Ness, J.L. Wolff, and T. Fried, "Family and other unpaid caregivers and older adults with and without dementia and disability", *J. Am. Geriatr. Soc.,* vol. 65, no. 8, pp. 1821-1828, 2017.
[http://dx.doi.org/10.1111/jgs.14910] [PMID: 28426910]

[59] B.L. Rollman, B.H. Belnap, S. Mazumdar, K.Z. Abebe, J.F. Karp, E.J. Lenze, and H.C. Schulberg, "Telephone-delivered stepped collaborative care for treating anxiety in primary care: a randomized controlled trial", *J. Gen. Intern. Med.,* vol. 32, no. 3, pp. 245-255, 2017.
[http://dx.doi.org/10.1007/s11606-016-3873-1] [PMID: 27714649]

[60] G.P. Cernada, *International Quarterly of Community Health Education,* Farmingdale, N. Y: Baywood., 1981.

[61] S.A. Carter, T. Gutman, C. Logeman, D. Cattran, L. Lightstone, A. Bagga, S.J. Barbour, J. Barratt, J. Boletis, D. Caster, R. Coppo, F.C. Fervenza, J. Floege, M. Hladunewich, J.J. Hogan, A.R. Kitching, R.A. Lafayette, A. Malvar, J. Radhakrishnan, B.H. Rovin, N. Scholes-Robertson, H. Trimarchi, H. Zhang, K. Azukaitis, Y. Cho, A.K. Viecelli, L. Dunn, D. Harris, D.W. Johnson, P.G. Kerr, P. Laboi, J. Ryan, J.I. Shen, L. Ruiz, A.Y.M. Wang, A.H.K. Lee, S. Fung, M.K.H. Tong, A. Teixeira-Pinto, M. Wilkie, S.I. Alexander, J.C. Craig, and A. Tong, "Identifying outcomes important to patients with glomerular disease and their caregivers", *Clin. J. Am. Soc. Nephrol.,* vol. 15, no. 5, pp. 673-684, 2020. [http://dx.doi.org/10.2215/CJN.13101019] [PMID: 32354728]

[62] P.K.T. Li, G. Garcia-Garcia, S.F. Lui, S. Andreoli, W.W.S. Fung, A. Hradsky, L. Kumaraswami, V. Liakopoulos, Z. Rakhimova, G. Saadi, L. Strani, I. Ulasi, and K. Kalantar-Zadeh, "Kidney health for everyone everywhere: From prevention to detection and equitable access to care", *Can. J. Kidney Health Dis.,* vol. 7, p. 2054358120910569, 2020. [http://dx.doi.org/10.1177/2054358120910569] [PMID: 32215213]

[63] L.P. Gregg, and S.S. Hedayati, "Pharmacologic and psychological interventions for depression treatment in patients with kidney disease", *Curr. Opin. Nephrol. Hypertens.,* vol. 29, no. 5, pp. 457-464, 2020. [http://dx.doi.org/10.1097/MNH.0000000000000629] [PMID: 32701597]

[64] S.N. Rai, N. Tiwari, P. Singh, D. Mishra, A.K. Singh, E. Hooshmandi, E. Vamanu, and M.P. Singh, "Therapeutic potential of vital transcription factors in alzheimer's and parkinson's disease with particular emphasis on transcription factor EB mediated autophagy", *Front. Neurosci.,* vol. 15, p. 777347, 2021. [http://dx.doi.org/10.3389/fnins.2021.777347] [PMID: 34970114]

[65] A. Kumar, M.K. Nandi, B. Kumar, A. Kumar, R. Kumar, A.K. Singh, and M. Singh, *Int. J. Pharm. Investig.,* vol. 11, no. 3, pp. 306-311, 2021. Antiurolithiasis, antioxidant, anti-inflammatory, analgesic, and diuretic activity of ethanolic extract of seeds of *Caesalpinia bonducella.* [http://dx.doi.org/10.5530/ijpi.2021.3.54]

[66] J. D. Smith, E. Fu, J. Rado, L. J. Rosenthal, A. J. Carroll, J. A. Atlas, A. D. Carlo, I. Burnett-Zeigler, N. Jordan, C. H. Brown, and J. Csernansky, "Collaborative care for depression management in primary care: a randomized roll-out trial using a type 2 hybrid effectiveness-implementation design", *Contemp Clin Trials Commun,* vol. 23, p. 100823, 2021. [http://dx.doi.org/10.1016/j.conctc.2021.100823]

[67] A. Kumar, B. Kumar, R. Kumar, A. Kumar, M. Singh, V. Tiwari, A. Trigunayat, P. Paul, and P. Singh, "Acute and subacute toxicity study of ethanolic extract of *Calotropis procera* (Aiton) Dryand flower in Swiss albino mice", Phytomed Plus, vol. 2, no. 2, p. 100224, 2022. [http://dx.doi.org/10.1016/j.phyplu.2022.100224]

[68] R. Garg, "COVID-19: A challenge towards the sustainability of health in platform era," *J. Pharm. Negat. Results,* pp. 1666–1678, 2022.

[69] A.K. Alzubaidi, W.J. Al-Kaabi, A.A. Ali, S. Albukhaty, H. Al-Karagoly, G.M. Sulaiman, M. Asiri, and Y. Khane, "Green synthesis and characterization of silver nanoparticles using flaxseed extract and evaluation of their antibacterial and antioxidant activities", *Appl. Sci. (Basel),* vol. 13, no. 4, p. 2182, 2023. [http://dx.doi.org/10.3390/app13042182]

[70] M. Singh, B. Kumar, A. K Singh, C. Shekhar Azad, M. K Yadav, A. Kumar, R. Kumar, A. Kumar, and P. Chaudhary, "Pharmacokinetic screening to Estimate the drug likeliness characteristics of selected Herbal Anticancer Drugs", *Res. J. Pharm. Technol.,* vol. 16, no. 7, pp. 3422-3426, 2023. [http://dx.doi.org/10.52711/0974-360X.2023.00566]

[71] A. Daneshvar, M. Farokhi, S. Bonakdar, and M. Vossoughi, "Synthesis and characterization of

injectable thermosensitive hydrogel based on Pluronic-grafted silk fibroin copolymer containing hydroxyapatite nanoparticles as potential for bone tissue engineering", *Int. J. Biol. Macromol.,* vol. 277, p. 134412, 2024.
[http://dx.doi.org/10.1016/j.ijbiomac.2024.134412] [PMID: 39097043]

[72] A. Singh, R. Kumar, R. Patel, Trishna, R.N. Gautam, M.K. Bharty, and L.B. Prasad, "Dithiocarbamate-based novel anti-histaminic agents: synthesis, characterization, crystal structure and thermal study", *Dalton Trans.,* vol. 53, no. 33, pp. 14077-14088, 2024.
[http://dx.doi.org/10.1039/D4DT01777C] [PMID: 39113544]

[73] A. Hamid, R. Kumar, V. K. Chaturvedi, S. Dutt, G. Sulabh, V. Kumar, and D. P. Yadav, "Artificial intelligence (AI) based colonoscopy," In: *Nanobiotechnology and Artificial Intelligence in Gastrointestinal Diseases,* 2024, pp. 12-1, 12-23.
[http://dx.doi.org/10.1088/978-0-7503-6134-7ch12]

[74] A.M. Kilbourne, K.M. Prenovost, C. Liebrecht, D. Eisenberg, H.M. Kim, H. Un, and M.S. Bauer, "Randomized controlled trial of a collaborative care intervention for mood disorders by a national commercial health plan", *Psychiatr. Serv.,* vol. 70, no. 3, pp. 219-224, 2019.
[http://dx.doi.org/10.1176/appi.ps.201800336] [PMID: 30602344]

[75] D.E. Goodrich, A.M. Kilbourne, K.M. Nord, and M.S. Bauer, "Mental health collaborative care and its role in primary care settings", *Curr. Psychiatry Rep.,* vol. 15, no. 8, p. 383, 2013.
[http://dx.doi.org/10.1007/s11920-013-0383-2] [PMID: 23881714]

[76] W.L. Miller, and J. Cohen-Katz, "Creating collaborative learning environments for transforming primary care practices now", *Fam. Syst. Health,* vol. 28, no. 4, pp. 334-347, 2010.
[http://dx.doi.org/10.1037/a0022001] [PMID: 21299280]

[77] M. Butler, R.L. Kane, D. McAlpine, R.G. Kathol, S.S. Fu, H. Hagedorn, and T.J. Wilt, "Integration of mental health/substance abuse and primary care", *Evid. Rep. Technol. Assess. (Full. Rep),* no. 173, pp. 1-362, 2008.
[PMID: 19408966]

[78] W. Katon, J. Unützer, K. Wells, and L. Jones, "Collaborative depression care: history, evolution and ways to enhance dissemination and sustainability", *Gen. Hosp. Psychiatry,* vol. 32, no. 5, pp. 456-464, 2010.
[http://dx.doi.org/10.1016/j.genhosppsych.2010.04.001] [PMID: 20851265]

[79] P. Varkey, A. Sathananthan, A. Scheifer, S. Bhagra, A. Fujiyoshi, A. Tom, and M.H. Murad, "Using quality-improvement techniques to enhance patient education and counselling of diagnosis and management", *Qual. Prim. Care,* vol. 17, no. 3, pp. 205-213, 2009.
[PMID: 19622271]

[80] E. B. Arkin, *Making Health Communication Programs Work. A Planner's Guide,* 1989.

[81] M.A. Hagbaghery, M. Salsali, and F. Ahmadi, "The factors facilitating and inhibiting effective clinical decision-making in nursing: a qualitative study", *BMC Nurs.,* vol. 3, no. 1, p. 2, 2004.
[http://dx.doi.org/10.1186/1472-6955-3-2] [PMID: 15068484]

[82] T. Fenwick, "Problem based learning, group process and the midcareer professional: implications for graduate education", *High Educ Res Dev,* vol. 21, 2002.
[http://dx.doi.org/10.1080/07294360220124620]

[83] C. G. Whiting, "Efforts by the national alliance for caregiving", *Innov Aging,* vol. 5, Suppl 1, p. 177, 2021.
[http://dx.doi.org/10.1093/geroni/igab046.675]

[84] S.C. Reinhard, "The work of caregiving: What do ADLs and IADLs tell us", In: *Family caregivers on the job*, 2004, pp. 37-65.

[85] S. Vahdat, L. Hamzehgardeshi, S. Hessam, and Z. Hamzehgardeshi, "Patient involvement in health care decision making: a review.," *Iran. Red Crescent Med.* J., vol. 16, no. 1, p. e12454, Jan. 2014.

[86] M. Gibson, K. Kelly, and A. Kaplan, "Family caregiving and transitional care: A critical review family care giver alliance", In: *National Centre on Care Giving*, 2012.

[87] V.S. Freimuth, and S.C. Quinn, "The contributions of health communication to eliminating health disparities", In: *Am. J. Public Health,* vol. 94, pp. 2053-2055, 2004. [http://dx.doi.org/10.2105/AJPH.94.12.2053]

[88] J.M. Bernhardt, "Communication at the core of effective public health", In: *Am. J. Public Health* vol. 94. , 2004, pp. 2051-2053. [http://dx.doi.org/10.2105/AJPH.94.12.2051]

[89] B.K. Rimer, K. Glanz, and G. Rasband, "Searching for evidence about health education and health behavior interventions", *Health Educ. Behav.,* vol. 28, no. 2, pp. 231-248, 2001. [http://dx.doi.org/10.1177/109019810102800208] [PMID: 11265831]

[90] M. Jenicek, "Epidemiology, evidenced-based medicine, and evidence-based public health", *J. Epidemiol.,* vol. 7, no. 4, pp. 187-197, 1997. [http://dx.doi.org/10.2188/jea.7.187] [PMID: 9465542]

[91] W. Rosenberg, and A. Donald, "Evidence based medicine: an approach to clinical problem-solving", *BMJ,* vol. 310, no. 6987, pp. 1122-1126, 1995. [http://dx.doi.org/10.1136/bmj.310.6987.1122] [PMID: 7742682]

[92] S. L. Lusk, "Health promotion planning: An educational and environmental approach: Lawrence W. Green and Marshall W. Kreuter Mayfield Publishing, Mountain View, California, 2nd edn.," *Patient Educ. Couns.,* vol. 19, no. 3, p. 298, 1992.

[93] W.H. Org., *Ottawa Charter for Health Promotion,* Geneva, WHO, 1986. AVailable from: https://www. who. int/teams/health-promotion/enhanced-wellbeing/first-glo-al-conference

[94] D. Nutbeam, "The WHO health promotion glossary", *Health Promot (Oxf),* vol. 1, pp. 113-127, 1986. [http://dx.doi.org/10.1093/heapro/13.4.349]

[95] S. Polgar, "Health and human behavior: areas of interest common to the social and medical sciences", *Curr. Anthropol.,* vol. 3, no. 2, pp. 159-205, 1962. [http://dx.doi.org/10.1086/200266] [PMID: 14487526]

[96] N. Tuyen and T. Nguyen, *Telemedicine - Technologies Information Technologies in Medicine and Telehealth,* John Wiley & Sons, 2011.

[97] P. N. Mechael, "Exploring health-related uses of mobile phones: An Egyptian case study," *London School of Hygiene and Tropical Medicine,* University of London, 2007.

[98] E. Eifert, B. Chaney, K. Redican, and J. Eddy, "Responsibilities and competencies for health education specialists: implications for research and practice in the American journal of health education", *Am. J. Health Educ.,* vol. 52, no. 1, pp. 1-7, 2021. [http://dx.doi.org/10.1080/19325037.2020.1854134]

[99] S.L. Lusk, *Health promotion planning: An educational and environmental approach: Lawrence W. Green and Marshall W.* Kreuter Mayfield Publishing: Mountain View, California, 1992.

[100] S.M. Pancer, and G. Nelson, "Community-based approaches to health promotion: guidelines for community mobilization", *Int. Q. Community Health Educ.,* vol. 10, no. 2, pp. 91-111, 1989. [http://dx.doi.org/10.2190/1RLP-F21M-7Y4T-33BB] [PMID: 20840923]

[101] J. Gordon, H. Britt, G.C. Miller, J. Henderson, A. Scott, and C. Harrison, "General practice statistics in Australia: pushing a round peg into a square hole", *Int. J. Environ. Res. Public Health,* vol. 19, no. 4, p. 1912, 2022. [http://dx.doi.org/10.3390/ijerph19041912] [PMID: 35206101]

[102] P. Crampton, and M.C. Brown, "General practitioner funding policy: from where to whither?", *N. Z. Med. J.,* vol. 111, no. 1071, pp. 302-304, 1998.

[PMID: 9760958]

[103] M.S. Mittelman, D.L. Roth, D.W. Coon, and W.E. Haley, "Sustained benefit of supportive intervention for depressive symptoms in caregivers of patients with Alzheimer's disease", *Am. J. Psychiatry,* vol. 161, no. 5, pp. 850-856, 2004. [http://dx.doi.org/10.1176/appi.ajp.161.5.850] [PMID: 15121650]

[104] S. Rivière, S. Gillette-Guyonnet, T. Voisin, E. Reynish, S. Andrieu, S. Lauque, A. Salva, G. Frisoni, F. Nourhashemi, M. Micas, and B. Vellas, "A nutritional education program could prevent weight loss and slow cognitive decline in Alzheimer's disease", *J. Nutr. Health Aging,* vol. 5, no. 4, pp. 295-299, 2001. [PMID: 11753499]

[105] M. Prince, "Care arrangements for people with dementia in developing countries", *Int. J. Geriatr. Psychiatry,* vol. 19, no. 2, pp. 170-177, 2004. [http://dx.doi.org/10.1002/gps.1046] [PMID: 14758582]

[106] C. Clausen, F.J. Strohschein, S. Faremo, D. Bateman, N. Posel, and D.M. Fleiszer, "Developing an interprofessional care plan for an older adult woman with breast cancer: from multiple voices to a shared vision", *Clin. J. Oncol. Nurs.,* vol. 16, no. 1, pp. E18-E25, 2012. [http://dx.doi.org/10.1188/12.CJON.E18-E25] [PMID: 22297017]

[107] R.M. Carney, K.E. Freedland, B.C. Steinmeyer, E.H. Rubin, and G. Ewald, "Collaborative care for depression symptoms in an outpatient cardiology setting: A randomized clinical trial", *Int. J. Cardiol.,* vol. 219, pp. 164-171, 2016. [http://dx.doi.org/10.1016/j.ijcard.2016.06.045] [PMID: 27327502]

[108] F. Taleghani, M. Ehsani, S. Farzi, S. Farzi, P. Adibi, A. Moladoost, M. Shahriari, and M. Tabakhan, "Challenges to family caregivers in caring for gastric cancer patients from perspectives of family caregivers, patients, and healthcare providers: a qualitative study," *Indian J. Palliat. Care,* vol. 27, no. 4, pp. 521–529, 2021. [http://dx.doi.org/10.25259/IJPC_98_21]

SUBJECT INDEX

A

B

C

D

E

F

G

H

I

K

L

M

N

O

P

R

S

T

U

V

W

X

www.ingramcontent.com/pod-product-compliance
Lightning Source LLC
LaVergne TN
LVHW070946210926
848482LV00002B/110
9798898812515